LOGISTICS OF FACILITY LOCATION AND ALLOCATION

INDUSTRIAL ENGINEERING

A Series of Reference Books and Textbooks

1. Optimization Algorithms for Networks and Graphs, *Edward Minieka*
2. Operations Research Support Methodology, *edited by Albert G. Holzman*
3. MOST Work Measurement Systems, *Kjell B. Zandin*
4. Optimization of Systems Reliability, *Frank A. Tillman, Ching-Lai Hwang, and Way Kuo*
5. Managing Work-In-Process Inventory, *Kenneth Kivenko*
6. Mathematical Programming for Operations Researchers and Computer Scientists, *edited by Albert G. Holzman*
7. Practical Quality Management in the Chemical Process Industry, *Morton E. Bader*
8. Quality Assurance in Research and Development, *George W. Roberts*
9. Computer-Aided Facilities Planning, *H. Lee Hales*
10. Quality Control, Reliability, and Engineering Design, *Balbir S. Dhillon*
11. Engineering Maintenance Management, *Benjamin W. Niebel*
12. Manufacturing Planning: Key to Improving Industrial Productivity, *Kelvin F. Cross*
13. Microcomputer-Aided Maintenance Management, *Kishan Bagadia*
14. Integrating Productivity and Quality Management, *Johnson Aimie Edosomwan*
15. Materials Handling, *Robert M. Eastman*
16. In-Process Quality Control for Manufacturing, *William E. Barkman*
17. MOST Work Measurement Systems: Second Edition, Revised and Expanded, *Kjell B. Zandin*
18. Engineering Maintenance Management: Second Edition, Revised and Expanded, *Benjamin W. Niebel*
19. Integrating Productivity and Quality Management: Second Edition, Revised and Expanded, *Johnson Aimie Edosomwan*
20. Mathematical Programming for Industrial Engineers, *edited by Mordecai Avriel and Boaz Golany*
21. Logistics of Facility Location and Allocation, *Dileep R. Sule*

LOGISTICS OF FACILITY LOCATION AND ALLOCATION

DILEEP R. SULE

Louisiana Tech University
Ruston, Louisiana

CRC Press
Taylor & Francis Group
Boca Raton London New York

CRC Press is an imprint of the
Taylor & Francis Group, an **informa** business

CRC Press
Taylor & Francis Group
6000 Broken Sound Parkway NW, Suite 300
Boca Raton, FL 33487-2742

First issued in paperback 2019

ISBN-13: 978-0-8247-0493-3 (hbk)
ISBN-13: 978-0-367-39751-7 (pbk)

Visit the Taylor & Francis Web site at
http://www.taylorandfrancis.com

and the CRC Press Web site at
http://www.crcpress.com

Preface

Facility location has long been a subject of interest among industrial engineers, transportation engineers, management scientists, operations researchers, and logistics personnel. Major contributors to the field have come from many sources, but perhaps the largest single source has been well-trained mathematicians. As such, most of the facility location research published in journals and books has been mathematical in nature. Although the theorems and proofs that go along with this research are very important for analyzing the subject matter, the associated derivations and mathematical rigor can be intimidating to practicing engineers and business executives. And the same is true with most undergraduate and first-year graduate students, who may not be so mathematically inclined. Yet facility location is an important subject with numerous practical applications, and a happy medium must thus be found between theory and practice. Procedures that can be easily understood have a higher probability of being used in real life.

This book outlines such procedures for various location and allocation objectives. To facilitate understanding of concepts, each procedure is illustrated by a problem and its solution. However, this is not a cookbook. There are mathematical and logical foundations for the methods; these become apparent as one follows the necessary steps of the procedures. The idea is to take out the needless complexity and convey the solution procedure through simple steps. It is helpful, but not necessary, for the reader to have had one course in operations research. Many models are formulated as linear programming (LP) models to illustrate the mathematical structure, but are solved by simpler, alternative methods. For those with access to a computer program to solve LP problems, the formulations may be used to verify the results obtained by these alternative methods. Operations research techniques using the branch-and-bound algorithm, transportation algorithm, assignment algorithm, and dynamic programming are illustrated before being used in location models.

The book is designed to cover most of the broad topics in location analysis and can be used as a textbook as well as a reference book. The course can be a one-semester course for advance undergraduate or early graduate students in industrial engineering, management science, transportation science, logistics, systems engineering, or related fields. The content of the book includes models in which facilities may be placed anywhere in the plane (continuous location theory), at some discrete locations (discrete models), or on a network (network analysis).

The text has 12 chapters. The first is an introductory chapter; it also presents an elementary but popular ranking method for location selection. Chapter 2 presents some of the recent applications of fuzzy logic and the analytical hierarchy procedure (AHP) in location selection. Some of these procedures are long, but they can be computerized once the fundamentals are understood.

Chapters 3 and 4 are associated with continuous location problems for a single facility. A facility can be located to optimize the number of different objectives; the optimal location in each case may not be the same. Chapter 3 addresses the objective of minimizing the travel cost, called the minisum problem. Based on the mathematical expression for travel cost, a number of different procedures are applicable. Chapter 4 incorporates objectives such as minimizing the maximum distance, the circle covering problem, working with an undesirable facility location, and linear path facility development.

Chapter 5 addresses placement of multiple facilities in a continuous location problem. Unfortunately, it is not a direct extension of a single facility location problem and requires some effort. This chapter also discusses the machine layout models for efficient material flow analysis.

Chapter 6 is a basic location–allocation model that initiates discrete location analysis. The objective is to select from among the known locations the required number of locations to place facilities, and then allocate customers to receive service from one of these facilities to minimize cost.

Chapter 7 describes facility location in network-based problems. These problems are typical in transportation planning and other such applications in which travel is permitted only by a path represented on the network. The chapter describes, for example, where to place a competitive facility or how to develop a transportation hub.

Chapter 8 describes the procedures in tour development. In many instances the objective is to develop efficient routes for deliveries and collections of customer orders. This is a logistical problem of connecting different customers in sequence to minimize transportation cost. The procedures illustrated in this chapter accommodate many different modes of operation.

Chapter 9 deals with data changes due to such factors as shifts in demand pattern or foreseen changes in the use of the facilities. Changes are time-dependent; we often have to decide the initial location of the facility, and then when and where to move the facility to respond to changing costs and demands.

Chapter 10 addresses simultaneous facility location or, as popularly called in the literature, a quadratic assignment problem. Besides the well-known branch-and-bound procedure, a few easy-to-apply heuristics are explained that lead to a good, often optimal, solution.

Chapter 11 introduces transportation network–related problems, as it mainly applies the transportation algorithm to minimize nonlinear transportation costs as well as the maximum response time from a source to a destination in a transportation network.

Chapter 12 describes new location–allocation modes in a production environment. It describes which locations to select if there is a fixed cost for a location, if the cost of production varies from location to location, or if there is an advantage associated with a large-scale production at one place. It also discusses the machine or facility capacity selection procedure based on the various costs associated with machines of different capacity. It is an interesting chapter, and although the procedures seem lengthy at first glance, they can be easily grasped if the example solutions are followed.

An instructor should have no problem in developing a facility location course by selecting appropriate chapters that he or she feels are suitable for the class. Chapters 1, 2, 7, and 11 are independent and require no previously acquired information from other chapters.

Dileep R. Sule

Acknowledgments

I would like to thank a number of students in my facilities location class who suffered through an incomplete manuscript while it was being developed and made some useful suggestions for improvement. My special thanks to Rahul Joshi and Kedar Panse, who spent many days (and nights) developing some of the topics. Their efforts are sincerely appreciated. My thanks also to Advait Damle, Vikram Patel, and Amol Damle for proofreading the final copy.

Important suggestions were also made by Horst Eiselt, Trevor Hale, and Vedat Verter, who served as reviewers. Although not all the suggestions could be incorporated, the book has benefited greatly from their comments. I thank these reviewers for their time and effort.

The staff of Marcel Dekker, Inc., especially acquisitions editor John Corrigan and production editor Michael Deters, were very helpful in production of the book, and I thank them for their support.

And finally, to my wife, Ulka, and my children, Sangeeta and Sandeep, my thanks for their support during this proejct.

Contents

PART I

Introduction and Qualitative Methods

1

Introduction and the Traditional Approach

1.1 NATURE AND SCOPE

The primary purpose of this book is to introduce and demonstrate pragmatic methods for solving complex problems in the broad area called facilities locations. Although industrial engineers, economists, logisticians, and management scientists are primarily concerned with the topics discussed in succeeding chapters, the subject matter is also addressed by many other professionals, as is evident by the articles appearing in professional journals, such as *International Journal of Production Research*, *Logistics and Transportation Review*, *Journal of Farm Economics*, *Geographical Analysis*, *Econometrics*, *Transportation Science*, *Operations Research*, and *SIAM Review*. This illustrates that a large number of problems in many different fields can be investigated and solved by facility location methods.

Facility location and allocation is an important topic in logistic management. *Logistics*, as defined by the Council of Logistics Management, is "that part of supply chain process that plans, implements and controls the efficient, effective flow and storage of goods, services, and related information from the point of origin to the point of consumption in order to meet customers' requirements." Facility location and allocation forms one of the core link of this chain. Location of a facility and allocation of customers to that facility determines the distribution pattern and associated characteristics, such as time, cost, and efficiency, of the distribution pattern. Placement of one or more facilities, each in optimum locations and assigning the customers to them in the best possible manner, not only improves flow of material and services offered by the facility to customers, but also utilizes the facilities in an optimum manner, thereby reducing a need for multiple duplicating or redundant facilities.

The basic question may be to decide how to choose from the known feasible locations or from an infinite number of locations described as an area, a location, or coordinates for a location, in which to place a "facility (facilities)" and how to assign the "customers" to this facility. But what is a facility, and what is a customer? The nature of the problem defines these terms. For example, in determining the suitable locations for industrial plants so they can best serve the demands from various regions in the country, the plants are the facilities, and the product users are the customers. In determining the market territories to assign to sales personnel, the territories are customers, and the latter are facilities. For supplying water to different farm patches, these farms are the customers, and where water wells are bored are the facility locations. In airline business, location of a hub is important to determine how to serve different cities (customers) that are connected by shuttle air traffic to the hub. Other examples include locations of hospitals, fire stations, electric power plants, equipment in machine shops, and even a component on a PCB board. In each case, the facility and corresponding customer can be easily identified.

The applications of facility location problems are not restricted to just locating facilities. Numerous other examples can also be suggested in areas ranging from selection of machine capacity to placement of a facility so that it can respond to an emergency as quickly as possible, The following list illustrates a few more applications for facility location analysis.

1. Selecting sites for emergency service facilities, such as hospitals and fire stations
2. Determining the "best" locations for tool rooms, machines, water fountains, wash areas, concession areas, and first aid stations in a manufacturing plant
3. Choosing sites for warehouses and distribution centers
4. Selecting subcontractors and assigning appropriate work to each
5. Picking vendors and determining items to purchase from each, with or without quantity discounts
6. Choosing sites for maintenance departments in plants or, on a larger scale, sites for state agencies, such as a highway department or garage
7. Selecting sites and capacities of machines to meet expected demands from customers distributed throughout a given area
8. Developing a layout for a machine shop or an instrument panel
9. Selection of site for an obnoxious facility

The broad scope of topics is not restricted to just the problems cited in the foregoing, and later in the text we see many additional applications that, at first glance, may not seem to be facility location problems. However, for the present, let us state briefly the basic ideas that are discussed in the following chapters. They are as follows:

1. Where should the facilities be located?
2. How should the customers be assigned to these facilities or which facility should serve which customer?
3. What should the capacity of each facility be?
4. How should vendors be selected?
5. How does a quantity discount affect vendor selection?
6. How is a facility location schedule developed that will allow for changes in production, transportation, and other costs relative time?
7. How are appropriate machines selected from an available set consisting of machines with different capacities and associated costs?
8. How are locations selected to minimize the maximum response time?
9. How are the effects of changes included in data from period to period?
10. How are items with similar group characteristics assigned for production in different plants to minimize production and transportation costs?
11. How are many facilities placed simultaneously to develop a plant layout or an intricate design?
12. How is an optimum placement selected for the location of a facility?

The book is divided into four parts. The first part, consisting of chapters 1 and 2, addresses the traditional and not so traditional methods of location analysis. These methods are mainly qualitative but are combined with quantitative analysis. The second part, which includes chapters 3 through 7, addresses continuous and discrete facility location problems. In continuous or universal facility location problems, it is possible to place a facility in any position, and the objective is to seek the optimum position. It may also involve representation and related problems. In discrete location problems, a few locations are selected for new facilities from a number of predetermined locations. The third part is on tour development, covered in Chapter 8. This chapter presents variant methods for developing efficient tours to visit different demand points. The fourth section covers topics such as multiperiod location problems, simultaneous location problems, and so forth. Though some of these topics involve rather specialized applications, they introduce interesting concepts in location analysis.

1.2 TRADITIONAL APPROACH

There are numerous approaches for facilities location problems. Broadly speaking, they can be divided into two classifications: qualitative and quantitative. Qualitative methods are more easily understood and, therefore, frequently followed. We will study some of the more popular qualitative methods in this and the next chapter.

The first concern in deciding location is to analyze the factors that might influence the decision. For example, one widely known problem in facility location is that of selecting a site for a new plant. Several factors affect this decision, and the analysis involves comparing these factors in a judicial manner. Because identification of those factors influencing a site selection is also a very challenging problem, we will discuss an example in some detail.

Generally, site selection is a collective decision. A group of executives may decide what factors are important. In our example, after consulting with engineers and managers, the team of analysts considers the following elements to be important factors:

1. Transportation facilities
2. Labor supply
3. Availability of land
4. Nearness to markets
5. Availability of suitable utilities
6. Proximity to raw materials
7. Geographic and weather characteristics
8. Taxes and other laws
9. Community attitudes
10. National security
11. Proximity to the company's existing plants

Some of these factors are interrelated; therefore, one could have easily produced a list with a slightly different breakdown. Continuing with the present list, however, the following explanation further examines the discussion and rational as to why the group thinks that these factors affect the decision on site selection for a manufacturing plant.

Transportation Facilities

Suitable transportation facilities must be available to move personnel, equipment, raw materials, and products to and from the plant. Highways, railways, waterways, and airways are commonly used to transport raw material and finished goods. The volumes and type of raw materials and products often determine the best-suited mode of transportation. Trains, automobiles, and bus services are generally preferred by workers as the transportation means to travel to and from the manufacturing plant.

Adequate Labor Supply

Even in the coming age of computers, automation, and robotics, no company is able to operate without employees. The plant location study must assure that the types and the number of employees who will be needed will be available. The interstate highway system has made the suburbs as easily accessible as most cities, and as a result a plant can draw its work force from an area easily as large as 75 miles in diameter. The following factors are important: prevailing wages, work-week restrictions, existence of competing companies that can cause high turnover or labor unrest, productivity level, labor problems, and the education and experience of available potential employees. Many semiskilled factory positions are filled by trainable unskilled workers. Some light industries have found that locating plants in smaller communities allows them to use lower-cost, and more-productive, country labor. Other companies have chosen to locate new plants in rural areas to avoid the labor difficulties generally associated with areas that are staunchly union. A university nearby can supply good and talented technical staff, and the convenience of having a university nearby is also an attraction for skilled workers.

Availability of Land

Communities that are attempting to attract new industries often provide land at a low cost, but the company must make sure of the suitability of the land. For example, the soil characteristics and topography of the location must be evaluated, because they can influence the building costs. Additional space for future expansion should either be available or should be accessible for future acquisition, if required.

Nearness to Markets

The location of a plant is very important in determining the overall costs of goods and services to the customers. For example, when the transportation cost is high, it is highly desirable that a plant be located in or near the market area. This is especially true for bulky items for which the cost of shipping is significant compared with the cost of materials and labor, such as with foundry items, fertilizer, and building materials. On the other hand, for products with high labor or material cost, such as jewelry, computers, and watches, proximity of the market is not that critical. Another consideration is the use of by-products. For example, sawdust from a lumber mill might be sold to a paper mill or a particle board manufacturer.

Suitable Utilities

Most industrial plants require electricity, heating and cooling, and compressed air and steam. Availability of a cheap fuel source can be very important factor in site selection. Some plants produce their own power and steam by burning oil, gas, wood, or coal. For these plants the availability of an inexpensive fuel supply is almost necessary. For others, an area of the country that has an ample supply of low-cost electricity becomes an attractive alternative.

The availability of an adequate water supply is also very important. Plants that consume small volumes of water often purchase it from local public utilities, but those that use large quantities may need a source, such as a large river or lake, or an area where a deep water well could be drilled. Water table information is available from the state geological survey and the U.S. Corps of Engineers, who should be checked for seasonal fluctuations of lake and river levels, and future plans for water sources.

Waste disposal is also a critical consideration. The proposed site should have adequate facilities for solid, liquid, and gas disposals. The laws of the state and municipality should be carefully examined for determining the economic aspect of the waste disposal.

Another consideration is the flood and fire damage. If the plant site under evaluation is exposed to potential flood damage, the availability of flood protection facilities and flood insurance should be reviewed. The closeness and efficiency and quality of local fire department should also be considered.

Proximity of Raw Materials

The cost of shipping raw materials and fuel to the plant site can be considerable. Plants that require perishable or bulky raw material, tend to locate near the source of the raw materials. Plants in which raw materials lose much of their weight during the manufacturing process, such as ore refineries, steel plants, and paper mills, also often locate as near their raw material sources as practicable.

Geographic and Weather Considerations

The geographic characteristics of the site can affect the building and operating costs. A severely cold climate necessitates additional sheltering for the equipment, whereas a very hot climate requires the plant to have air-conditioning for personnel comfort, and an additional cooling tower for process equipment. Thus, factors such as altitude, temperature, humidity, average wind speed, annual rain fall, and terrain, become important.

Taxes and Legal Considerations

Because taxes form a significant part of operating cost, the types, bases and rates of taxes charged by the state and local governments must be considered. To attract new industries, many cities and states offer tax incentives. Some taxes to be evaluated are property, income, and sales. Unemployment compensation taxes also vary from state to state. In terms of legal considerations, local regulations concerning real estate, health and safety codes, truck transportation, roads, acquisition of easements and rights-of-way, zoning, building codes, and labor codes also influence the site selection.

Community Considerations

The community (both local authorities and the people) under consideration should welcome the placement of the plant within its area. Successful operation of the plant will require essential services, such as police and fire protection, street maintenance, and trash and garbage collection, from the community. Good living conditions that include cultural facilities, churches, libraries, parks, good schools, community theaters and symphonies, and recreational facilities are needed to attract and maintain motivated workers.

National Security

The U.S. government sometimes encourages companies that supply strategic materials to locate such that the sources for such products are widely dispersed. A company may increase its probability of being awarded a government contract if it locates its manufacturing facilities where the government wishes. In other cases the government may desire to locate a very security-sensitive manufacturing facility within the limits of a government installation.

Proximity to an Existing Plant

Some companies prefer to locate a new satellite facility in the general area of an existing major plant. Doing so facilitates direct supervision by upper-level management of both plants. Executives and consultants are able to minimize travel time between plants because they are near to each other. We may, however, also want to consider locating the plants far enough apart so they do not have to compete with each other for the same labor force.

1.3 RANKING PROCEDURES FOR SITE SELECTION

From the preceding discussion, it is evident that a large amount of information must be accumulated to aid in the decision-making process; federal, state, and local governmental agencies can provide very useful information. The Department of Labor, the Federal Communications Commission, the Federal Power Commission, and the Federal Trade Commissions are other good sources to gather some facts. Local agencies, such as industrial commissions and chambers of commerce, can provide details more specific to a particular location, such as labor, utility, land and housing costs, and availability of transportation facilities, such a railroad, trucking, and airports.

Once the necessary information about alternative locations has been gathered, the advantages and disadvantages of one location over another must be evaluated. One may have the lowest raw material cost, another the lowest utility cost, and yet another the best labor supply. A commonly used aid in

TABLE 1.1 Sample Location Rating Procedure

Considerations	Maximum weight	Loc X	Loc Y	Loc Z
Transportation facilities	100	80	90	90
Labor supply	100	75	80	90
Land	100	80	75	70
Markets	100	60	70	80
Utilities	75	70	70	65
Raw materials	75	50	75	60
Geographic/weather	50	40	40	50
Taxes and legal	50	30	30	40
Community	40	20	40	35
National security	15	10	5	15
Proximity to existing plant	40	30	10	25
TOTAL	745	545	585	620

selecting from alternative sites is the use of a rating procedure. Each of the major factors is rated from 0 to 100 relative to its importance. Each individual location is then rated from 0 to the maximum for each factor. The scores for the locations will determine the final ranking. An example of this procedure is shown in Table 1.1. Location Z would be selected, based on its maximum score.

It is clear that the ranking obtained by this method is largely subjective. When the factors are evaluated, they express an analyst's feelings that are measured in terms of assigned weights. It is quite possible that different analysts might choose varying weights for the same physical conditions, leading to entirely diverse site selections. Such variations in judgment can be eliminated if all the locations that meet certain minimum requirements are made candidates, and the procedures, presented later text, are then applied based on the cost data that can be collected for each site. We have seen in our discussion that in most factors (e.g., land, labor, utility, and taxes), cost plays a very dominant role in the analyst's subjective rating. Because the ultimate objective of any business is to provide goods or services at a profit, the site that can minimize the operational cost and, at the same time, satisfy other subjective criteria (e.g., size of the land, community feeling) would be a natural choice. The models in the latter chapters would evaluate each site in a quantitative manner.

The minimum score requirements for screening purposes may be set in many different ways. For example, for each factor, a minimum necessary value may be set, and the site that fails to meet the minimum is then eliminated from any future consideration; or a minimum could be set just for the more important factors, such as the first four considered in Table 1.1; or an overall minimum may be set, for instance at 560, eliminating location X from further examination.

Another approach is to allow all the sites to be considered for objective evaluation by first applying the methods presented in later chapters and then checking to see if the best site selected also meets the minimum requirements of the foregoing subjective factors. If it does not, and an alternative site is to be chosen, management will at least understand the dollar cost associated with that decision.

1.4 ANOTHER EXAMPLE

As noted earlier, facility location problems are not restricted to placement of plants or warehouses. Even within a plant, location of machines and placement of offices follow the same basic rules. For example, some factors that may influence selection of possible sites for a machine are the following:

1. Required floor space and floor space available
2. Cost of installation
3. Availability of auxiliary equipment
4. Travel time and distance between existing and new machines or offices
5. Work flow between existing and new machines and offices
6. Weight or physical size of the units that are processed
7. In-process inventory

With so many different factors affecting location and allocation decisions, careful analysis, similar to that for the plant location problem, is essential if we are to make the "right" decision. Again the combination of qualitative and quantitative approaches is of major benefit.

1.5 ECONOMIC MODELS

In some location problems, the site can be selected by using traditional economic models, if we can estimate the cost and benefits associated with each site. Listed in the following are the few well-known economic models (for more details refer to an engineering economy book):

1. Payback period: Determine the number of years it will take to recover the invested capital at each site. This is equal to

   Initial investment for the site/expected annual profit from the site

   Choose a site with least payback period.

2. Rate of return method: Determine the rate of return from each site and select the site with maximum rate of return. Rate of return is the interest rate at which the present worth of all expenditures are equal to the present worth of all earnings.

3. Benefit/cost ratio: Calculate the present worth of benefits and present worth of cost. Take the ratio and select the site that gives maximum value.

1.6 MULTINATIONAL CORPORATIONS

Although not a major thrust of this book, the following discussion presents some of the considerations given in location analysis by multinational corporations (MNC). These are major investment decisions, and added political, social, and local factors must also be considered. That industries had been protected in the third world and developing countries earlier, which are now opening up their economies, makes these countries more interesting to research for investment and locations by MNC.

The MNC seeking expansion may be from developed countries going into business in developing or underdeveloped countries or vice versa. The factors consider in each case in making decisions may be distinct. As an example, we might consider the United States as the developed world, countries such as South Korea as developing countries, and countries such as India, as an underdeveloped country. MNC from India will have different criteria for making an investment decision in the United States than a U.S.-based MNC trying to set a plant in India. Firm-specific advantages that are required to operate in a country, largely depend on the economic development in that country. As a result, firm-specific advantages to operate in a less-developed country and developed country are different. MNCs make substantial investment in terms of time and money before a project is begun, and usually the gestation period is long, meaning that a huge outflow of cash has to be managed for long time periods, with no immediate return on the money. As a result, an MNC makes a careful study of the socio-economic–political conditions before going into a new venture in an alien land.

The factors usually considered by an MNC before making an investment decision follow:

1. The political stability of the country
2. The commitment of the government to the opening up of the economy
3. The advantages of introducing a new technology and the possibility of adapting the new technology to the host country
4. The influences of the host's economy on the regional economy and vice versa
5. The product life cycle of the local manufacturers (if the product to be introduced is similar to that which is already being manufactured at the host country)
6. The possibility of exporting the products from the host country (or importing)

7. Government restrictions on import of raw materials, knocked-down kits and such
8. The variability of the market size

Political Stability

If we consider the huge investment required, any MNC is skeptical about the commitment of the local rulers to liberalization. Investments in countries that do not have a stable political setup are usually considered to be vulnerable. To avoid a possible loss of investment, MNCs usually prefer countries where all major parties (in case of democracies) are committed to liberalization. In Asia, for example, it is apparent that U.S. MNCs prefer investing in China, a communist country, with a stable political scenario, vis-a-vis India, which has a multiparty democracy.

Commitment of the Local Government

The commitment of the local government toward liberalization becomes another important factor. A government that is usually more populist could at any time roll back on its liberalization plans. For example, many MNC prefer Japan as an investment destination because most of the important economic and trade decisions in Japan are made by the bureaucrats in the ministry of international trade and industry (MITI), rather than politicians, which means that any change in the government will not affect the liberalization process.

Adaptability of Technology

Any MNC that plans to invest capital in new projects at a new country has to make sure that it can adapt its technology to the new country. There have been cases of good technologies failing in new markets as a result of the inability of the local populace to adapt itself to new technologies. A typical example would be the case of Rover Motors' introduction of its car in India. The car, albeit had a good technology, could not perform satisfactorily because Indian servicemen were not used to the new type of carburetor used in these cars and, hence, could not fix any problems that arose in the same.

Influence of the Local Economy

Despite that liberalizing economies integrate themselves with the global economy, there are still clusters of regional economies operating in the world, such as NAFTA, ASEAN, SAARC, and others. As a result of the regionalization of the economies, any investment decision is influenced by the effect of the local economy on the regional economy. Any decision made without considering this could be disastrous. One example for this would be the failure of the Southeast Asian (Pacific) rim countries economies in the 1990s, which was largely triggered by the failure of some banks in Japan and, hence, the imminent depression in the local economy (Japan).

Product Life Cycle of Competitors

Companies, which derive leverage in developed countries as a result of their technological capabilities, usually have large investments in research and development (R&D) and, as a result, in time, make their own product obsolete. If such companies are to make investments in a country where the product life cycles of (already existing) competitors are long, there is an advantage.. The investing company could generate larger revenues out of a product in these countries and, thereby, make up for short life cycles in their home (developed) countries. This entails that, for the investment already made in R&D to develop the product (in the form of sunk cost) could make more revenue by making investment in such a country.

Export and Reimport Possibility

The MNCs usually operate in a number of economies, and there will be bilateral and multilateral treaties between these economies. An MNC that manufactures labor-intensive products might usually be interested in investing and creating new manufacturing facilities in countries that do not have such high labor costs. However, these decisions are also based on the tariff rates that prevail between the host and home countries. If the home country has high import tariff rates, then the MNC is at a disadvantage because reimporting into the home country (to avoid high labor costs) would entail huge tariffs in the form of import duties.

Government Tariffs for Import of Raw Materials

Any MNC that is operating in technology-intensive business will have to consider this factor in making an investment decision. That such MNCs will usually be unwilling to part with their technology would largely force them to manufacture the core items at home and manufacture the rest of the assemblage in the new country. In some cases, governments charge large import tariffs on such items. In addition, some governments may even force an MNC to make a timetable to completely localize production (100%).

Variability of the Market Size

Variability of the market size becomes an important criterion, for the cash flows are usually considered in advance before venturing into a new project, and they are projected using an assumed market size.

There are also other aspects of investment that must be decided, and this is the investment pattern or the ownership percentage and the investment partner, if the ownership is not 100%. Firm, specific advantages are those leverages obtained by a firm, such as low-interest source of capital and technological advantage.

Source of Funds

Multinational companies usually enjoy high leverage of funds, as a result of their clout, and have the ability to source funds from different parts of the world (in the economies in which they operate). This ability to secure funds from low-interest country and use it in a high-interest market usually gives MNCs an advantage.

Technological Intensity

Companies that are in technology-intensive businesses tend to patent their technologies to safeguard their interests commercially. In such cases, they would like to maximize their revenue within the 15–25 years operating window allowed by the Intellectual Property Rights. MNCs tend to prefer places where a competitor's technology is not by any means better than their own.

Investment Pattern Decision

When a multinational corporation makes a decision to invest in a country, it is faced with the problem of determining the mode of investment. For example, different modes of investment include the following: total ownership of a subsidiary; partnership with a local firm in which the multinational owns a major part of the subsidiary; partnership with a local firm in which the local firm owns a major chunk in the subsidiary; or just giving technology to the local firm. The factors that are to be considered before making an investment decision are

1. Existence and strength of the local players
2. Market size
3. Cultures prevalent in the host country
4. Government regulations

Existence and Strength of the Local Players

When an MNC makes an investment in which there are local players, it has to consider the strength of the local players. For example, if a company such as Daimler-Benz, which makes luxury cars, decides to invest in a third-world country, the road conditions and the infrastructure would not be very well developed to exploit the luxury built into the car. To avoid this problem, the company would like to go for a partnership with a local car manufacturer, which would obviate the need to study the road conditions and adapt the vehicle for that country's conditions. Also, for an MNC that decides to go it alone, it has to consider the difficulties in marketing the product all by itself in a new country. To take advantage of the existing dealership network, an MNC would choose to go for a partnership with a local manufacturer.

Market Size

It would be more difficult for an MNC to estimate the market size and type under various conditions of the economy. So they usually tend to go for partnerships to avoid any possible overestimation of the size of the market.

Cultures Prevalent in the Country

In some third-world countries, the cultures are very complex, and there are different languages spoken and different religions being practiced. In such countries, it may be better for an MNC to go for partnerships with a local organization than to go for full ownership.

Government Regulations

When an MNC comes into a new country, it is competing for the local market, without bringing the nation any value other than employment for local people. To take care of this problem, governments decree regulations on total ownership of subsidiaries. For example, there are countries in which, if an MNC goes for total ownership of a local subsidiary, it has to meet some export obligations spread over a period of years, say a decade. This helps the local government obtain valuable foreign exchange. In such cases, if the company is not sure of the possibility of exporting, it must go for a partnership.

Thus, while making an investment decision, an MNC goes through two stages of decision; namely,

1. Location of investment
2. Pattern of investment in the decided location

Although at the outset, these two steps seem to be in sequence, they are, in fact, intertwined in a complex manner. A company cannot decide a location without making a pattern decision and vice versa. However, each of these factors are important in every sense of the word, and they can make or break a fortune for any corporation.

Similar analysis can also lead to development of qualitative factors for location analysis within a country. For example, people may differ in their work habits, because of religious beliefs or social structure. Language variations within different parts of a country may cause additional problems, and so on.

1.7 TYPE OF FACILITY LOCATION PROBLEMS

Use of location analysis is illustrated in multiple disciplines in many different scenarios. Here we will review some of the commonly used terminology in location literature. The problems of location analysis are generally categorized into one of the following broad classes.

1. p-Median problem (Weber problem)
2. p-Center problem
3. Uncapacitated facility location problem
4. Capacitated facility location problem
5. Quadratic assignment problem.

The p-median problem, or p-MP, deals with placement of p facilities (in p locations) to minimize a cost criteria. If p is one, the problem is 1-MP and so on. The cost may be defined in terms of time, dollars, number of trips, total distance, or any such measure. Because the objective is to minimize the total cost, it is also referred to as minisum problem or Weber problem. Some locations in the plan may not be available for a facility placement, in which case it is called a closed location. Otherwise the location is an open location. If there is a "weight" associated with the demand point, such that every demand point does not contribute equally to the objective function, the problem is called p-MP weighted.

The p-center problem, or p-CP, deals with placement of p facilities to minimize the maximum distance from any facility to the demand point it is assigned to serve. This is a typical objective for placement of emergency services in the community, such as fire stations, ambulance services, and police stations.

Uncapacitated facility location problem or UFLP is one in which the objective is that of minisum, but the cost expression contains a fixed cost, depending on the location in which a facility is placed. The number of facilities to be located are not prespecified, but are determined such as to minimize the cost. Because the capacity of each facility is unlimited, it is never profitable to assign a demand to more than one supply point.

Capacitated facility location problem (CFLP) is similar to the UFLP, except the capacity of each facility is limited. The optimum solution may, in such case, require that a customer be supplied from more than one source.

Quadratic assignment problem (QAP) defines a problem where in n facilities, such as n machines having flow among themselves, are to be placed in n locations simultaneously to minimize the total cost, which is measured as flow times the distance. If we have four machines to locate, we obtain 4! combinations for possible solutions. For a 20-machine problem, 20! possible solutions will require about 2×10^{18} evaluations—an impossible task even for today's high-speed computers to complete in a reasonable time (less than 1 year).

Even though this gives a short summary, we have addressed these and some more topics in Chapters 4 through 11 in detail. As we go through these chapters, we will see some more variations of the facility location problems.

Exercises

1.1 What is logistics? How do location–allocation decisions play a major role in effective flow of goods and services?

1.2 What are the four most important factors to consider in each of the following location problems:

a. A hospital
b. A chemical plant
c. A shopping center
d. A school
e. An auto service center
f. Airport

Are the factors and their ranking the same? If not discuss why the variations.

1.3 We want to set up an Italian restaurant in a city of Heavenly Living. As with any other city the population it is somewhat segregated based on ethnic background. In four major counties the population distribution (in percentage) is as follows:

County	German	Italian	Indian	Chinese	Polish	Turkish	English	Population
A	20	40	3	4	7	6	20	3,000
B	3	3	60	14	10	10	0	4,000
C	10	5	10	15	20	20	15	15,000
D	0	5	5	55	10	10	15	8,000

Of course, everyone loves Italian food, except Italians love it more than others. Suppose average frequency of weekly visits from a person from a group of ten person, in an ethnic group can be estimated. If the restaurant is within one county (block) from their residence, the number of persons visiting from group of ten is displayed in the following table. If the travel required is more than one block, then the number of people visiting the restaurant decreases with the following relationship (people within block/ number of blocks of travel required):

German	Italian	Indian	Chinese	Polish	Turkish	English
2	4	1	1	2	1	2

If a suitable site is available in each county, which location should be selected to have maximum customers? Assume that the distance between each county is as follows

$$A-B = 1;\ A-C = 2;\ \text{and } A-D = 3.$$

1.4 Compare three locations, one in a city, one in a small suburb or nearby town, and one in a surrounding country to locate a water-bottling plant.

1.5 We want to locate a plant to produce sugar. There are three alternatives available for us, for which comparison is given in the following table. Weigh them giving 0–100 points and select the best alternative. Note the difference in selection among your friends.

Point	Alternative 1	Alternative 2	Alternative 3
1. Transportation facilities	Can be connected by water, highways	Can be connected by highways	Can be connected by all modes of transport
2. Labor supply	Readily available at lower cost	Available at moderate cost	Is available, but expensive
3. Availability of land	Available at lower cost, but building costs are higher owing to soil quality	Available at lower cost but constraint on additional space if required	Available at moderate cost
4. Markets	To be distributed all over United States	To be distributed all over United States	To be distributed all over United States
5. Utilities (plant can produce its own electricity from the waste)	Abundance of water	Abundance of water	Moderate availability of water
6. Raw material	Has to be brought from long distance	Has to be brought from moderate distance	Is readily available
7. Geographic–weather conditions	Pleasant to work for 9 months, hot for 3 months	Cold climate	Pleasant to work
8. Taxes and legal information	Taxes are moderate	Taxes are moderate with state offering incentives for new plants	Taxes are moderate

(continued)

Point	Alternative 1	Alternative 2	Alternative 3
9. Community consideration	Is neutral about the plant	Not happy with the plant coming	Welcomes the plant
10. Proximity to existing plant (no existing plant)	—	—	—

1.6 When are the application of economic models, such as payback period or rate of return method appropriate and when are the applications not appropriate or suspect?

1.7 What are the type of problems that can be addressed by facility location–allocation methods?

1.8 Explain the difference between the p-median problem and a p-center problem.

2

Fuzzy Logic and the Analytical Hierarchy Procedure

In this chapter we illustrate recent concepts of fuzzy logic and analytical hierarchy procedure (AHP) as they are applied to facility locations problems. Both concepts combine subjective and objective evaluation procedures, and are illustrated by examples.

2.1 FUZZY LOGIC

Classic logical systems are usually based on Boolean two-valued logic (i.e., true or false). Frequently, such as in determining temperature, this is too restrictive. Most people would consider 32°F cold and 100°F hot. Now how do we classify 60° or 70°F? In Boolean logic we could define 70°–80°F as high-comfort zone (and 60–69° as low-comfort zone), but humans feel very little difference from 69°–71°F, yet 69°F would be considered low-comfort zone and 71°F as high-comfort zone. In fuzzy logic systems a factor between 0 and 1 is assigned to each level of a parameter, 0 being the lack of the variable and 1 being the maximum of that variable. In this example of comfort zone, a value of 60°F could be assigned a value of 0.1 and 80°F could be assigned a value of 1.0. Consequently, temperatures in the 70s could be assigned various values between 0.4 and 0.8 (or whatever weighting value is assigned by an expert in comfort zones).

Let us consider an application of fuzzy set theory in location analysis. Ideally, the location of a facility should be such that it maximizes the resource utilization, while minimizing the overall cost. As we have seen in Chapter 1, many factors are taken into consideration in choosing a site for a new facility. These attributes can be classified into three classes: critical, objective, and subjective. The critical criteria are the attributes that each location must have for further evaluation. The objective attributes can be defined in monetary terms such as the investment cost, labor cost, and land cost. The subjective attributes are qualitative. Climatic conditions and availability of skilled labor are subjective in judgment and, hence, are classified as the subjective attributes.

Objective factors can be easily expressed in terms of numbers; hence, they do not pose many problems in decision making. Subjective attributes, however, are expressed in qualitative terms; therefore, they are harder to incorporate in an analysis. The fuzzy set theory can convert these qualitative evaluations into quantitative ones, allowing an effective way to measure the contributions of the subjective factors.

Subjective criteria are generally expressed in terms of "very low," "low," "good," "very good," "medium," "high," and the like. With the fuzzy set theory, these are converted to quantifiable evaluations that generally take a triangle or a trapezoidal shape with different weights, given in Figure 2.1 and Table 2.1.

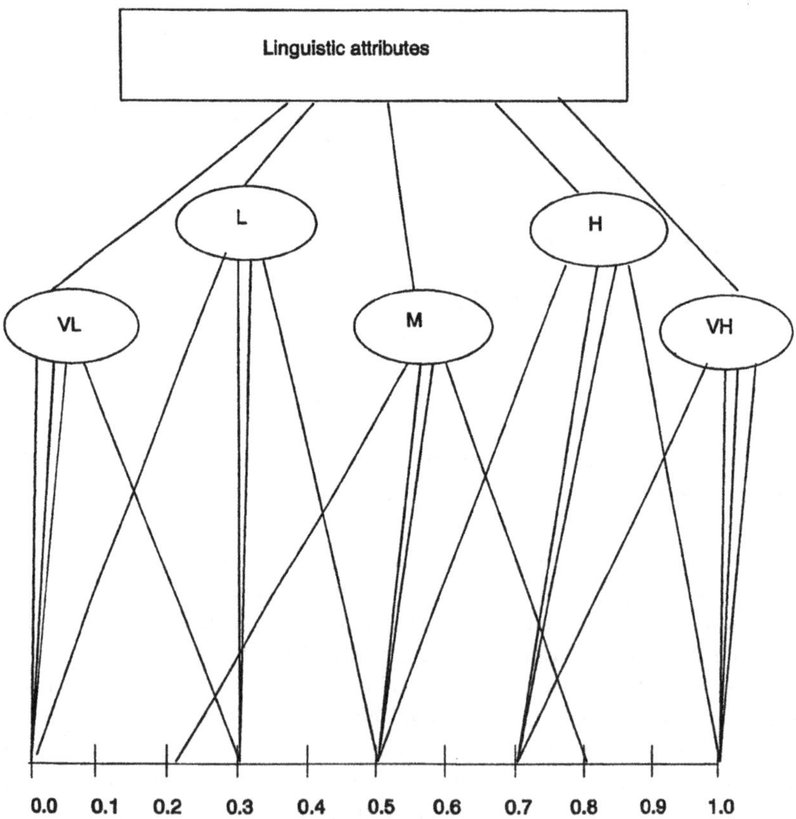

FIGURE 2.1 Weights of the linguistic attributes for criteria.

TABLE 2.1 Attributes and Four-point Grading Weight

Linguistic attribute	Numerical weight
VL	(0, 0, 0, 0.3)
L	(0, 0.3, 0.3, 0.5)
M	(0.2, 0.5, 0.5, 0.8)
H	(0.5, 0.7, 0.7, 1)
VH	(0.7, 1, 1, 1)

For example, from Table 2.1, "low rating" (L) has 0–0.5 weight in decision making. So if four payoff values for the low rating are 200, 250, 300, and 450 then in trapezoidal terms it is interpreted as, low of 200 with the weight of 0, first midpoint of 250 with the weight of 0.3, the second midpoint with the payoff of 300 with the weight of 0.3, and the maximum payoff of 450 with the weight of 0.5. The trapezoidal fuzzy set numbers are easy to interpret. For example "approximately equal to $400" may be represented as (395, 400, 400, 405), "approximately between $360 and $400" may be represented as (355, 360, 400, 405), and the nonfuzzy numbers such as $500 can be represented as (500, 500, 500, 500). Thus, in linguistic description, each attribute—such as "low rating"—has a four-point trapezoidal distribution. The weight distribution is 0, 0.3, 0.3, and 0.5. Ultimately, we plan to develop a composite trapezoidal distribution that represents all subjective criteria.

2.2 AN EXAMPLE OF FUZZY THEORY IN LOCATION ANALYSIS

Consider a problem of a team of four physicians trying to set up a new clinic. They have predetermined three cities, Shreveport, Monroe, and Ruston, from which they want to select one city where their new clinic could be located. However, there are several factors they would like to consider. They are

1. Commuting services and infrastructure in the city
2. Responsiveness of emergency services
3. Need for physicians of their expertise
4. Whether people can afford health insurance and medical expenses
5. Climatic conditions
6. Investment required to set up a clinic and employ the necessary personnel

In applying the fuzzy logic technique, the first step is to divide the criteria (denoted by Ci) into subjective and objective categories. In our example the physicians divide them as follows:

Subjective

1. Commuting services and infrastructure in the city (C1)
2. Availability of emergency services (C2)
3. Need for physicians of their expertise (C3)
4. Whether people can afford health insurance and medical expenses (C4)
5. Climatic conditions (C5)
6. Labor cost (C6)

Objective

1. Investment required to set up a clinic and employ the necessary personnel (C6). This comprises
   a. Cost of purchasing land
   b. Medical equipment cost

Evaluation of the Relative Weights of Criteria

The next step is to assign weight: very high (VH), high (H), medium (M), low (L), or very low (VL) to each site criteria. Each physician is considered as an expert decision maker (denoted by D1, D2, D3, and D4) and asked to assign a linguistic rating for each factor, both for subjective and objective factors, and to display their subjective evaluation of importance of each factor. Results are shown in Table 2.2.

The linguistic rating from each expert is aggregated for each criterion. We will use the numerical scale shown in Figure 2.2 and Table 2.3 for this purpose (these distributions are fairly standard in the fuzzy theory literature).

Again, the figures assume a trapezoidal or triangular distribution. The minimum, maximum, and two values (same as for triangular distribution) within which probability of finding the attribute are illustrated in these figures. These estimations, give a fuzzy or linguistic attribute a numerical weight distribution.

TABLE 2.2 Four Decision Makers Weight Assignment to Each Criterion

	D1	D2	D3	D4
C1	VH	L	M	H
C2	M	VL	H	L
C3	H	M	H	L
C4	M	H	L	M
C5	L	H	L	M
C6	H	M	VH	H

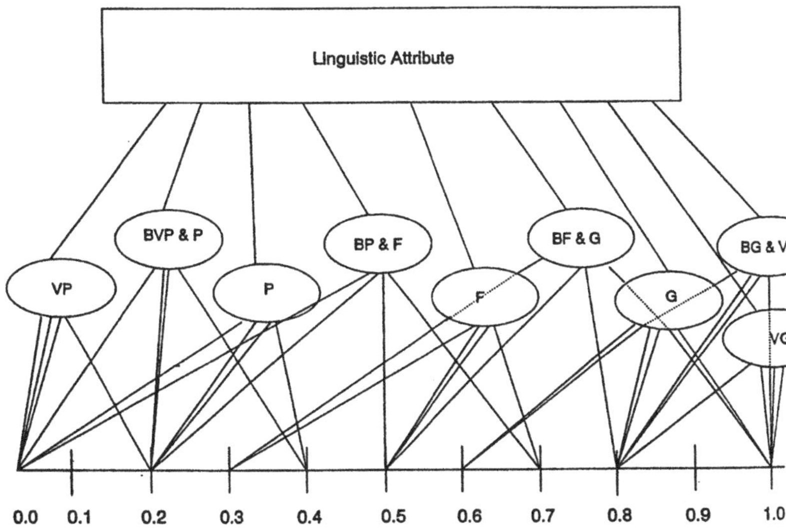

FIGURE 2.2 Weights of the linguistic attributes for alternative sites.

TABLE 2.3 Attributes and Four-point–Grading Weight

Linguistic attribute	Numerical weights
Very poor	(0, 0, 0, 0.2)
Between poor and very poor	(0, 0.2, 0.2, 0.4)
Poor	(0, 0.2, 0.2, 0.4)
Between poor and fair	(0, 0.2, 0.5, 0.7)
Fair	(0.3, 0.5, 0.5, 0.7)
Between fair and good	(0.3, 0.5, 0.8, 1)
Good	(0.6, 0.8, 0.8, 1)
Between good and very good	(0.6, 0.8, 0.8, 1)
Very good	(0.8, 1, 1, 1)

The linguistic attributes and associate four-point numerical weights are represented in the figures, but can also be easily shown in a chart (see Table 2.3).

For each criteria in Table 2.2, calculate the aggregate of all expert's weight assignment to form a new trapezoidal distribution with a minimum, two-modal value and a maximum weight (value). For example, for criterion 1, C1, the composite lower bound of all decision makers based on Table 2.1 linguistic assignments and their corresponding weights from Table 2.1 (or from the foregoing chart is):

$$(VH + L + M + H)/4 = (0.7 + 0 + 0.2 + 0.5)/4 = 0.35$$

Similarly, for the same decision string, taking the average of the middle (because it is a triangular distribution, both modal weights are the same) and the upper bounds are

First middle weight $= (1 + 0.3 + 0.5 + 0.7)/4 = 0.625$
Second middle weight $= (1 + 0.3 + 0.5 + 0.7)/4 = 0.625$
Upper bound $= (1 + 0.5 + 0.8 + 1)/4 = 0.825$

Thus, the weight distribution, considering the opinions of all four experts, of the new trapezoidal distribution for C1 is

$$w_1 = (0.35,\ 0.625,\ 0.625,\ 0.825)$$

Similarly, the weights for the other criteria are

$$w_2 = (0.175,\ 0.375,\ 0.375,\ 0.65)$$
$$w_3 = (0.30,\ 0.55,\ 0.55,\ 0.825)$$
$$w_4 = (0.225,\ 0.5,\ 0.5,\ 0.775)$$
$$w_5 = (0.175,\ 0.45,\ 0.45,\ 0.70)$$
$$w_6 = (0.475,\ 0.725,\ 0.725,\ 0.95)$$

Evaluation of Relative Weight of Each City Relative to Each Subjective Criterion

Next, the decision makers should evaluate each criterion, relative to each possible location. In our example, we have three available sites: A1, Shreveport; A2, Monroe; or A3, Ruston.

The evaluations are shown in Table 2.4 through 2.8, each for a specific criterion. For example, Table 2.4 shows the evaluation of each city for the criterion C1, community services and infrastructure in the city. Note that it is possible in these linguistic ratings to give a rating in between two major weights

TABLE 2.4 Evaluation of Alternative Cities Under Criterion C1

	D1	D2	D3	D4
A1 Shreveport	VG	F	BG and VG	BP and VP
A2 Monroe	G	BP and F	BF and G	F
A3 Ruston	BG and VG	P	BF and G	BP and F

TABLE 2.5 Evaluation of Alternative Cities Under Criterion C2

	D1	D2	D3	D4
A1 Shreveport	G	F	F	P
A2 Monroe	BF and G	BP and F	BF and G	VG
A3 Ruston	G	G	G	BVP and P

TABLE 2.6 Evaluation of Alternative Cities Under Criterion C3

	D1	D2	D3	D4
A1 Shreveport	VG	F	VG	BG and VG
A2 Monroe	BP and F	G	G	F
A3 Ruston	P	P	BF and G	P

TABLE 2.7 Evaluation of Alternative Cities Under Criterion C4

	D1	D2	D3	D4
A1 Shreveport	G	VP	G	BP and F
A2 Monroe	BF and G	F	P	BF and G
A3 Ruston	BP and F	BF and G	F	G

TABLE 2.8 Evaluation of Alternative Cities Under Criterion C5

	D1	D2	D3	D4
A1 Shreveport	VP	P	VG	G
A2 Monroe	VG	BVP and P	F	BF and G
A3 Ruston	BF and G	F	BF and G	P

such as BP and VP meaning between poor and very poor. The categories and associated four-point numeric values are shown in Figure 2.3 and are represented in the chart.

Following the same procedure as before, determine the lower, mid-two, and upper bound-weight values for each location and for each associated criterion. For example, for criterion 1 (Table 2.4) for Shreveport, the lower bound $(0.8 + 0.3 + 0.6 + 0)/4 = 0.425$, the middle two bounds are $(1 + 0.5 + 0.8 + 0.2)/4 = 0.625$, and the upper bound is $(1 + 0.7 + 1 + 0.4)/4 = 0.775$. Thus, if we define S_{ij} as weight for site i and criteria j then performing similar calculations leads to,

$$\text{Criterion 1: } S_{11} = (0.425,\ 0.625,\ 0.625,\ 0.775)$$
$$S_{21} = (0.30,\ 0.50,\ 0.65,\ 0.85)$$
$$S_{31} = (0.225,\ 0.425,\ 0.575,\ 0.775)$$

$$\text{Criterion 2: } S_{12} = (0.30,\ 0.50,\ 0.50,\ 0.70)$$
$$S_{22} = (0.35,\ 0.55,\ 0.775,\ 0.925)$$
$$S_{32} = (0.45,\ 0.65,\ 0.65,\ 0.85)$$

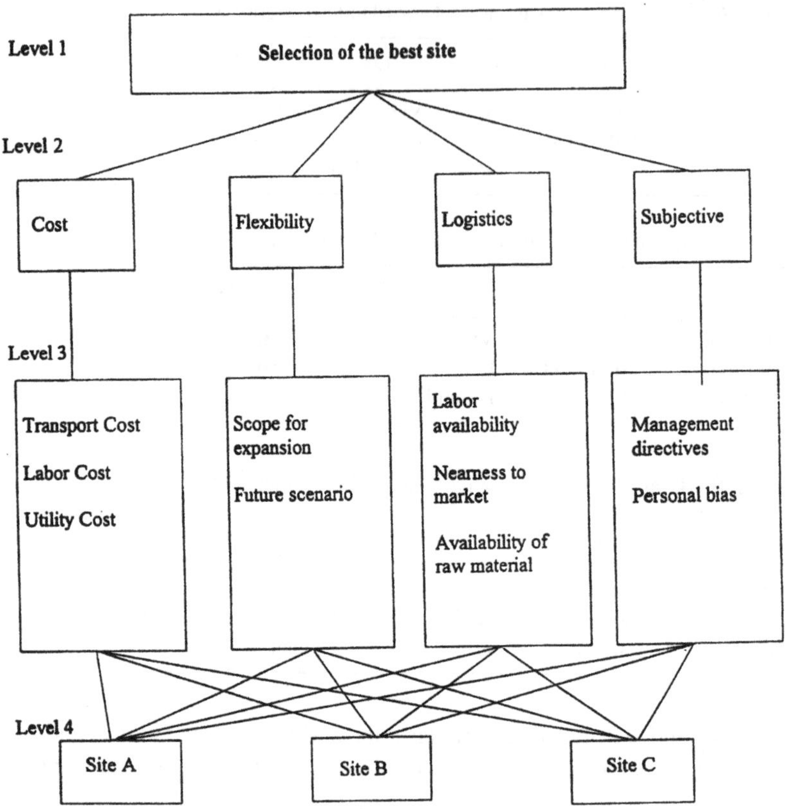

FIGURE 2.3 Hierarchy levels of the site selection.

$$
\begin{aligned}
\text{Criterion 3: } S_{13} &= (0.625,\ 0.825,\ 0.825,\ 0.925)\\
S_{23} &= (0.375,\ 0.575,\ 0.65,\ 0.85)\\
S_{33} &= (0.075,\ 0.275,\ 0.35,\ 0.55)
\end{aligned}
$$

$$
\begin{aligned}
\text{Criterion 4: } S_{14} &= (0.30,\ 0.45,\ 0.525,\ 0.725)\\
S_{24} &= (0.225,\ 0.425,\ 0.575,\ 0.775)\\
S_{34} &= (0.30,\ 0.50,\ 0.65,\ 0.85)
\end{aligned}
$$

$$
\begin{aligned}
\text{Criterion 5: } S_{15} &= (0.35,\ 0.50,\ 0.50,\ 0.65)\\
S_{25} &= (0.35,\ 0.55,\ 0.625,\ 0.775)\\
S_{35} &= (0.225,\ 0.425,\ 0.575,\ 0.775)
\end{aligned}
$$

Objective Criteria

Now, let us consider the objective criterion of cost. It can be evaluated independently of decision makers, because the values can be estimated based on market and economic research. Again four cost estimates for each site are obtained in terms of their numeric value. For example, land cost in Shreveport would be between 28,000 and 42,000 dollars with 30,000 and 40,000 dollars also likely values. To ensure compatibility between the objective criteria and rating of the subjective criteria, the total cost must be converted into a dimensionless form. The evaluation is needed in terms of cost distribution for each site (in Table 2.9, the row values) and comparing the sites with each other (the column values). In addition, the alternative with the minimum cost should have the maximum rating. To achieve these goals, the following transformation of the data for each site i is suggested.

$$R_{ti} = \{t_i * [t1^{-1} + t2^{-1} \cdots t_m^{-1}]\}^{-1}$$

where $t1$, $t2$, $t3$ ... are the cost factors, 1, 2, 3, ..., each with four estimates, and t_i is the sum of the costs for the factors at appropriate levels. For example, the total lower bound for the cost (sum of the costs at lower estimates) for Shreveport is $28 + 15 + 35 = 78$.

In calculating R_{ti} (relative total cost), each sum is subjected to the same transformation and, therefore, the relative weights between them do not change. However, because the cost is to be converted into relative ranking, the high-cost option should be ranked lower than the low-cost option. It is easier to compare numbers that are small yet greater than zero. Hence, we select an arbitrary number that is greater than the largest number in the cost distribution, and divide it by each cost. The resulting values have ranking such that higher the value, the better off we are. In this example, to keep the numbers small, we divide 100 (an arbitrary number) by each cost value. The values for each site are tabulated in column 2 of Table 2.10. Again note that $100/78 = 1.28$ is greater than $100/92 = 1.08$. Thus, the ranking in column 2 for Shreveport is $100/92 = 1.08$, $100/90 = 1.11$, $100/80 = 1.25$, and $100/78 = 1.28$. Add the lowest values for each site, the midvalues for each site, and so on to obtain the

TABLE 2.9 Calculation of *ti* (Total Cost) for Each Alternative Site

Alternatives	Purchase of land ($t1$)	Cost of equipment ($t2$)	Labor cost ($t3$)	Total *ti*
A1 Shreveport	(28, 30, 40, 42)	(15, 15, 15, 15)	(35, 35, 35, 35)	(78, 80, 90, 92)
A2 Monroe	(18, 20, 25, 27)	(16, 18, 20, 22)	(24, 25, 25, 26)	(58, 63, 70, 75)
A3 Ruston	(32, 32, 32, 32)	(16, 18, 20, 22)	(24, 26, 28, 30)	(72, 76, 80, 84)

TABLE 2.10 Calculations of *Rti* (Relative Total Cost) for Each Site

	$1/ti * 10^{-2}$	$ti/\text{sumi} * 10^{-2}$	$Rti = S_{i6}$
A1 Shreveport	(1.08, 1.11, 1.25, 1.28)	(2.74, 3.02, 3.72, 4.02)	(0.248, 0.26, 0.33, 0.36)
A2 Monroe	(1.33, 1.42, 1.58, 1.72)	(2.04, 2.38, 2.89, 3.28)	(0.304, 0.34, 0.42, 0.49)
A3 Ruston	(1.11, 1.25, 1.31, 1.38)	(2.53, 2.87, 3.31, 3.67)	(0.272, 0.30, 0.348, 0.395)
Total	(3.52, 3.78, 4.14, 4.38)		

distribution for the objective criteria. For example, $1.08 + 1.33 + 1.11 = 3.52$, gives the lowest value. To convert this back into site value, say for Shreveport, multiply 3.52 by 78 (the lowest cost value for Shreveport from Table 2.9) and divide by 100, the constant we had used in the previous calculations. The resulting rating of 2.74 is then associated with Shreveport lower value in column 3. Inverse of 2.74 is 0.36 which is the top value in the forth column associated with Shreveport. Similar calculations for other sites give the final rating distribution for criterion 6, S_{16} by corresponding R_{ti}.

To find the fuzzy suitability index F_I for site I we find the average of the product of weight for each criterion times the relative weight of that criterion for site I. This is given by:

$$F_i = 1/k * [(S_{i1} * w_1) + (S_{i2} * w_2) + (S_{i3} * w_3) + (S_{i4} * w_4) * (S_{i5} * w_5) + (S_{i6} * w_6)]$$

where $k = 6$, is the number of factors that are being evaluated.

For example, for alternative one we have,

$$F_1 = 1/6 * [(S_{11} * w_1) + (S_{12} * w_2) + (S_{13} * w_3) + (S_{14} * w_4) + (S_{15} * w_5) + (S_{16} * w_6)]$$

$$\begin{aligned} F_1 = 1/6 * [&(0.425, 0.625, 0.625, 0.775) * (0.35, 0.625, 0.625, 0.825) \\ &+ (0.3, 0.5, 0.5, 0.7) * (0.175, 0.375, 0.375, 0.65) \\ &+ (0.625, 0.825, 0.825, 0.925) * (0.30, 0.55, 0.55, 0.825) \\ &+ (0.3, 0.45, 0.525, 0.725) * (0.225, 0.5, 0.5, 0.775) \\ &+ (0.35, 0.50, 0.50, 0.65) * (0.175, 0.45, 0.45, 0.70) \\ &+ (0.248, 0.26, 0.33, 0.36) * (0.475, 0.725, 0.725, 0.95)] \end{aligned}$$

Thus, from F_1, the value of the first element for the first site F_{11} (Shreveport) is equal to:

$$F_{11} = 1/6 * [(0.425 * 0.35 + 0.30 * 0.175 + 0.625 * 0.30 + 0.30 * 0.225 + 0.35 * 0.175 + 0.248 * 0.47)] = 0.105$$

Performing similar calculations, we obtain the values for the fuzzy suitability index shown in Table 2.11. For deciding the final rating for each location (see Table 2.12), the fuzzy set index is added. For instance, for A1, the final rating is:

$$A1 = 0.105 + 0.278 + 0.293 + 0.536 = 1.212$$

The highest value is for Monroe and, therefore, the physician would set up his practice in Monroe.

Thus, the steps in fuzzy analysis can be summarized as follows:

Choose facility site selection criteria and identify facility alternative sites.
Choose preference ratings for criteria weight
Choose preference ratings for alternative sites for subjective and objective criteria.
Tabulate the weighing of criteria and pool them to obtain the aggregate weight, for each subjective criterion.
Calculate the fuzzy ratings of alternative sites for subjective criteria.
Calculate the fuzzy ratings of alternative sites for objective criteria.
Aggregate the weightings of criteria and fuzzy ratings of alternatives relative to all criteria and obtain the fuzzy suitability indices.
Calculate the ranking value of each site and choose the highest-rank site.

2.3 THE ANALYTIC HIERARCHY PROCESS

The analytic hierarchy process (AHP) was developed by Thomas L. Saaty (1980) in the 1970s. Since then, it has proved to be a useful method for decision making where the objective is to select the best alternative. The analytic hierarchy process acts as an aid to the process by sorting out the inconsistencies in making subjective judgments. AHP is a multicriteria decision-making method that uses hierarchic or network structures to represent a decision problem and then

TABLE 2.11 Fuzzy Suitability Index

Alternatives	Fuzzy suitability index
A1 Shreveport	(0.105, 0.278, 0.293, 0.536)
A2 Monroe	(0.089, 0.256, 0.321, 0.602)
A3 Ruston	(0.069, 0.219, 0.271, 0.537)

TABLE 2.12 Final Ranking

Alternatives	A1 Shreveport	A2 Monroe	A3 Ruston
Ranking	1.212	1.268	1.096

develops priorities for alternatives based on the decision maker's judgments throughout the system. It can be used in both subjective and objective evaluation criteria. It obtains not only the rank order of the alternatives, but also their relative standing measured on a ratio scale.

The AHP framework is called "analytical" and "hierarchical" because it organizes decision variables into successive levels of importance. This framework is then processed for evaluating the interrelation among parts. As a result, it drastically improves the decision-making process, giving numbers of benefits. The stated benefits for AHP are as follows:

1. AHP allows the systematic consideration and evaluation of multiple decision criteria. These criteria could be financial, nonfinancial, quantitative, or qualitative (intangible).
2. AHP allows managerial judgments to be included formally and systematically in the decision-making process.
3. As a process AHP enables managers to focus on those aspects of the decision that need refinement or have the highest degree of uncertainty. It accomplishes this through the use of matrix-weighing techniques.
4. Finally the AHP is well suited to support the growing movement toward group decision making in business. This can be especially important in a decision involving site selection because it involves consultants, managers, engineers, and other functional persons of an organization.

The Working Of Analytic Hierarchy Process

The use of AHP in solving a decision problem involves the following four steps:

Step 1 Setup the decision hierarchy by breaking down the decision problem into a hierarchy of interrelated decision elements.

Step 2 Collect input data by pairwise comparison of decision elements.

Step 3 Use the eigenvalue method to estimate the relative weights of decision elements.

Step 4 Aggregate the relative weights of decision elements to arrive at a set of ratings for the decision alternatives (or outcomes).

Step 1. Structuring any decision problem hierarchically is an efficient way of dealing with its complexity and identifying the major components of the problem. In step 1, the decision maker should breakdown the decision problem into a hierarchy of interrelated decision elements. Each level of the hierarchy consists of some important elements, and each element, in turn, is further divided into another set of subelements. The process continues down to the most detailed elements of the problem (see Figure 2.3).

Step 2. In step 2, the input data for the problem is collected. It consists of matrices of pairwise comparison of elements of one level that contribute to achieving (or satisfying) the objectives of the next higher level. This is accomplished by asking the managers (or decision makers) to evaluate each set of elements in the pairwise fashion. The diagonal elements of the matrix are comparisons of the decision elements with themselves; therefore, the associated values should be 1.

Each comparison in the pair is made to evaluate the importance of one factor over another relative to the criteria to be evaluated at that point. In typical analytic hierarchy studies a nine-point scale is used as explained in Table 2.13.

Consider the following example. Suppose the site in Shreveport is evaluated relative to flexibility, logistics, and cost. When we compare flexibility with flexibility the weight is 1. Comparing flexibility with logistics, for Shreveport, the analyst estimates the logistics to be five times more important than flexibility. Then for the matrix to be consistent (or the analyst judgment to be consistent) logistics when compared with flexibility should be 1/5. This ratio is automatically entered in the logistic–flexibility comparison as shown in the following chart. But we may still unwittingly develop inconsistencies in weight assignments. The next example illustrates that.

Inconsistency

Suppose an analyst has given the following rating for three factors when they are compared with each other relative to the site in Shreveport.

	Flexibility (x)	Logistics (y)	Labor cost (z)
Flexibility (x)	1	5	3
Logistics (y)	1/5	1	6
Labor cost (z)	1/3	1/6	1

Now from the matrix, if flexibility, logistics, and labor cost are denoted by x, y, and z, respectively, then the stated relations are as follows:

$$y = 5x \text{ and } z = 3x$$

which means $1/5y = 1/3z$, or $z = 3/5y$. This is not true when z and y comparison was made in the foregoing chart. Thus, we can say that the judgment of the analyst is inconsistent. To correct this problem the eigen vector method forms an integral part of AHP.

Why the Eigenvector Method Works

Any $n \times n$ square matrix has a property that the sum of its eigenvalues (λ) is equal to its trace, trace being the sum of the diagonal elements of the square

TABLE 2.13 The Nine-Point Scale Used by the AHP

Intensity of importance		Definition explanation
1	Equal importance	Two activities contribute equally to the objective
3	Weak importance of one	Experience and judgment over the other slightly favor one activity over another
5	Essential or strong	Experience and judgment importance strongly favor one activity over another
7	Demonstrated importance	An activity is strongly favored and its dominance is demonstrated in practice
9	Absolute importance	The evidence favoring one activity over another is of the highest possible order
2,4,6,8	Intermediate values	When compromise is between two adjacent needed judgments

matrix. In the pairwise comparison matrix all the judgments have positive values; hence, only the positive eigenvalues provide solutions. If the matrix is totally consistent, then there will be only one eigenvalue, which will be equal to the trace of the matrix, and all other eigenvalues will be equal to zero. Because the diagonal element represents comparison of each factor with itself, it will always be equal to 1, giving the sum of diagonal elements to be equal to n. Therefore, in the totally consistent matrix,

$$\lambda_1 = \lambda_2 = \cdots \lambda_{n-1} = 0, \quad \lambda_n = n$$

For example, consider the following matrix **P** where three factors **A**, **B**, and **C** are judged. This is a 3 × 3 matrix with trace = n = 3.

	A	**B**	**C**
A	1	5	3
B	1/5	1	3/5
C	1/3	5/3	1

To obtain the eigenvalues, solve for λ in the following determinant equation. $|\mathbf{P} - \lambda\mathbf{I}| = 0$ where **I** is the identity matrix

$$\begin{vmatrix} 1-\lambda & 5 & 3 \\ 1/5 & 1-\lambda & 3/5 \\ 1/3 & 5/3 & 1-\lambda \end{vmatrix} = 0$$

Solving the foregoing determinant we obtain

$$\lambda^2(\lambda - 3) = 0$$

so we have $\lambda = 0$ or $\lambda = 3$. Here, the eigenvalue $\lambda = 3$ is equal to the trace of the matrix. The matrix is therefore consistent.

When the matrix is somewhat inconsistent, however, we obtain a number of eigenvalues that sum up to the value that is equal to the trace of the matrix, but the sum of positive eigenvalues will not be equal to the trace. In this case, we select the highest positive eigenvalue to calculate the relative weights, for that value is the closest to n.

Suppose, for the foregoing example, if now the judgments were such that we have an inconsistent matrix as shown

	A	B	C
A	1	5	3
B	1/5	1	1/4
C	1/3	4	1

Then solving for the eigenvalues we obtain

$$\begin{vmatrix} 1-\lambda & 5 & 3 \\ 1/5 & 1-\lambda & 1/4 \\ 1/3 & 4 & 1-\lambda \end{vmatrix} = 0$$

$$\lambda = 3.25 \text{ or } \lambda = -0.251$$

Now, because we do not consider negative values in this case, the largest positive eigenvalue of 3.25 is selected. Note the trace of the matrix is still 3. The closer the eigenvalue is to the trace of the matrix, the more consistent the matrix is.

The closer the value of computed λ_{max} is to n, the more consistent are the observed values of **A**. This property forms the basis for constructing the consistency index (CI) and the consistency ratio (CR), where $CI = (\lambda_{max} - n)/(n - 1)$ and

$$CR = (CI/ACI) * 100$$

ACI is the average index of randomly generated weights. According to Satty (1986), the values for ACI depended on the order of the matrix and are as follows (first row is the order of the matrix, second row is the ACI value).

1	2	3	4	5	6	7	8	9	10	11	12	13	14	15
0.00	0.00	0.58	0.90	1.12	1.24	1.32	1.41	1.45	1.49	1.51	1.48	1.56	1.57	1.59

As a working rule of AHP, a CR value of 10% or less is considered acceptable. Otherwise it is recommended that **A** be reevaluated to resolve inconsistencies in the pairwise comparisons.

In our case $CI = (3.25 - 3)/2 = 0.125$ and $CR = 0.125/0.58 \times 100 = 21.5\%$

Because the value of CR is greater than 10%, the present matrix is inconstant.

Step 3. In step 3, we take the input of the foregoing pairwise comparisons and produce the relative weights of the elements as the output. Supposing we have n criterion $P1, P2 \ldots Pn$, and they have corresponding weights of $W = (W1, W2, \ldots Wn)$. Based on the matrix theory, if matrix **A** is consistent then $PW = nW$ where, **P** is the judgment matrix that we obtained by pairwise comparisons. If the matrix is somewhat inconsistent then $PW = \lambda_{\max} W$ where, $\lambda_{\max}$ is the maximum positive eigenvalue.

Since, $PW = nW$, it also means $|P - nI||W| = 0$. And since $n = \lambda_{\max}$, the values of weights are obtained by solving $|P - \lambda_{\max}||W| = 0$. The details are illustrated in an example latter in the chapter.

Step 4. The final step is to calculate relative weight or priority for each alternative. This can be achieved by taking the expected value or weighted arithmetic average of all factors that influence the decision.

2.4 HOW INCONSISTENCIES CAN LEAD TO RANK REVERSAL

This section explains why AHP method of weight allocation using eigenvalues of step 3, is important. As we will see in the discussion, inconsistent evaluations and weight assignments can lead to misleading selection of an alternative.

Consider one simple procedure of allocation of weights to different factors, the largest-column total method. This method involves totaling the largest weight associated with each row and selecting this value to normalize all factors. For instance, in the following problem there are three alternatives A, B, C that are compared against three criteria C1, C2, and C3, and the corresponding judgments are given in the following matrices. By applying the method for weight determination, for example, for criteria 1, the largest sum of the columns is 10 $(1 + 8 + 1)$; therefore, each largest element of the row is divided by the total 10, giving weights for each factor in that criteria. Is this a good method? Not really. We will show an instance of a rank reversal as a result of weight calculations using this method in the following data.

Comparison with three alternatives relative to criterion 1

C1	**A**	**B**	**C**	**W** (calculated)
A	1	1/7	1	1/10
B	7	1	8	8/10
C	1	1/8	1	1/10

Note, that if the matrix is consistent (i.e., if A is 1/7th of B), then B is 7 times better than A, which should be indicated in appropriate comparative values.

Now, consider additional criteria 2 and 3, with associated weight distribution and comparative evaluations of the criteria relative to the sites as shown in the following charts.

Comparison with three alternatives relative to criterion 2

C2	**A**	**B**	**C**	**W** (calculated)
A	1	8	9	9/11
B	1/8	1	1	1/11
C	1/9	1	1	1/11

Comparison with three alternatives relative to criterion 3

C3	**A**	**B**	**C**	**W** (calculated)
A	1	6/7	6	6/14
B	7/6	1	7	7/14
C	1/6	1/7	1	1/14

Criterion	**A**	**B**	**C**	**W** (calculated)
C1	1	1	1	1/3
C2	1	1	1	1/3
C3	1	1	1	1/3

By applying the largest-column total method and finding the arithmetic average relative to each criterion, the overall weights for **A**, **B**, and **C** are as follows:

$$W_A = (1/3)(1/10) + (1/3)(9/11) + (1/3)(6/14) = 0.4489$$
$$W_B = (1/3)(8/10) + (1/3)(1/11) + (1/3)(7/14) = 0.4636$$
$$W_C = (1/3)(1/10) + (1/3)(1/11) + (1/3)(1/14) = 0.0874$$

Giving ranking of B, A, and C.

Now suppose we add an alternative **D**, keeping the same judgments for **A**, **B**, and **C**, and we obtain the following comparisons:

Criterion 1 comparison with four factors

C1	A	B	C	D	W
A	1	1/7	1	1/8	1/18
B	7	1	8	1	8/18
C	1	1/8	1	1/7	1/18
D	8	1	7	1	8/18

Criterion 2 comparison with four factors

C2	A	B	C	D	W
A	1	8	9	1/7	9/12
B	1/8	1	1	1	1/12
C	1/9	1	1	1	1/12
D	1/7	1	1	1	1/12

Criterion 3

C3	A	B	C	D	W
A	1	6/7	6	7/8	6/16.14
B	7/6	1	7	8	8/16.14
C	1/6	1/7	1	1	1/16.14
D	8/7	1/8	1	1	(8/7)/16.4

By using the same weight distribution for each criterion as in the previous evaluation (i.e., 1/3), the comparison of four factors gives the following rankings,

$$W_{\mathrm{A}} = 0.39, \quad W_{\mathrm{B}} = 0.34, \quad W_{\mathrm{C}} = 0.07, \quad W_{\mathrm{D}} = 0.19$$

thus reversing the rank between A and B. Such reversal of ranks will not occur when the weights are calculated based on the eigenvalue method.

2.5 APPLYING THE AHP TO SITE SELECTION

Consider that a manufacturer X is planning to open a new plant for making a product Y. After initial deliberations the choice of alternatives for a plant site were narrowed down to three locations, namely, A, B, and C. To decide among these three selection alternatives a committee of five members was formed to choose the best suitable location. The members of the decision-making committee were picked from different functions, such a finance, production/engineering, marketing/sales, management, and research. To follow the standard AHP procedure it is necessary to define the problem and its variables. The objective of the problem is to "select the best suitable site/location for the production plant." As described before, there are four basic steps for development of the AHP. The first step involved model specification; that is, identifying the major categories, criterion, and alternatives in the problem. The criteria represents specific factors against which the location alternatives are to be evaluated. The criteria, categories, and alternatives are also individually referred to as the decision elements. Once this is done, the problem needs to be broken down into different levels of hierarchy. The first level of the hierarchy consists of the ultimate objective; that is, to select the most suitable site for the plant (see Figure 2.3). The subsequent levels of the hierarchy consist of the various decision elements. The last level consists of the site selection alternatives, A, B, and C. The factors affecting the site selection decision are broadly classified into four categories

1. Overall cost: This can be further divided into transportation costs, labor wages and training, land cost, raw materials, local taxes, and utility costs.
2. Flexibility: This constitutes scope for future expansion, availability of land, and facilities for expansion.
3. Logistics or operation: This includes the linguistic variables, such as availability of labor and transportation, community considerations, and proximity to the market.
4. Subjective: These are the factors that cannot be quantified, but nevertheless, play an important role in making the choice. They include management and personal bias toward a particular site, and management directives that are directly associated with the location alternative. These factors will be referred to as I, II, III, and IV for the purpose of our calculations.

The second step involves a series of pairwise comparisons to establish relative importance (weights) of the categories and criteria in the model. These weights will ultimately be used to develop the overall ratings of each alternative. The decision elements are compared, one at a time, with all the other decision elements at the same level of hierarchy, relative to the decision elements at the immediately higher level.

Table 2.14 shows the matrix representation of these comparisons for the elements at level 2 relative to the elements at level 1. In this case, the decision elements at level 2 are the factors influencing site selection; namely, I, cost; II, flexibility; III, logistics; and IV, subjective. These factors ultimately influence the objective in the first level that of the site selection. It is conventional to compare the factors in the rows with those in the columns using the 9-point rating scale of Table 2.13. In this case, factor I was strongly preferred over II and, therefore, was assigned the rating 5. Because IV was moderately preferred over II, the element II was given a rating of 1/3. When n factors are being compared, only $n(n-1)/2$ rating inputs are necessary to fill the matrix. The elements in the lower triangle are reciprocal of associated elements above the diagonal.

The next step is to calculate the eigenvalues and eigenvectors. The procedure is as follows: Let

$$\mathbf{A} = \begin{bmatrix} 1 & 5 & 3 & 4 \\ 1/5 & 1 & 1/6 & 1/3 \\ 1/3 & 6 & 1 & 4 \\ 1/4 & 3 & 1/4 & 1 \end{bmatrix}$$

Then,

$$\mathbf{A} - \lambda\mathbf{I} = \begin{bmatrix} 1 & 5 & 3 & 4 \\ 1/5 & 1 & 1/6 & 1/3 \\ 1/3 & 6 & 1 & 4 \\ 1/4 & 3 & 1/4 & 1 \end{bmatrix} - \lambda \begin{bmatrix} 1 & 0 & 0 & 0 \\ 0 & 1 & 0 & 0 \\ 0 & 0 & 1 & 0 \\ 0 & 0 & 0 & 1 \end{bmatrix}$$

$$= \begin{bmatrix} 1-\lambda & 5 & 3 & 4 \\ 1/5 & 1-\lambda & 1/6 & 1/3 \\ 1/3 & 6 & 1-\lambda & 4 \\ 1/4 & 3 & 1/4 & 1-\lambda \end{bmatrix}$$

TABLE 2.14 Pairwise Comparison Matrix for Categories

	Cost	Flexibility	Logistics	Subjective
Cost	1	5	3	4
Flexibility	1/5	1	1/6	1/3
Logistics	1/3	6	1	4
Subjective	1/4	3	1/4	1

Solving for the det$(\mathbf{A} - \lambda\mathbf{I}) = 0$, We obtain

$$\lambda^4 - 4\lambda^3 - 163/36\lambda - 91/90 = 0$$

Which gives us the eigenvalues for the matrix **A**. The largest eigenvalue is called λ_{max} and is equal to 4.27. The value can also be obtained by using software such as Mathcad.

Is this a consistent matrix? Once the pairwise comparison is completed, it is necessary to calculate the consistency index CI, to establish the validity of these ratings. Here the consistency index is

$$\begin{aligned} \mathrm{CI} &= \lambda_{max} - n/n - 1 \\ &= (4.27 - 4)/4 - 1 \\ &= .09 \end{aligned}$$

and consistency ratio CR $= 0.09/0.90 = 0.10$ which is within the acceptable range, less than or equal to 10%.

The next step is to calculate the relative weights for each criterion. Substitute the value of λ_{max} in the matrix $|\mathbf{A} - \lambda\mathbf{I}||\mathbf{W}| = 0$ and solve for $\mathbf{W}$ to obtain the appropriate weights. These weights are then normalized to 1. The details of the calculation are as follows:

$$\mathbf{A} - \lambda\mathbf{I} = \begin{bmatrix} -3.27 & 5 & 3 & 4 \\ 1/5 & -3.27 & 1/6 & 1/3 \\ 1/3 & 6 & -3.27 & 4 \\ 1/4 & 3 & 1/4 & -3/27 \end{bmatrix}$$

and $[\mathbf{A} - \lambda\mathbf{I}]\ [\mathbf{W}] = 0$ leads to

$$\begin{bmatrix} -3.27 & 5 & 3 & 4 \\ 1/5 & -3.27 & 1/6 & 3 \\ 1/3 & 6 & -3.27 & 4 \\ 1/4 & 3 & 1/4 & -3/27 \end{bmatrix} \begin{bmatrix} \mathbf{W1} \\ \mathbf{W2} \\ \mathbf{W3} \\ \mathbf{W4} \end{bmatrix} = \begin{bmatrix} 0 \\ 0 \\ 0 \\ 0 \end{bmatrix}$$

The foregoing can be written as:

$$-3.27W_1 + 5W_2 + 3W_3 + 4W_4 = 0 \quad (1)$$

$$0.2W_1 - 3.27W_2 + 0.16W_3 + 3W_4 = 0 \quad (2)$$

$$0.33W_1 + 6W_2 - 3.27W_3 + 4W_4 = 0 \quad (3)$$

$$0.25W_1 + 3W_2 + 0.25W_3 - 3.27W_4 = 0 \quad (4)$$

Likewise,

$$3.27W_1 = 5W_2 + 3W_3 + 4W_4 \quad (1)$$
$$3.27W_2 = 0.2W_1 + 0.16W_3 + 3W_4 \quad (2)$$
$$3.27W_3 = 0.33W_1 + 6W_2 + 4W_4 \quad (3)$$
$$3.27W_4 = 0.25W_1 + 3W_2 + 0.25W_3 \quad (4)$$

Solving Eqs. (3) (4):

$$3.27W_3 + 4W_4 = 0.33W_1 + 6W_2$$
$$-0.25W_3 + 3.27W_4 = 0.25W_1 + 3W_2$$

we obtain,

$$3.77W_3 + 10.54W_4 = -0.17W_1$$

Substituting the value of W_1 from Eq. (2)

$$3.27W_2 = 0.2(-22.17W_3 + 62W_4) + 0.16W_3 + 3W_4$$
$$3.27W_2 = 12.74W_4 - 4.264W_3$$

Substituting the value of W_2 and W_1 in the equation for W_3

$$3.27W_3 = 0.33(-22.17W_3 + 62W_4) + 6(-1.3W_3 + 3.89W_4) + 4W_4$$

Solving,

$$W_3 = 2.6W_4$$
$$3.27W_2 = 12.74W_4 - 4.264W_3$$

Substituting value W_3 in the foregoing

$$W_2 = 0.5W_4$$

Substituting the value of W_3 in

$$-0.17W_1 = 3.77W_3 - 10.54W_4$$

$$W_1 = 4.34W_4$$

Therefore,

$$W_1 = 4.34$$
$$W_2 = 0.5$$
$$W_3 = 2.6$$
$$W_4 = 1.0$$

Total 8.44

By normalizing the values to the scale of 1 we obtain

$$W_1 = 4.34/8.44 = 0.517$$

$$W_2 = 0.5/8.44 = 0.059$$

$$W_3 = 0.306$$

$$W_4 = 0.118$$

The result is Table 2.15.

Next, go to level 3. The pairwise comparison process described in the foregoing is repeated for criteria within each category to establish the relative importance of each subcriterion. For example, for cost, the subcriteria are transportation cost, labor cost, and utility cost. These comparisons are shown in Tables 2.16–2.19. The respective values of relative weights for these criterion are also included in the tables. Remember that for each data table the consistency CR is calculated and consistency is checked. If the data is not consistent it must

TABLE 2.15 Weight Assignments

	Cost	Flexibility	Logistics	Subjective	Relative weights
Cost	1	5	3	4	0.517
Flexibility	1/5	1	1/6	1/3	0.059
Logistics	1/3	6	1	4	0.306
Subjective	1/4	3	1/4	1	0.118

TABLE 2.16 Pairwise Comparison Matrix for Costs

Overall cost	Transport cost	Labor cost	Utility cost	Relative weights
Transport cost	1	3	5	0.637
Labor cost		1	3	0.258
Utility cost			1	0.105

TABLE 2.17 Pairwise Comparison Matrix for Logistics

Logistics	Labor availability	Nearness to market	Raw material availability	Relative weights
Labor availability	1	5	1	0.455
Nearness to market		1	1/5	0.091
Raw material availability			1	0.455

TABLE 2.18 Pairwise Comparison Matrix for Subjective Factors

Subjective	Management directive	Personal bias	Relative weights
Management directive	1	5	0.833
Personal bias		1	0.167

TABLE 2.19 Pairwise Comparison Matrix for Flexibility

Flexibility	Scope of expansion	Future scenario	Relative weights
Scope of expansion	1	1	0.5
Future scenario		1	0.5

be reevaluated. For brevity we will not show these calculations or the values any further.

After the decision criteria at level 3 have been compared relative to the categories at level 2, the elements of the final level, level 4 are compared with all the elements at the same level relative to all the elements of the previous level, level 3. The decision elements of the fourth level of hierarchy are the three location alternatives, A, B, and C. So, these alternatives A, B, and C are compared pairwise, with one another relative to the ten criteria from level 3, as shown in Tables 2.20–2.29. Again the consistency is important.

TABLE 2.20 Comparisons of Sites Relative to Transport Cost

Transport cost	A	B	C	Relative weights
A	1	3	1/3	0.258
B		1	1/5	0.105
C			1	0.637

TABLE 2.21 Comparison of Sites Relative to Labor Cost

Labor cost	A	B	C	Relative weights
A	1	5	1/3	0.272
B		1	1/8	0.067
C			1	0.661

TABLE 2.22 Comparisons of Sites Relative to Utility Cost

Utility cost	A	B	C	Relative weights
A	1	3	1	0.429
B		1	1/3	0.143
C				0.429

TABLE 2.23 Comparisons of Sites Relative to Labor Availability

Labor availability	A	B	C	Relative weights
A	1	3	1/5	0.183
B		1	1/8	0.075
C			1	0.742

TABLE 2.24 Comparisons of Sites Relative to Nearness to Market

Nearness to market	A	B	C	Relative weights
A	1	1/5	3	0.183
B		1	8	0.744
C			1	0.075

TABLE 2.25 Comparisons of Sites Relative to Raw Material Availability

Raw material availability	A	B	C	Relative weights
A	1	5	3	0.637
B		1	1/3	0.105
C			1	0.258

TABLE 2.26 Comparisons of Sites Relative to Personal Bias

Personal bias	A	B	C	Relative weights
A	1	3	1	0.429
B		1	1/3	0.143
C			1	0.429

TABLE 2.27 Comparisons of Sites Relative to Management Directive

Management directive	A	B	C	Relative weights
A	1	3	1/3	0.258
B		1	1/5	0.105
C			1	0.637

TABLE 2.28 Comparisons of Sites Relative to Scope for Expansion

Scope for expansion	A	B	C	Relative weights
A	1	1	1/3	0.2
B		1	1/3	0.2
C			1	0.6

TABLE 2.29 Comparisons of Sites Relative to Future Scenario

Future scenario	A	B	C	Relative weights
A	1	1	1/3	0.2
B		1	1/3	0.2
C			1	0.6

These results are summarized in Figure 2.4. Note the numbers at level 4 represent the converging factors from level 3 and because there are ten such factors, they are evaluated for every site.

The last step calculated the overall rankings for the alternatives. This is done by finding the global priority of each alternative relative to the ultimate objective or goal, which is at the first level of the hierarchy. The global priority of an alternative is the composite vector of relative weights established at each level. These are found by respectively multiplying the values of the local priorities by the local priorities of the elements at the next higher level and then taking the sum. For example for site A, based on Figure 2.4 the global weight (GP) or priority is

$$\begin{aligned} GP = {} & 0.517[0.637 * 0.258 + 0.258 * 0.272 + 0.105 * 0.429] \\ & + 0.059[0.5 * 0.2 + 0.5 * 0.2] + 0.306[0.455 * 0.183 \\ & + 0.091 * 0.183 + 0.455 * 0.637] + 0.118[0.833 * 0.429 \\ & + 0.167 * 0.258] = 0.322 \end{aligned}$$

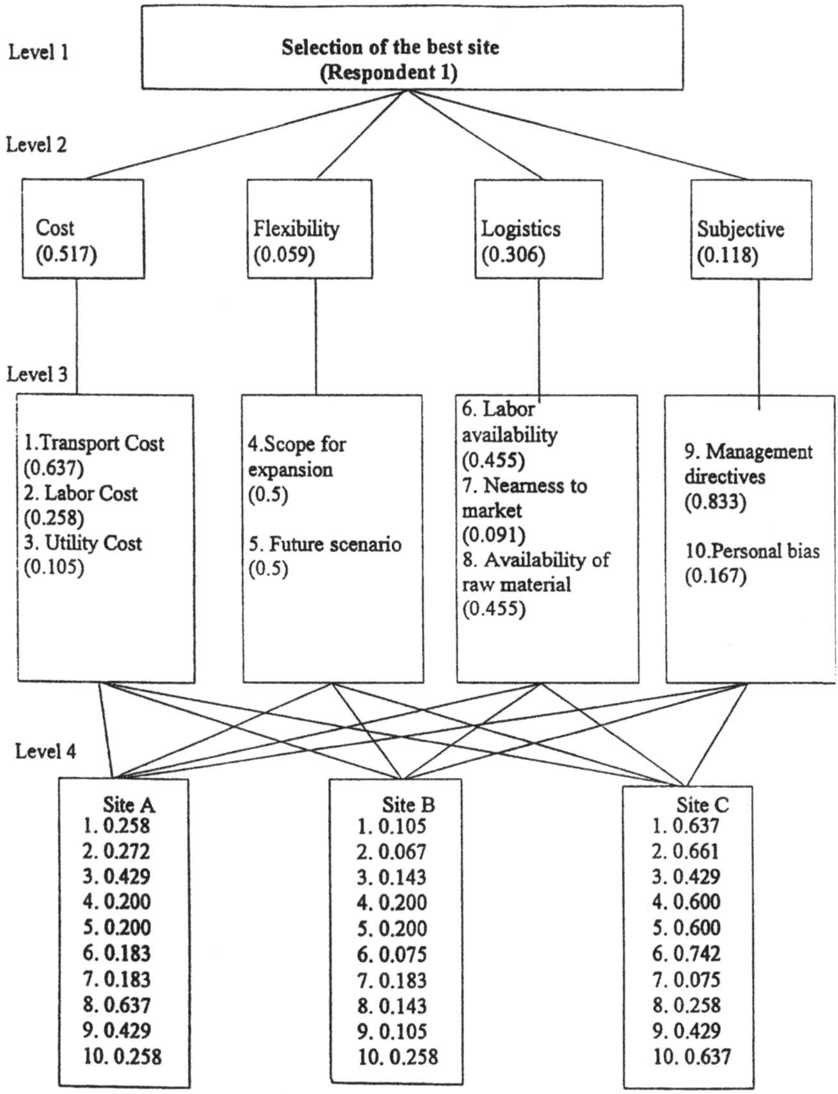

FIGURE 2.4 Weight distribution of respondent 1.

So far, we have determined the weight distribution for three sites by the first respondent. Because we have five evaluators, similar calculations must be performed on all five respondents. Suppose Table 2.30 displaying each respondents weight for each site.

TABLE 2.30 Overall Priorities for Location of Alternatives for all Five Respondents

Alternative	Respondent 1	Respondent 2	Respondent 3	Respondent 4	Respondent 5
A	0.322	0.298	0.282	0.246	0.280
B	0.114	0.207	0.124	0.233	0.209
C	0.553	0.495	0.594	0.521	0.511

The overall rating of an alternative gives us the respective ranking. This can be done by taking a simple arithmetic mean. The alternative with the highest overall priority is ranked first and so on. These rankings are shown in Table 2.31.

2.6 COMPARISONS OF RATING, FUZZY LOGIC, AND AHP

The Rating Method

The rating method is the simplest and provides the crudest approach to solve the facility location issue. Following are the advantages and disadvantages of this method:

Advantages

- Simplicity of application, is the major advantage of the rating method. The rating procedure requires numerical-rating values as an input from the evaluators and then a simple sum or arithmetic mean is all that is needed for assigning the final ranking. There are no particular restriction on weight and ratings assignments to various criteria and theoretically each evaluator may subjectively assign different weight and rating.
- There is no training required to perform the decision-making process. One can simply use personal judgment and the knowledge of the problem to come to a conclusion.

TABLE 2.31 The Final Global Priorities and Rankings of the Sites

Alternative	Arithmetic mean	Rankings
A	0.286	2
B	0.178	3
C	0.535	1

Disadvantages

- The rating method is very basic and subjective and can resolve only those problems that are direct and simple.
- If there are too many subjective factors in the decision process, the method may not accurately assess the interactions between these factors.
- In a group decision making where more than one decision maker is involved, each rating may vary considerably and the method does not provide any way of checking the inconsistency of these ratings.
- On occasion, there might be some outliners in ranking, and such judgment might be critical. In this method, there is no means to incorporate or evaluate these outliners because only the mean value is of importance.

Fuzzy Set Theory

Fuzzy set theory is capable of handling situations that involve decision criteria that are of an intangible nature. The concepts of fuzzy numbers and logistic variables are used to evaluate the objective and subjective factors in such a manner that the viewpoints of an entire decision-making body can be expressed without any constraints. The fuzzy set of decision algorithm used for site selection can also be computerized to make the implementation a little easier.

Advantages

- There is no limit to the number of factors that can be considered in a problem because the number of factors does not greatly effect the complexity of analysis.
- There is a large selection of everyday words to express the exact rating feeling so it gives the evaluator a large range of scales from which to choose.
- The concept of membership sets allows a large amount of information to be represented in a single construct for logical or algebraic manipulation.

Disadvantages

- The method does not provide a means of detecting inconsistency in rating.
- If there are multiple decision makers, the individual inconsistencies contribute toward the total inconsistency, which is the geometric function of individual consistencies. The application of fuzzy set theory to a problem, however simple, requires a complete knowledge of the theory

itself. This overshadows the capability of the process to solve complex problems, because an evaluator or decision maker must be trained to use this method.

- If the decision makers or the evaluator is to give only the rankings and weights to various entries, an expert is required to implement these input values to calculate the final rankings of the alternatives or to obtain the ultimate solution.
- Once the algorithm for the fuzzy set problem has been developed, it can be computerized. The input values can be obtained and the computer code can be utilized for obtaining the desired results. This is sometimes very expensive.

The Analytic Hierarchy Process

The analytic hierarchy process has been successfully used in several areas, and it has been well established as one of the most powerful and capable decision-making tools available today.

Advantages

- The AHP breaks down the problem into hierarchy, and this helps in simplifying the problem to a great extent.
- The various levels of hierarchy can be structured to suit a problem and because there are no rigid rules for doing so, this enables a decision maker to address problems in many domains.
- The analytic hierarchy process involves pairwise comparisons of the decision elements. This helps the evaluator focus solely on the two factors that are under consideration; therefore, the judgments that are made are usually objective and consistent.
- The pairwise comparisons are done for decision elements at one level relative to each decision element at the adjacent higher level, separately. When an evaluator is making the judgments, he or she takes into consideration the relative importance or weights of the two factors being compared, only in the view of the decision element they are being evaluated against. This avoids the unnecessary speculation of what global influence these factors might have on the ultimate decision, and the decision maker can assign the corresponding weights without being intimidated.
- The pairwise comparison matrices require only one-half of the judgment values as input data. The other half are the reciprocal of these inputs. This reduces the work of an evaluator.
- An inherent advantage of the AHP is its flexibility and ability to adjust and accommodate any problems. If, for any reason, there are changes in

the parameters or factors of the problem, AHP can accommodate those changes with considerable ease.

- The subjective factors influencing a decision can be blended with the objective ones.
- The initial decision comparisons of the factors, being made in comparing two factors at a time is not very difficult and does not require much training by the evaluator.
- There are software programs available that can assist the evaluators in performing the mathematical operations needed—that is, calculations of eigenvalues and eigenvectors.
- A major advantage of the analytic hierarchy process is its ability to highlight the inconsistencies in the rating procedure, if there are any.
- The pairwise comparison process forces the evaluator to think about comparisons that might otherwise have been thought of as insignificant.

Disadvantages

- Implementation of the procedure to obtain the final ranking is tedious and time-consuming. Checking for consistency at each stage requires considerable effort if the software for determining eigenvalues is not available.
- The nine-point rating scale does not always translate well the marginal differences of importance. Several other rating scales have been proposed that may be more applicable.

Exercises

2.1 The 6th grade soccer team at Ruston Junior High is trying to decide what T-shirt company they want to buy their new soccer shirts from. The coach and two soccer moms have made a list of factors to consider when choosing a company. In the following table, their opinions about the importance of these factors are shown. From this information, determine the four-point weight distribution for each of the subjective factors based on the linguistic weights from Table 2.1.

	Coach (D1)	Mom 1 (D2)	Mom 2 (D3)
Availability of size (C1)	M	VL	L
Availability of colors (C2)	H	M	VH
Delivery time (C3)	VL	L	M
Design features (C4)	L	H	VH

2.2 The decision makers in problem number 1 must decide between two T-shirt companies. They have evaluated both companies on the criteria from problem 1. From the linguistic ratings from Table 2.3 and relative to objective costs (in terms of the expected cost to the school), the decision is as follows:

Alternatives	Expected cost to the school
A1 T-Shirts Plus	(300, 325, 350, 400)
A2 T's-R-Us	(350, 375, 400, 425)

Criteria 1	Coach (D1)	Mom 1 (D2)	Mom 2(D3)
A1 T-Shirts Plus	G	BP and F	BF and G
A2 T's-R-Us	VG	BP and F	BG and VG

Criteria 2	Coach (D1)	Mom 1 (D2)	Mom 2 (D3)
A1 T-Shirts Plus	BP and F	G	G
A2 T's-R-Us	P	BF and G	F

Criteria 3	Coach (D1)	Mom 1 (D2)	Mom 2 (D3)
A1 T-Shirts Plus	VP	P	VG
A2 T's-R-Us	BP and F	BF and G	G

Criteria 4	Coach (D1)	Mom 1 (D2)	Mom 2 (D3)
A1 T-Shirts Plus	VP	VG	G
A2 T's-R-Us	F	BF and G	P

Determine whether A1 or A2 should be selected as the supplier.

2.3 A manufacturing firm requires new machines to improve its production schedule. Three alternative machines have been decided on. The firm wants to select a machine among the alternatives based on the following objective and subjective criteria:

1. Quality: measured in defectives/million	Objective in type
2. Cost of M/c	Objective in type
3. Capacity: machining time per piece	Objective in type
4. Space occupied	Objective in type
5. Ergonomics	Subjective

6. Aesthetics Subjective
7. Flexibility with different product designs Subjective
8. Degree of unattended operation Subjective

The decision has to be made by departments of Design, Manufacturing, Finance, and Industrial Engineering and the labor union. The linguistic weights assigned for these criteria by personnel from various departments is given in the following chart.

Criteria	Design	Manufacturing	Finance	Ind eng	Labor union
Quality	VH	M	VH	VH	M
Cost	M	M	VH	VH	L
Capacity	M	VH	VH	VH	L
Space	M	VH	VH	VH	L
Ergonomics	M	M	M	VH	VH
Aesthetics	M	VH	M	M	M
Flexibility	VH	VH	VH	VH	L
Independence	M	M	H	VH	L

The individual ratings for the criteria given by these departments are given in the following tables:

Criteria 5

Machine	Design	Manufacturing	Finance	End eng	Labor union
Machine 1	VH	H	H	VH	VH
Machine 2	H	H	H	H	H
Machine 3	H	H	M	M	M

Criteria 6

Machine	Design	Manufacfturing	Finance	End eng	Labor union
Machine 1	H	H	M	H	M
Machine 2	M	M	H	M	M
Machine 3	M	H	M	L	M

Criteria 7

Machine	Design	Manufacturing	Finance	Ind eng	Labor union
Machine 1	H	H	M	H	M
Machine 2	M	H	M	M	L
Machine 3	M	H	H	VH	M

Criteria 8

Machine	Design	Manufacturing	Finance	Ind eng	Labor union
Machine 1	VH	VH	H	H	VH
Machine 2	H	M	M	H	M
Machine 3	L	M	H	H	M

Objective Criteria

Criteria 1 Cost of bad quality is directly proportional to number of defectives per million.

Machine	Defects/million
Machine 1	28, 30, 40, 42
Machine 2	18, 20, 20, 23
Machine 3	30, 26, 36, 40

If a machine produces more defectives products per million, it should have less likelihood of being selected. Hence, it is required that we rescale the criteria results as follows: The rescaling is done by dividing 100 by the defective rates. We obtain the relative total cost as follows:

Machine 1 (3.57, 3.33, 2.50, 2.38)
Machine 2 (5.55, 5.00, 5.00, 4.35)
Machine 3 (3.33, 2.77, 2.77, 2.50)

Criteria C2 Cost of machine is known in thousands of dollars.

Machine 1 (10.0, 8.33, 7.14, 6.25)
Machine 2 (16.0, 5.55, 5.55, 5.0)
Machine 3 (14, 14, 14, 14)

The relative total cost then would be

Machine 1 (10.0, 10.0, 10.0, 10.0)
Machine 2 (6.25, 6.25, 6.25, 6.25)
Machine 3 (7.14, 7.14, 7.14, 7.14)

Criteria C3 Machining time is directly proportional to cost of product

Machine 1 (10, 12, 12, 14)
Machine 2 (8, 9, 9, 10)
Machine 3 (8, 10, 10, 12)

The relative cost index for this is

Machine 1 (10, 8.33, 8.33, 7.14)

Machine 2 (12.5, 12.5, 12.5, 10)
Machine 3 (12.5, 10. 10, 8.3)

Criteria C4 Space occupied is directly proportional to cost of product

Machine	Space occupied in square meters
Machine 1	(12, 14, 14, 16)
Machine 2	(16, 18, 18, 20)
Machine 3	(8, 10, 10, 12)

The relative total cost indices for this are as follows:

Machine 1 (8.33, 7.14, 7.14, 6.25)
Machine 2 (6.25, 5.55, 5.55, 5.0)
Machine 3 (12.5, 10.0, 10.0, 8.33)

Decide which machine the firm should select. Refer to Tables 2.1 and 2.3 for linguistic ratings.

2.4 Which factor in the following matrix accounts for more costs to the company and by what degree?

Company costs	*Capital cost*	*Training cost*	*Operating cost*
Capital cost	1	5	7
Training cost	1/5	1	4
Operating cost	1/7	1/4	1

Is the matrix consistent? Explain your answer.

2.5 Discuss two situations besides facility location where fuzzy logic or AHP may be applied.

2.6 Pairwise comparison of three factors is given in the matrix. Determine the relative weight of each factor.

	Cost	*Flexibility*	*Logistics*
Cost	1	3	2
Flexibility	1/3	1	3
Logistics	1/2	1/3	1

2.7 Determine if the following matrix is consistent?

	A	B	C
A	1	3	2
B	1/3	1	4
C	1/2	1/4	1

Using the largest column total method, what is the weight for each factor?

REFERENCE

Saaty, TL. The Analytical Hierarchy Process. New York: McGraw-Hill, 1980.

Part II

Basic Quantitative Models

3

Single-Facility Minisum Location

3.1 INTRODUCTION

In the previous chapters we discussed the qualitative techniques for site selection. A single best site was selected from among several possible locations. These qualitative techniques, however, are largely influenced by subjective and personal opinions and, therefore, may not always result in an optimum site selection relative to a single dominating quantifiable objective. In this and the following chapters we will study different quantitative methods. Those methods select the best location based on optimization of a mathematically formed objective that describes quantifiable costs of the problem.

A class of problems discussed in this chapter is one for which we are not restricted to just a few locations, but are free to choose any site for the new facility. It is called the continuous or universal facility location problem. This method may be used to determine the ideal location when the sole consideration is the cost. The term *ideal* is used to stress that on such a site, construction or placement of a facility may or may not be physically possible. It does, however, guide us to the location(s) that would result in the minimum costs, and if it should prove infeasible to utilize such a location, the "penalty" incurred by selecting an alternative location is better understood.

In this chapter we study a few variations of the single-facility location problem, based on the way the travel cost is measured. We assume that the demand points (or the customers), their x–y coordinates, and the necessary number or trips from each customer to the new facility are known. The cost is measured in terms of the number of trips multiplied by the distance a customer has to travel to reach the facility. For example, in manufacturing plants, the travel of men and material is on aisles; therefore, the distance is measured in rectilinear fashion. In the event that direct travel by the shortest path is possible, the cost can be measured as a function of direct distance. For example, it can be linearly proportional, called euclidean distance, or proportional in quadratic fashion. Figure 3.1, illustrates some examples of rectilinear and euclidean measured distances.

Quadratic cost implies that a customer farther away contributes much more to the cost than the one nearby. This is appropriate when one also desires to implicitly include other parameters, such as time, in the cost. For example, for placing a public facility, such as a hospital, within a community, it might be important that the hospital not be very far from any center of population it is expected to serve: timely treatment of emergency cases being a critical criterion.

Both rectilinear and euclidean costs, on the other hand, maintain the direct, proportionality of the cost with distance and are appropriate when only the travel distance and not time, is worthy of attention. Some examples are the following:

1. Location of a tool crib in a manufacturing plant: In a plant, movements are allowed only on the aisles, which are positioned in a grid fashion; therefore, rectilinear distances between the new location for the tool crib and existing machines within the plant may be appropriate to measure the cost of travel.
2. Location of sewer treatment plant: Generally, it is desirable to connect a treatment plant directly with storage tanks by pipe lines, if, physically possible. The euclidean distance model is then appropriate to determine the cost of the pipe line.
3. Other calculation of cost may include a power p of the distance, where exponent p has some positive value. For, example, a long pipe, line carrying gas may require frequent booster pumps, making the cost proportional to a power of the length of the pipeline between two points. In another example, in going from point A to B in a city, many persons prefer freeway than traveling through a downtown, even though a downtown route may be shorter. This preference for other factors beside distance can be included in the model by raising the cost

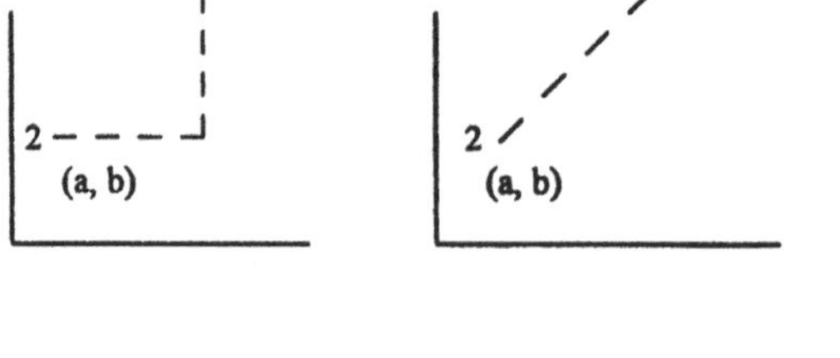

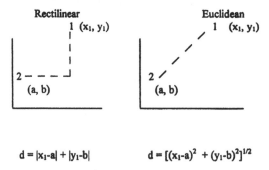

$d = |x_1 - a| + |y_1 - b|$ $\qquad$ $d = [(x_1 - a)^2 + (y_1 - b)^2]^{1/2}$

FIGURE 3.1 Rectilinear and euclidean distance.

of downtown route by a power larger than that associated with the freeway.

3.2 RECTILINEAR DISTANCE: SINGLE-FACILITY LOCATIONS

Suppose we want to install a new machine in a plant that will process output of two existing work centers. For the sake of simplicity, in this example, assume that these work centers are on the same aisle, the first being 10 ft from the entrance door and the other being 450 ft from the same door as shown in Figure 3.2. The number of units transported from these work centers to the new machine are 1000 and 5000, respectively. It is required that we find the optimum location for the new machine.

The first thought that comes to mind is why not locate the new facility at the center of gravity. After all the center of gravity should equal or balance the costs from A and B and, therefore, should perhaps also lead to the best location. Suppose we follow our intuition and decide to minimize the weighted cost of travel by locating the new machine at the center of gravity of the existing facilities. To find the center of gravity, (CG) we take the movements along the door, which results in:

$$CG = \frac{1000^{*}10 + 5000^{*}450}{1000 + 5000} = 376.66$$

This is called center of gravity, or gravity (CG), solution for short. If the new machine is placed in this location the total cost, of travel is

$$(376.66 - 10)^{*}1000 + (450 - 376.66)^{*}5000 = 733.360$$

Let us investigate further and determine what would have been the cost if we had placed the new machine at some location other than the CG. Table 3.1 shows the points investigated and associated costs.

It is obvious that placing the new machine at location 450, the existing location of work center 2, gives the minimum cost and not the location calculated using center of gravity.

```
x-----------A---------------------------------------------------------------------B

Door

<---10'---->

<---------------------------------------------450'-------------------------------------------->
```

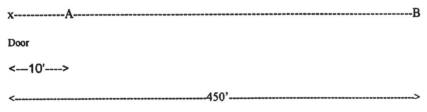

FIGURE 3.2 Location of existing facilities.

TABLE 3.1 Assorted Locations for New Facility and Associated Costs (rounded to closest 000).

Location of new facility	10	100	200	300	400	450	CG
Cost in 0000	220	184	144	104	64	44	73

Suppose we expand the problem by considering a third existing work center, located 200 ft from the door, that may use the new machine for 6000 parts. The cost at different locations for the new facilities are as shown in Table 3.2.

The minimum cost location again is not the center of gravity solution, but the location at 200, the present location of work center 2 (or the existing facility 2).

A major observation that can be made from these calculations is that whenever demand from one work center is equal to or greater than the sum of demands from other work centers, the new facility location coincides with the existing facility with the largest demand. This observation is called majority theorem (Witzgall, 1964). It is valid for the cost calculations that are based on linear distances (l), and furthermore, it can also be shown that the same observation is also valid if the costs are a function of some power (p) of distances (l^p).

What if a demand from one source does not dominate the demands from other sources? Let us examine the three work centers problem again, except change the demand from source 3 from 6000 to 5500 parts. The costs are now as shown in Table 3.3.

Again we see that the optimum location for the facility is not the center of gravity location, but at a location of one of the existing facilities. However, because there is no dominating demand, it is not possible to predict which existing facility location will provide the minimum cost solution. So, to obtain the optimum location for the new facility one needs to examine only the locations of existing facilities.

TABLE 3.2 Costs Associated with the New Setup (rounded to 000)

Location of the new facility	10	100	200	288.33(cg)	300	400	450
Cost in 0000	334	244	144	161.1	164	184	194

TABLE 3.3 Costs with New Demands

Location of the new facility	10	100	200	292.17(cg)	300	400	450
Cost in 0000	269.5	239	144	157.8	159	174	330.5

However, rather than checking the cost associated with each location, a shortcut could be devised that gives the same results. This method is call 50th percentile method. The major advantage of the method is that it reduces the calculations considerably.

50th-Percentile Method

List the existing facilities in order of increasing values of their abscissa, and the corresponding trips. The optimum x coordinate for the new facility is the one associated with the 50th percentile, or the median value of the cumulative number of trips. It should be understood that the trips originate from discrete points and, therefore, the median is associated with one of those points. For example, if the cumulative trips are as shown in Table 3.4, the median is the 30th trip, which is made by customer 3.

We leave it to the reader to check if applying the foregoing procedure leads to the same optimum locations for the new facility in each of the three examples, illustrated earlier.

Two-Dimensional Problem

Now consider a two-dimensional problem for which each location i, has a_i and b_i coordinates. Let w_i be the demand from location i. There are n existing locations from which the demand for the new machine may originate. Let (x,y) be the location of new facility. The cost of transportation, which we want to minimize, is directly proportional to the distance traveled and is given by:

$$d = \sum w_i[|(x - a_i)| + |(y - b_i)|]$$

The absolute value associated with each term assures us that the distance is always positive. The function is linear; hence, it is not possible to obtain the optimum by applying the usual method from calculus of taking the first derivative, setting it equal to zero, and solving the resulting expression.

The function can be optimized by following a different approach. To demonstrate this concept, suppose n is equal to 2 (i.e., there are only two

TABLE 3.4 Cumulative Trips

Customer	Abscissa	No. of trips	Cumulative trips
1	13	10	10
2	21	15	25
3	24	35	60

customers). They are located at (a_1, b_1) and (a_2, b_2), respectively, and are making w_1 and w_2 trips to the new facility. The cost of travel is given by:

$$w_1|(x - a_1)| + w_1|(y - b_1)| + w_2|(x - a_2)| + w_2|(y - b_2)|$$

the value of which is to be minimized. This expression can be separated into two parts:

$$\text{min: } w_1|(x - a_1)| + w_2|(x - a_2)|,$$

and

$$\text{min: } w_1|(y_1 - b_1)| + w_2|(y - b_2)|$$

These are two one-dimensional problems.

We have seen that in one-dimensional problems the optimum location for the new facility is one of the existing facility's location. Thus, the x coordinate for the new facility is the x coordinate of one of the existing facility, and the y coordinate is also the y coordinate of one of the existing facility. These two coordinates, however, may not be associated with the same existing facility. Indeed, as a result, the location of the new facility would be on one of the points that are obtained on the intersections of the lines drawn horizontally and vertically from the existing facilities.

For illustration of this optimization procedure, suppose the introduction of a new product will require a new molding machine, along with six existing machines, and the parts from the new molding machine need to be transported to these machines in the plant. The coordinates of the existing machines (a, b) and the number of trips that would be made each day from the new machine to the existing machines (w) are given in Table 3.5. The problem is to determine the appropriate location for the new molding machine.

To determine the best abscissa x for the new machine, list the existing machines (facilities) in order of increasing values of the abscissas, the corresponding trips. Also determine the cumulative trips value for each machine as listed in the table. The optimum x coordinate is the one associated with the 50th

TABLE 3.5 Coordinates of Existing Machines

Machine	Coordinates (a, b)	Trips/day(w)
1	20, 46	20
2	15, 28	15
3	26, 35	30
4	50, 20	18
5	45, 15	20
6	1, 6	15

TABLE 3.6 x-Axis (Abscissa) Determination

Machine	Abscissa	No. of trips	Cumulative trips
6	1	15	15
2	15	15	30
1	20	20	50
3	26	30	80
5	45	20	100
4	50	18	118

percentile, or the median value of the cumulative trips. Table 3.6 lists the machine arrangement based on the abscissa values.

The 50th percentile would be the 59th trip (118/2). That trip occurs at machine 3, with an abscissa of 26. Therefore, the optimum x-axis value for the new molding machine is 26.

To obtain the optimum y coordinate, arrange the machines in ascending order of ordinate values as shown in Table 3.7. The 50th-percentile trip (i.e., the 59th trip), occurs with machine 2, the ordinate of which is 28. This becomes the optimum y axis value of the new facility. Therefore, the location for the new molding machine is (26, 28). The optimum cost is 2776.

Now, suppose the demand from machine 1 is 42. The best location would then be (26, 35). Because a facility already exists in this location, it cannot be used for the new facility. Therefore, we want to find a nearby location that is usable, knowing it will have a higher cost than the optimum location.

To do this, we construct the contour sets. The associated contour line has the property that every point on that contour line has the same value of the objective function. The cost function, if the new facility is located at (x, y), is the sum of cost from each existing machine to the new facility in the x and y direction. The cost of travel in x direction for machine i is

$$(\#\text{ of trips for machine } i)(|x - \text{abscissa of machine } i|)$$

TABLE 3.7 y-Axis (Ordinate) Determination

Machine	Ordinate	No. of Trips	Cumulative Trips
6	6	15	15
5	15	20	35
4	20	18	53
2	28	15	68
3	35	30	98
1	46	20	118

If $F(x)$ and $F(y)$ denote the total cost in x and y direction travel, respectively, then in our case,

$$F(x) = 15|x-1| + 15|x-15| + 20|x-20| + 30|x-26| + 20|x-45| + 18|x-50|$$

and

$$F(y) = 15|y-6| + 15|y-28| + 20|y-46| + 30|y-35| + 20|y-15| + 18|y-20|$$

In each expression, the coordinates of the existing facilities are sorted in the increasing order.

The function $F(x)$ is linear between adjacent points and the slope increases from one interval of adjacent points to the next interval of adjacent points. To compute the slope in any interval requires that positive and negative slopes should the properly identified (we want to eliminate the need for absolute value). For example, consider the interval A (20, 26), that is, $20 < x < 26$, for any x between this range we have,

$$|x-1| = x-1,\ |x-15| = x-15,\ |x-20| = x-20,$$
$$|x-26| = 26-x,\ |x-45| = 45-x,\ |x-50| = 50-x$$

Substituting these values in the cost expression, we obtain for $20 < x < 26$,

$$F(x) = (15)(x-1) + (15)(x-15) + (20)(x-20) + (30)(26-x) + (20)(45-x) + (18)(50-x)$$

or

$$F(x) = (-18)x + 1940$$

Note that the slope of x, (-18), is the sum of weights of xs below the lower limit of the range of x (below and including 20) minus the sum of slopes above the upper limit of x (above and including 26). The total weight below x equal to 20 is $(15 + 15 + 20) = 50$, and the total weight above x equal to 26 is $(30 + 20 + 18) = 68$, giving -18.
These values are noted in Figure 3.3 as interval coefficients.

Similarly for all values of y in the interval (35,46), we obtain

$$|y-6| = y-6,\ |y-15| = y-15,\ |y-20| = y-20,$$
$$|y-28| = y-28,\ |y-35| = y-35,\ |y-46| = 46-y$$

or

$$F(y) = (78)y - 1300$$

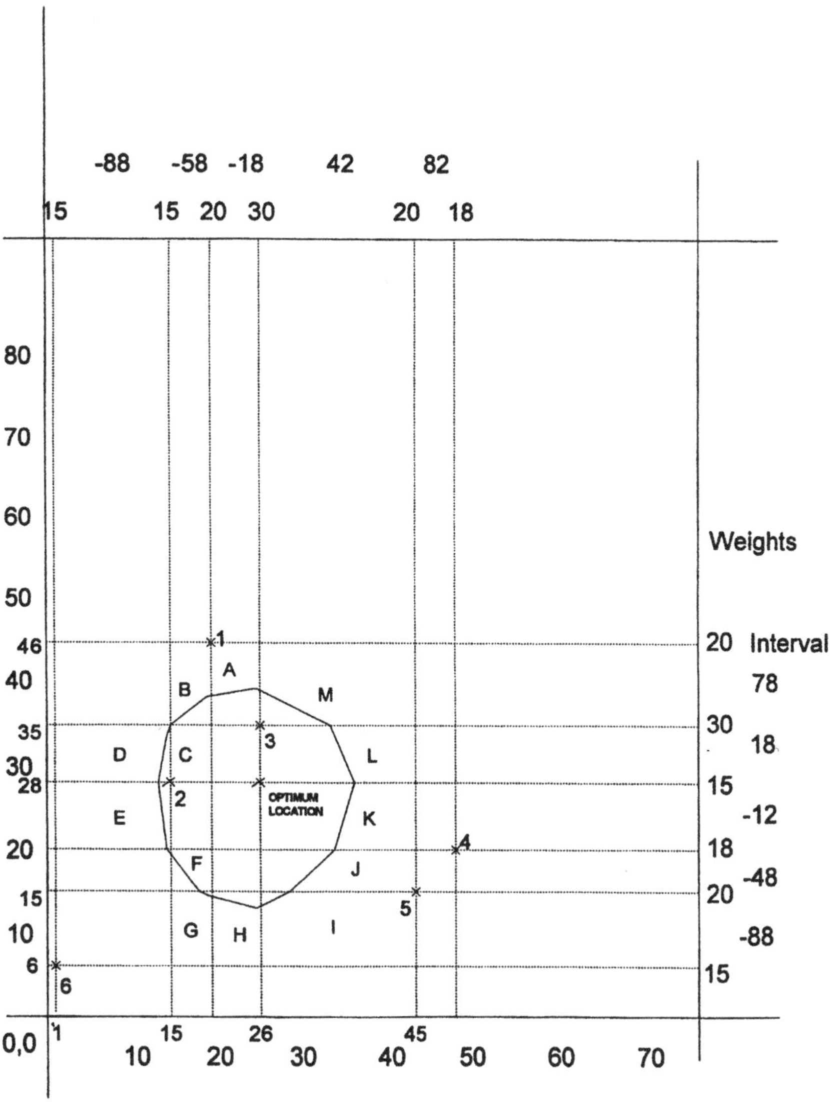

FIGURE 3.3 Contour for $3276.

Again, the same rule can be applied to obtain the slope of y (78), the interval coefficient.

Thus, we can calculate the slopes of $F(x)$ and $F(y)$ for different intervals of x and y and obtain the interval coefficients. This is easily done by using the following procedure.

Plot the existing locations and draw vertical and horizontal line through each location and attach the associated facility weight to that line. For example, Figure 3.3, the vertical and horizontal lines drawn from machine 1 has associated weight of 20 attached to them (both on x andy y axes). This creates several rectangles, each enclosed by two vertical parallel lines drawn through two adjacent facilities on the x axis and the other two horizontal parallel lines drawn through two adjacent facilities on the y axis. For example, consider a block A in this case formed by facilities 1 and 3 (these two are adjacent facilities in both the x and y direction). To calculate the slope of the objective function in rectangle A, simply take the ratio of the associated interval coefficients, -18 for x and 78 for y.

For any point P in this box we can verify that

$$F(x, y) = F(x) + F(y) = (-18x + 1940) + (78y - 1300) = k' \qquad (1)$$

Now we are ready to draw the contour lines. Start with a value of the objective function k' for which we want to draw a contour. This value must be greater than the optimum solution value. For a given value of x, determine the corresponding y value from expression such as Eq. (1) for a rectangle associated with these values of x and y. Draw a line passing through this point with the slope associated with that of the rectangle. At the intersection of the rectangle boundary, determine the slope of the contour in the next rectangle and draw a line with that slope passing through the intersection point. Continue the procedure till we obtain the close loop. This is the contour with the selected value of the objective function.

For drawing a contour line that has a value of 500 more than the optimum (which is 3276 in this case), let us start with some block, say block A.

For block *A*

x coordinates are (20 and 26)
y coordinates are (35 and 46)
$F(x) = -18x + 1940$
$F(y) = 78y - 1300$

The contour line is for 3276 + 500 or

$$F(x, y) = -18x + 78y + 640 = 3276 + 500$$
$$78y = 18x + 3136$$

For some value of x in the box (for convenience we will take the boundary point), say $x = 26$,

$$y = 39.8$$
$$\text{Slope} = (18/78)$$

Draw the line through point (26, 39.8) with the slope of 18/78.

Continue on to block B. From the point at which the contour line first intersects the boundary of block B, draw a line with the slope of 58/78. Continue to block C. The process is repeated and displayed in the following with first description of the block boundaries and then indicating the slope of the contour line in that box, obtained from the graph.

Block *B*

x coordinates: (15 and 20)
y coordinates: (35 and 46)
Slope = (58/78)

Block *D*

x coordinates: (01 and 15)
y coordinates: (28 and 35)
Slope = (88/18)

Block *E*

x coordinates: (01 and 15)
y coordinates: (20 and 28)
Slope = −(88/12)

Block *F*

x coordinates: (15 and 20)
y coordinates: (15 and 20)
Slope = −(58/48)

Block *G*

x coordinates: (15 and 20)
y coordinates: (06 and 15)
Slope = −(58/88)

Block *H*

x coordinates: (20 and 26)
y coordinates: (06 and 15)
Slope = −(18/ 88)

Block *I*

x coordinates: (26 and 45)
y coordinates: (06 and 15)
Slope = (42/ 88)

Block *J*

x coordinates: (26 and 45)
y coordinates: (15 and 20)
Slope = (42/ 48)

Block *K*

x coordinates: (26 and 45)
y coordinates: (20 and 28)
Slope = (42/12)

Block *L*

x coordinates: (26 and 45)
y coordinates: (28 and 35)
Slope = −(42/18)

Block *M*

x coordinates: (26 and 45)
y coordinates: (35 and 46)
Slope = −(42/78)

The contour plot is as shown in Figure 3.4. Contours with different objective values give us some insight into the problem. These can be drawn for different objective values, by just drawing parallel lines to the existing lines, as long as the new contour is passing through the same blocks. Recall that the optimum location for the new facility in our problem is the location of facility 3. If we have to deviate front that point, then it seems like we have much more lead way to go down on the vertical scale than to go up. Similarly, perhaps a location to the right of 3 on the x axis has somewhat more lead way than to the left.

The cost of placing the facility in any other location increases as shown by the counter lines. The inner most lines have the smallest values, and as we progress outward, the cost will increase. Hence, in determining the possible locations for a new facility we start with the innermost contours. If no suitable location can be found then move outward and so on.

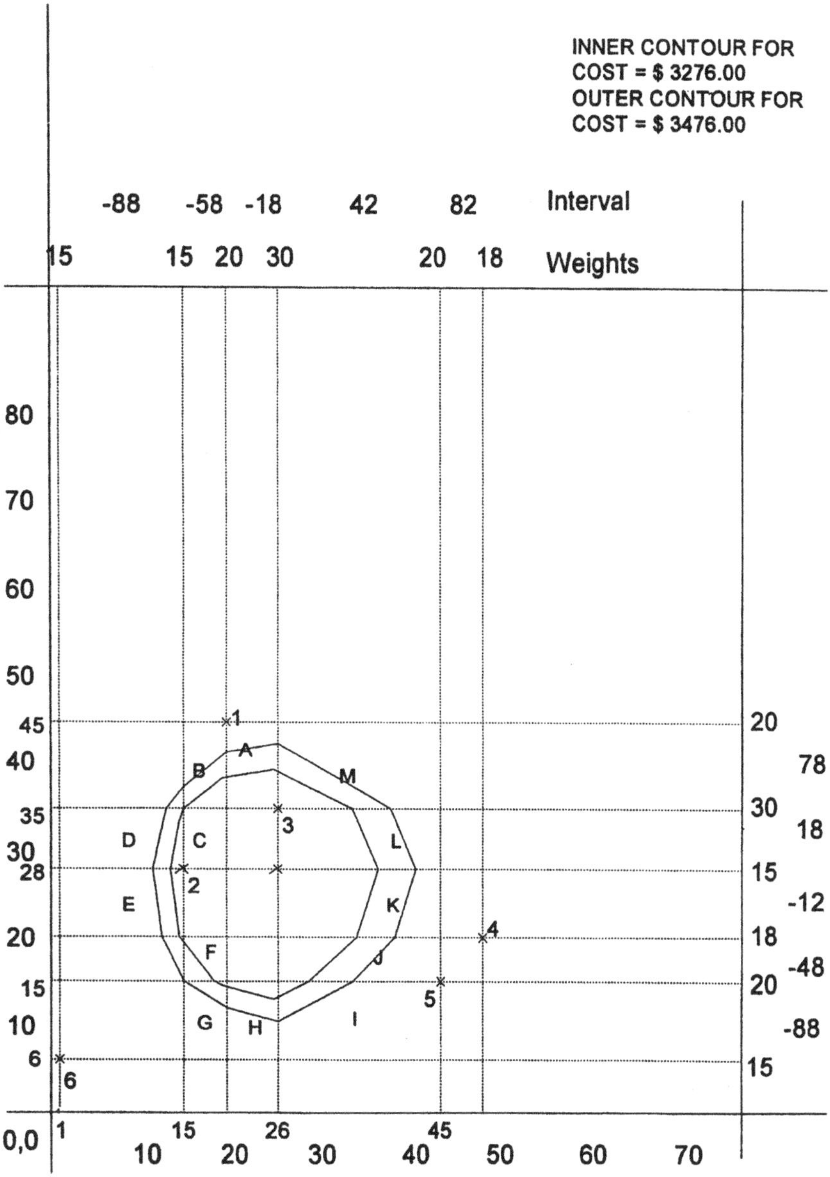

FIGURE 3.4 Multiple contours.

Three-Dimensional Problem

Now suppose we have a problem in three dimensions; that is, height is also an issue because the facility can be installed in a multistory building. To find the optimum location apply the same 50th percentile procedure that we have seen in the two-dimensional problem. The following example illustrates the point.

There are five departments of the college of engineering and science located in the university building. These departments would like to set up a coffee lounge so that it services the faculty of all the departments. The location of the departments is as shown in Table 3.8. Depending on the number of faculty, there is a weight associated with every department representing the usage of the coffee lounge. *Z* dimension indicates, the first or second floor of the building.

We use the method of the 50th-percentile. Because there are *x*, *y* and *z* directions we need to use the procedure for each direction separately.

For the *x* location, arranging, the *x* coordinates in the nondescending order, as shown in Table 3.9 (if coordinate is the same, the order is chosen in random). Next, divide the final cumulative weight by 2 to find the 50th percentile. The associated *x* coordinate is 0 associated with location 5. Perform similar calculations on the *y* and the *z* directions (Tables 3.10 and 3.11).

TABLE 3.8 Data for the Example

No.	Location	Weight	Location
1	1	4	(0, 1, 0)
2	2	2	(1, 0, 1)
3	3	1	(0, 0, 1)
4	4	3	(1, 1, 0)
5	5	2	(0, 1, 1)

TABLE 3.9 *x*-Axis (Abscissa) Determination

No.	Location	*x* Coord	Weight	Cumulative wt.
1	1	0	4	4
2	3	0	1	5
3	5	0	2	7
4	2	1	2	9
5	4	1	3	12

TABLE 3.10 y-Axis (Ordinate) Determination

No.	Location	y Coord	Weight	Cumulative wt.
1	2	0	2	2
2	3	0	1	3
3	1	1	4	7
4	4	1	3	10
5	5	1	2	12

TABLE 3.11 z-Axis (Ordinate) Determination

No.	Location	z Coord	Weight	Cumulative wt.
1	1	0	4	4
2	4	0	3	7
3	2	1	2	9
4	3	1	1	10
5	5	1	2	12

Half of the cumulative weight is 6. The location corresponding to the weight of 6 is 1 with the y coordinate of 1.

In the z direction we find the location is 4 with the z coordinate of 0.

Hence, the optimum location of the coffee lounge is (0, 1, 0). The cost of locating this new facility at (0, 1, 0) is

$$\text{Cost} = \Sigma w_o[|x - a_i| + |y - b_i| + |z - c_i|]$$

w_i: weight assigned to existing facility
z, y, z: coordinates of the new facility
a_i, b_I, c_I: coordinates of existing location

For the new coordinates the cost is

$$\begin{aligned}\text{Cost} = {} & 4\{|0-0| + |1-1| + |0-0|\} + 2\{|0-1| + |1-0| + |0-1|\} \\ & + 1\{|0-0| + |1-0| + |0-1|\} + 3\{|1-0| + |1-1| + |0-0|\} \\ & + 2\{|0-0| + |1-1| + |0-1|\}\end{aligned}$$

$$\text{Cost} = 13.$$

As a check, some other existing coordinate locations as possible sites for the new facility are evaluated. The results are shown in Table 3.12.

Each cost is higher. The least cost is obtained when we locate the new facility at the location (0, 1, 0) with a cost of 13 units.

TABLE 3.12 Possible Locations for New Facility

No.	New site	Cost
1	(1, 0, 1)	23
2	(0, 0, 1)	21
3	(1, 1, 0)	15
4	(0, 1, 1)	15

3.3 SINGLE FACILITY LOCATION: QUADRATIC COST

Suppose the cost is proportional to the square of the distance traveled and not linearly proportional, as in the previous case. Now the objective is to minimize the function:

$$\mathrm{d} = \Sigma w_i[(x - a_i)^2 + (y - b_i)^2]$$

By taking partial derivatives relative to x and y, equating them to 0 and solving the resultant expressions leads to the following solution:

$$x = \frac{\Sigma_{i=1}^{n} w_i a_I}{\Sigma_{i=1}^{n} w_I} \qquad \text{and} \qquad y = \frac{\Sigma_{i=1}^{n} w_i b_I}{\Sigma_{i=1}^{n} w_I}$$

The solution is referred to as a centroid or the center-of-gravity solution. Referring to the example in Section 3.2 (Tables 3.6 and 3.7), the centroid solution is:

$$x = \frac{(20*20 + 15*15 + 26*30 + 50*18 + 45*20 + 1*15)}{(20 + 15 + 30 + 18 + 20 + 15)} = 27.3$$

$$y = \frac{(20*46 + 15*28 + 30*35 + 18*20 + 20*15 + 15*6)}{(20 + 15 + 30 + 18 + 20 + 15)} = 26.6$$

Thus, the location for the new molding machine is (27.3, 26.6).

3.4 EUCLIDEAN DISTANCE COST

The euclidean distance between two points T and S with coordinates (x, y) and (a, b), respectively, is defined as

$$\mathrm{d} = [(x - a)^2 + (y - b)^2]^{1/2}$$

When the cost is proportional to euclidean distance, the objective is to minimize the resulting cost expression:

$$\mathrm{d} = \Sigma_{i=1}^{n} w_i[(x - a_i)^2 + (y - b_1)^2]^{1/2}$$

Following a procedure similar to that in Section 3.3, that is, taking partial derivatives relative to x and y and setting them equal to zero, leads to

$$\frac{\delta}{\delta x} = 0 = \Sigma_{i=1}^{n} w_i(x - a_i)/[(x - a_i)^2 + (y - b_i]^{1/2} \quad (1)$$

and

$$\frac{\delta}{\delta y} = 0 = \Sigma_{i=1}^{n} w_i(y - b_i)/[(w - a_i)^2 + (y - b_i)]^{1/2} \quad (2)$$

The optimum location should be obvious if we could solve the resulting expressions. However, if the new machine could be placed, at least mathematically, at the same coordinates as one of the old machines (i.e., for some i, $X = a_i$ and $Y = b_i$) both Eqs. (1) and (2) are undefined, and great difficulty is experienced in trying to obtain a unique solution.

An alternative is to engage in an iterative solution procedure. It can be proved mathematically that the objective function is convex, as shown in Figure 3.5.

One can begin with a centroid solution, and then perform a search relative to each dimension until the optimum solution is obtained. The procedure is easy to perform for small problems.

Another, and perhaps a better, approach to obtain the optimum location is suggested by Kuhn and also by Weiszfeld is known as the modified Kuhn's method or Weiszfeld method. If each of the foregoing derivatives is made equal to zero the subsequent expression would be

$$x^* \Sigma w_i/[(x - a_i)^2 + (y - b_i)^2]^{1/2} - \Sigma(w_i^* a_i)/[(x - a_i)^2 + (y - b_i)^2]^{1/2} = 0$$

Therefore,

$$x^* \Sigma\{w_i/[(x - a_i)^2 + (y - b_i)^2]^{1/2}\} = \Sigma(w_i^* a_i)/[(x - a_i)^2 +)y - b_i)^2]^{1/2}$$

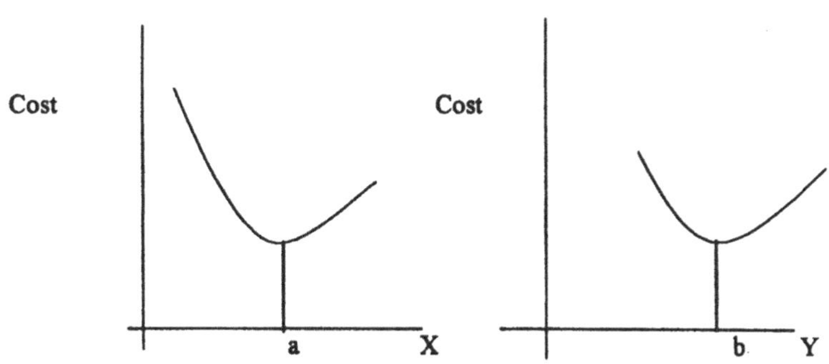

FIGURE 3.5 Typical objective function.

If we define, $\phi_i(x, y) = \{w_i/[(x - a_i)^2 + (y - b_i)^2]^{1/2}\}$

for $I = 1 \cdots n$,

We obtain

$$x = \Sigma a_I \phi_i(x, y)/\Sigma \phi_i(x, y)$$

and similarly

$$y = \Sigma b_I \phi_i(x, y)/\Sigma \phi_i(x, y)$$

We can solve these equations by iterative methods. The generalized equation to perform iterations is,

$$x^{(k)} = \Sigma a_I \phi_i(x^{(k-1)}, y^{(k-1)})/\Sigma \phi_i(x^{(k-1)}, y^{(k-1)})$$

$$y^{(9k)} = \Sigma b_I \phi_i(x^{(k-1)}, y^{(k-1)})/\Sigma \phi_i(x^{(k-1)}, y^{(k-1)})$$

where k denotes the iteration number. This iterative procedure is carried out until there is no appreciable change in x and y values from the previous iteration. The final value gives the least-cost optimum location for placing the new facility.

This procedure is illustrated by an example. Consider the coordinates and their respective weights given in Table 3.5 For convenience the table is reproduced as follows:

Coordinates of Existing Facilities

No.	x Coord (a_I)	y Coord (b_I)	Trips/day (Weight w_i)
1	20	46	20
2	15	28	15
3	26	35	30
4	50	20	18
5	45	15	20
6	1	6	15

To find the optimum location for a new facility using euclidean distance, start by choosing the initial coordinates for the facility (x^*, y^*) by the gravity method:

$$x^* = \Sigma w_i a_i/\Sigma w_i$$

$$y^* = \Sigma w_i b_i/\Sigma w_i$$

For our data,

$$x^0 = \{20{*}20 + 15{*}15 + 26{*}30 + 50{*}18 + 45{*}20 + 1{*}15\}/\{20 + 15 + 30 + 18 + 20 + 15\}$$

$$x^0 = 27.28$$

$$y^0 = \{46{*}20 + 28{*}15 + 35{*}30 + 20{*}18 + 15{*}20 + 6{*}15\}/\{20 + 15 + 30 + 18 + 20 + 15\}$$

$$y^0 = 26.61$$

With these as initial values we substitute in the first iteration,

$$X^{(1)} = \Sigma a_i \phi_i(x^{(0)}, y^{(0)})/\Sigma\phi_i(x^{(0)}, y^{(0)})$$

$$Y^{(1)} = \Sigma b_i \phi_i(x^{(0)}, y^{(0)})/\Sigma\phi_i(x^{(0)}, y^{(0)})$$

and in the second iteration,

$$X^{(2)} = \Sigma a_i \phi_i(x^{(1)}, y^{(1)})/\Sigma\phi_i(x^{(1)}, y^{(1)})$$

$$Y^{(2)} = \Sigma b_i \phi_i(x^{(1)}, y^{(1)})/\Sigma\phi_i(x^{(1), y^{(1)})}$$

and so on.

In the first iteration, we calculate

$$\phi_i(x, y) = \{w_i/[(x - a_i)^2 + (y - b_i)^2]^{1/2}\}\text{for every coordinate}$$

Thus, for the coordinates (20, 46)

$$\begin{aligned}\phi_1(x, y) &= \{w_1/[(x^0 - a_1)^2 + (y^0 - b_1)^2]^{1/2}\}\\ &= 20/[(27.28 - 20)^2 + (26.61 - 46)^2]^{1/2}\\ &= 0.9655\end{aligned}$$

Similarly,

$$\phi_2(x, y) = 1.212$$

$$\phi_3(x, y) = 3.534$$

$$\phi_4(x, y) = 0.760$$

$$\phi_5(x, y) = 0.944$$

$$\phi_6(x, y) = 0.449$$

The new coordinates after the first iteration (x^*, y^*) are given by

$$x^* = X^{(1)} = \Sigma a_i \phi_i(x^{(0)}, t^{(0)}) / \Sigma \phi_i(x^{(0)}, y^{(0)})$$

$$X^{(1)} = \frac{\{20(0.9665) + 15(1.212) + 26(3.534) + 50(0.7609) + 45(0.944) + 1(0.449)\}}{(0.9665 + 1.212 + 3.534 + 0.7609 + 0.944 + 0.449)}$$

$$X^{(1)} = 26.77$$

$$y^* = Y^{(1)} = \Sigma b_i \phi_i(x^{(0)}, y_/^{(0)} \Sigma \phi_i(x^{(0)}, t^{(0)})$$

$$Y^{(1)} = \frac{\{46(0.9665) + 28(1.212) + 35(3.534) + 20(0.7609) + 15(0.944) + 6(0.449)\}}{(0.9665 + 1.212 + 3.534 + 0.7609 + 0.944 + 0.449)}$$

$$Y^{(1)} = 29.763$$

Next we calculate the cost of locating these facilities based on our objective function,

$$\text{Minimize } f(x, y) = \Sigma w_i[(x - a_i)^2 + (y - b_i)^2]^{1/2}$$

$$\begin{aligned} f(x, y) = {} & 20[(26.77 - 20)^2 + (29.73 - 46)^2]^{1/2} + 15[(26.77 - 15)^2 + \\ & (29.76 - 28)^2]^{1/2} + 30[(26.77 - 26)^2 + (29.76 - 35)^2]^{1/2} + \\ & 18[(26.77 - 50)^2 + (29.76 - 20)^2]^{1/2} + 20[(26.77 - 45)^2 + \\ & (29.76 - 15)^2]^{1/2} + 15[(26.77 - 1)^2 + (29.76 - 6)^2]^{1/2} \end{aligned}$$

$$f(x, y) = 2137.52$$

For the second iteration, we have

$$X^{(2)} = \Sigma a_i \phi_i(x^{(1)}, y^{(1)}) / \Sigma \phi_i(x^{(1)}, y^{(1)})$$
$$Y^{(2)} = \Sigma b_i \phi_i(x^{(1)}, y^{(1)}) / \Sigma \phi_i(x^{(1)}, y^{(1)})$$

again we calculate, $\phi_i(x, y) = \{w_i/[(x - a_i)^2 + (y - b_i)^2]^{1/2}\}$ using the most recent values of x, y [i.e., $(x^{(1)}, y^{(1)})$]. Thus we have,

$$\begin{aligned} \phi_1(X^{(2)}, Y^{(2)}) &= 20/[(26.77 - 20)^2 + (29.76 - 46)^2]^{1/2} \\ &= 1.136 \end{aligned}$$

and the remaining values are:

$$\phi_2(X^{(2)}, Y^{(2)}) = 1.260$$
$$\phi_3(X^{(2)}, Y^{(2)} = 15.61$$
$$\phi_4(X^{(2)}, Y^{(2)} = 0.714$$
$$\phi_5(X^{(2)}, Y^{(2)} = 0.852$$
$$\phi_6(X^{(2)}, Y^{(2)} = 0.427$$

Because, $X^{(2)} = \Sigma a_i \phi_i(x^{(1)}, y^{(1)})/\Sigma\phi_i(x^{(1)}, y^{(1)})$

We have,

$$X^{(2)} = \frac{\{20(1.13) + 15(1.26) + 26(15.61) + 50(0.71) + 45(0.85) + 1(0.42)\}}{(1.13 + 1.26 + 15.61 + 0.71 + 0.85 + 0.42)}$$
$$= 26.1$$

and

$$y^* = Y^{(2)} = \Sigma b_i \phi_i(x^{(1)}, y^{(1)})/\Sigma\phi_i(x^{(1)}, y^{(1)})$$

$$Y^{(2)} = \frac{\{46(1.13) + 28(1.26) + 35(15.61) + 20(0.71) + 15(0.85) + 6(0.42)\}}{(1.13 + 1.26 + 15.61 + 0.71 + 0.85 + 0.42)}$$
$$= 31.37$$

The value of the objective function is,

$$f(X^{(2)}, Y^{(2)}) = \Sigma w_i[(X^{(2)} - a_i)^2 + (Y^{(2)} - b_i)^2]^{1/2}$$

$$\begin{aligned} f(x, y) = {} & 20[(26.15 - 20)^2 + (31.4 - 46)^2]^{1/2} + 15[(26.15 - 15)^2 + \\ & (31.4 - 28)^2]^{1/2} + 30[(26.15 - 26)^2 + (31.4 - 35)^2]^{1/2} + \\ & 18[(26.15 - 50)^2 + (31.4 - 20)^2]^{1/2} + 20[(26.15 - 45)^2 \\ & + (31.4 - 15)^2]^{1/2} + 15[(26.15 - 1)^2 + (31.4 - 6)^2]^{1/2} \end{aligned}$$
$$f(x, y) = 2111.9$$

By performing similar iterations we arrive at the series of results shown in the following. The iterative process is terminated when the difference between the successive cost is less than \$0.10, a figure chosen somewhat arbitrarily.

	Iteration coordinates	Cost
1.	(26.7429, 29.7638)	2137.52
2.	(26.1874, 31.3757)	2111.91

3.	(25.956, 32.3597)	2100.26
4.	(25.8951, 33.0029)	2094.18
5.	(25.8905, 33.4456)	2090.60
6.	(25.9002, 33.7639)	2088.31
7.	(25.9124, 34.0009)	2086.76
8.	(25.9241, 34.1821)	2085.67
9.	(25.9346, 34.3238)	2084.86
10.	(25.9437, 34.4364)	2084.26
11.	(25.9516, 34.5272)	2083.79
12.	(25.9584, 34.6012)	2083.42
13.	(25.9642, 34.6621)	2083.13
14.	(25.9691, 34.7126)	2082.89
15.	(25.9734, 34.7549)	2082.70
16.	(25.977, 34.7903)	2082.54
17.	(25.9802, 34.8203)	2082.41
18.	(25.9828, 34.8456)	2082.30
19.	(25.9852, 34.8672)	2082.21
20.	(25.9872, 34.8856)	2082.13

We can observe that the new facility tends to stabilize as close as possible to the existing facility or facilities having the largest weight assigned.

For large problems we can develop a computer program that performs the iterative process. For our example, the development of cost function after a number of iterations is shown in Figure 3.6. We see that the cost converges to a value. The associated location is the optimum location using euclidean cost function.

Similar to the rectilinear cost objective, we can also draw quasi-cost lines on the basis of the 110%, 120%, or 130% of the least cost associated with

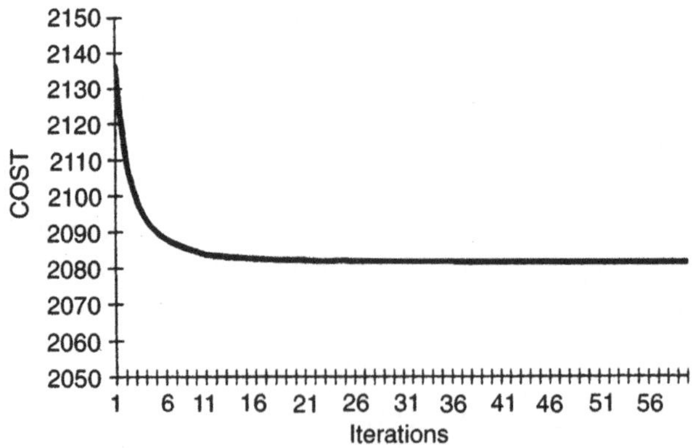

FIGURE 3.6 Cost curve.

the new facility location. We plot these cost values on the graph as shown in Figure 3.7. To draw this graph we have used the software, such as Mathcad, that gives sufficiently accurate results. The central almost circular region denotes the least-cost region. It is recommended to locate the new facility in that region. The outer curves show the increasing trend in the cost function. Thus locating the new facility in the other outer circular region would only add to the cost.

Details of Contour Development for Euclidean Cost Problem

The contour of plots for the euclidean cost is shown in Figure 3.7. These plots are obtained with the help of the software Mathcad. This software has an ability to draw a variety of plots, and we have used the "contour plot" option in the graphics mode of the main menu.

We can generate a multidimensional array of cost values. In our problem these cost values were obtained with the help of a C + + program. We calculated

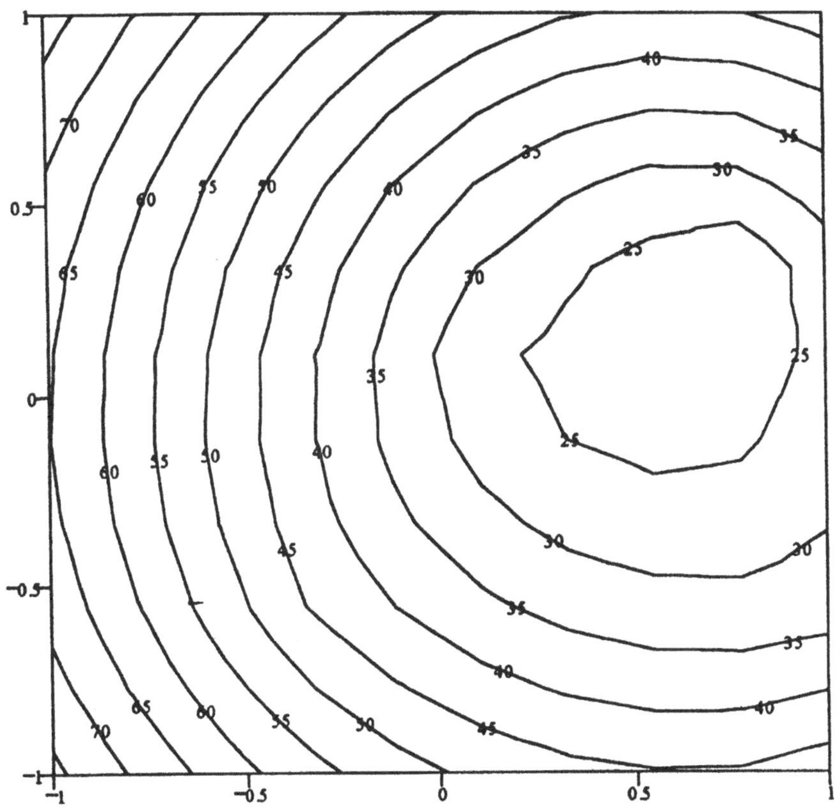

FIGURE 3.7 Contour plot for euclidean distance.

these cost values for the coordinates range of $0 \leq x \leq 50$ and $0 \leq y \leq 50$ with the increments of 1 for both x and y values. These cost values are entered in a matrix. On choosing the contour plot option in Mathcad and referencing it with the matrix of the cost values the plot is generated. The contours denote the equicost lines. The density of the lines (number of contours in a cost range) can be varied depending on the requirement.

3.5 DISTANCE MEASURED AS *p*-POWER

Now suppose, the objective function is neither rectilinear nor euclidean, but a power ***p*** of the direct distance. To find the optimum location, an iterative solution procedure used is adopted from the iterative procedure illustrated in the euclidean distance problem.

Mathematically the new problem can be stated as follows:

$$\text{Minimize} f(x, y) = \Sigma w_i[(x - a_i)^p + (y - b_i)^p]^{1/p}$$

Where p is other than 1 or 2.

To find the optimum point we first differentiate the foregoing equation relative to x and then with y and equate each derivative to zero resulting in:

$$\delta f(x, y)/\delta x = \Sigma w_i(x - a_i)/[(x - a_i)^p + (y - b_i)^p]^{1/p} = 0 \tag{3}$$

and

$$\delta f(x, y)/\delta y = \Sigma w_i(y - b_i)/[(x - a_i)^p + (y - b_i)^p]^{1/p} = 0 \tag{4}$$

Simplifying Eq. (3) further

$$x^*\Sigma w_i/[(x - a_i)^1 + (y - b_i)^p]^{1/p} - \Sigma(w_i^* a_i)/[(x - a_i)^p + (y - b_i)^p]^{1/p} = 0$$

Therefore,

$$x^*\Sigma\{w_i/[(x - a_i)^p + (y - b_i)^p]^{1/p}\} = \Sigma(w_i^* a_i)/[(x - a_i)^p + (y - b_i)^p]^{1/p}$$

$$\text{If} \phi_i(x, y) = \{w_i/[(x - a_i)^p + (y - b_i)^p]^{1/p}$$

For $i = 1 \cdots n$, we obtain

$$X = \Sigma a_i \phi_i(x, y)/\Sigma \phi_i(x, y)$$

Similarly

$$Y = \Sigma b_i \phi_i(x, y)/\Sigma \phi_i(x, y)$$

Similar to euclidean distance problem, writing a generalized equation to perform iterations

$$X^{(k)} = \Sigma a_i \phi_i(x^{(k-1)}, y^{(k-1)})/\Sigma \phi_i(x^{(k-1)}, y^{(k-1)})$$

$$Y^{(k)} = \Sigma b_i \phi_i(c^{(k-1)}, y^{(k-1)})/\Sigma \phi_i(x^{(k-1)}, y^{(k-1)})$$

K denotes the iteration number. This iterative procedure is carried out until no appreciable change occurs in the x and y values. This value gives the least cost for placing the facility.

The same procedure gives a good approximate solution even if p is not the same for each distance from the existing facility to the new facility, as long as variation in p is not very significant. The procedure is illustrated by following example:

Consider the coordinates and their respective weights as given in Table 3.13. Thus, for the starting values (x^*, y^*), to begin this iterative procedure, we use the values obtained by the gravity method. This method provides a simple solution and is, in fact, the weighted average of the x and y values.

Writing them empirically,

$$x^* = \Sigma w_i a_i/\Sigma w_i$$

$$y^* = \Sigma w_i b_i/\Sigma w_i$$

Where $a_i b_i$ are the existing coordinates.

TABLE 3.13 Coordinates of Existing Facilities

No.	x Coord	y Coord	Trips/day (weight)	Power p
1	20	46	20	1.7
2	15	28	15	1.3
3	26	35	30	2.1
4	50	20	18	2.3
5	45	15	20	2.2
6	1	6	15	1.9

This for the data of our example,

$$x^0 = \{20^*20 + 15^*15 + 26^*30 + 50^*18 + 45^*20 + 1^*15\}/$$
$$\{20 + 15 + 30 + 18 + 20 + 15\}$$
$$= 27.28$$

$$y^0 = \{46^*20 + 28^*15 + 35^*30 + 20^*18 + 15^*20 + 6^*15\}/$$
$$\{20 + 15 + 30 + 18 + 20 + 15\}$$
$$= 26.61$$

Using the recursive formula

$$X^{(k)} = \Sigma a_i \phi_i(x^{(k-1)}, y^{(k-1)})/\Sigma \phi_i(x^{(k-1)}, y^{(k-1)})$$

$$Y^{(k)} = \Sigma b_i \phi_i(x^{(k-1)}, y^{(k-1)})/\Sigma \phi_i(x^{(k-1)}, y^{(k-1)})$$

and starting with

$$X^{(1)} = \Sigma a_i \phi_i(x^{(0)}, y^{(0)})/\Sigma \phi_i(x^{(0)}, y^{(0)})$$

$$Y^{(1)} = \Sigma b_i \phi_i(x^{(0)}, y^{(0)})/\Sigma \phi_i(x^{(0)}, y^{(0)})$$

We obtain

$$\phi_i(x, y) = \{w_1/[(x - a_1)^{1.7} + (y - b_1)^{1.7}]^{1/1.7}\} + \{w_2/[(x - a_2)^{1.3} +$$
$$(y - b_2)^{1.3}]^{1/1.3}\} + \{w_3/[(x - a_3)^{2.1} + (y - b_3)^{2.1}]^{1/2.1}\} +$$
$$\{w_4/[(x - a_4)^{2.3} + (y - b_4)^{2.3}]^{1/2.3}\} + \{w_5/[(x - a_5)^{2.2} +$$
$$(y - b_5)^{2.2}]^{1/2.2}\} + \{w_6/[(x - a_6)^{1.9} + (y - b_6)^{1.9}]^{1/1.9}\}$$

or for $x^0 y^0$ of (20, 46)

$$\phi_1(x, y) = \{w_1/[(x - a_1)^{1.7} + (y - b_1)^{1.7}]^{1/1.7}\}$$
$$= 20/[(27.28 - 20)^{1.7} + (26.61 - 46)^{1.7}]^{1/1.7}$$
$$= 0.95$$

Similarly,

$$\phi_2(x, y) = 1.2$$
$$\phi_3(x, y) = 3.41$$
$$\phi_4(x, y) = 0.79$$
$$\phi_5(x, y) = 1.00$$
$$\phi_6(x, y) = 0.46$$

The new coordinates (x^*, y^*) are given by

$$x^* = X^{(1)} = \Sigma a_i \phi_i(x^{(0)}, y^{(0)}) / \Sigma \phi_i(x^{(0)}, y^{(0)})$$

$$X^{(1)} = \frac{\{20(0.95) + 15(1.2) + 26(3.41) + 50(0.79) + 45(1.0) + 1(0.46)\}}{(0.95 + 1.2 + 3.41 + 0.79 + 1.0 + 0.46)}$$

$$X^{(1)} = 26.9$$

$$y^* = Y^{(1)} = \Sigma b_i \phi_i(x^{(0)}, y^{(0)}) / \Sigma \phi_i(x^{(0)}, y^{(0)})$$

$$Y^{(1)} = \frac{\{46(0.95) + 28(1.2) + 35(3.41) + 20(0.79) + 15(1.0) + 6(0.46)\}}{(0.95 + 1.2 + 3.41 + 0.79 + 1.0 + 0.46)}$$

$$Y^{(1)} = 29.4$$

Next we calculate the cost of locating these facilities. This is our objective function. It is given by,

$$\text{Minimize} f(x, y) = \Sigma w_1[(x - a_1)^{1.7} + (y - b_1)^{1.7}]^{1/1.7}$$

$$\begin{aligned} f(x, y) = & \ 20[(26.9 - 20)^{1.7} + (29.4 - 46)^{1.7}]^{1/1.7} \\ & + 15[(26.9 - 15)^{1.3} + (29.4 - 28)^{1.3}]^{1/1.3} \\ & + 30[(26.9 - 26)^{2.1} + (29.4 - 35)^{2.1}]^{1/2.1} \\ & + 18[(26.9 - 50)^{2.3} + (29.4 - 20^{2.3}]^{1/2.3} \\ & + 20[(26.9 - 45)^{2.2} + (29.4 - 15)^{2.2}]^{1/2.2} \\ & + 15[(26.9 - 1)^{1.9} + (29.4 - 6)^{1.9}]^{1/1.9} \\ f(x, y) = & \ 2150.56 \end{aligned}$$

Other iterations are similarly evaluated; the results are displayed in Table 3.14.

Here, we find that the optimum location of the new facility relative to the existing locations is at (26.15, 32.05) with the cost of location the facility being 2119.14.

TABLE 3.14 Other Iterations

Iteration number	Coordinates	Cost
2	(26.2, 30.9)	2129.15
3	(26.18, 31.46)	2123.95
4	(26.15, 32.05)	2119.14
5	(26.15, 32.05)	2119.14

3.6 AREAS AS THE DEMAND POINTS

Suppose a hospital is to be located to serve several rural areas. Factually, the demands for the hospital services are generated from each of the households in each of the rural areas. If we conceive each household as the demand point, then there are an almost infinite number of demand points that we will have to consider. Fortunately, it is not necessary to go into such a minute degree of accuracy in formulating the models. Instead, we can divide each area into a number of homogenous segments, and use the center of gravity of each area segment to represent the area with the appropriate weight, based on the number and type of households that are attached to that center of gravity point. For example, on average, a healthy household may generate one trip per year, whereas a nursing home may generate one trip per bed per week. An area A, with 500 healthy households, will have a demand of 500 per year attached to its physical center of gravity, whereas an area B with 300 households and one 60-bed nursing home will have $300 \times 1 + 60 \times 1 \times 52 = 3420$ as the demand associated with its physical central of gravity.

If, on the other hand, the nursing homes are located in a small zoned area, then perhaps dividing area B into two parts, one associated with homes, and other with the nursing homes would be more appropriate. This will make each area more homogeneous. Again the appropriate demand should be attached to the appropriate area.

With this assumption, and modification for areas, the models developed in the previous sections, as well as those developed in latter chapters, can be appropriately applied.

Exercises

3.1 Define the continuous facility location problem.

3.2 Distinguish between rectilinear and euclidean distances.

3.3

a. Calculate the distances between department centers in Figure 3.8 if the material handling is by hand carts and must use aisles whenever possible.
b. Develop the shortest path for overhead conveyors that will connect each department center to all other department centers.

3.4 A water cooler is to be placed on the 15th floor of an office building. It has five offices on the that floor. A study was done to see how many trips per day, on the average, were made from each office. The data collected is shown in the following table along with the location of the five offices. Determine the location of water cooler.

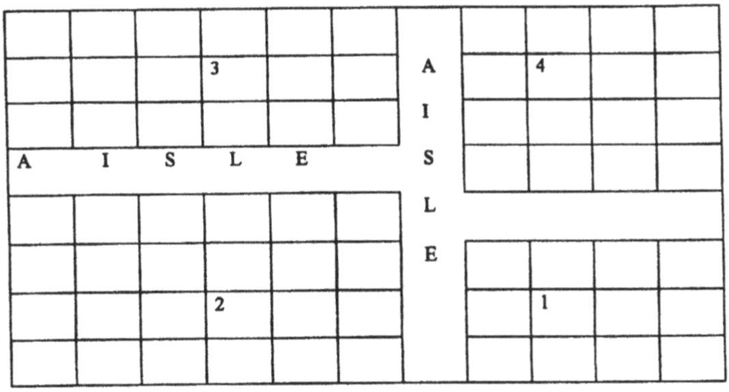

FIGURE 3.8 Layout of plant shown in Ex. 3.3.

Customer	Coordinates	Trips/day
1	5, 5	3
2	10, 40	5
3	70, 45	2
4	65, 20	1
5	25, 25	3

3.5 In the Gulf of Mexico, Tell Oil has 4 rigs. The location of each is given on the chart below. Food for all the rigs is brought out once a month by ships to a central warehouse. From here, the food is distributed according to the following data. Where should the warehouse be located?

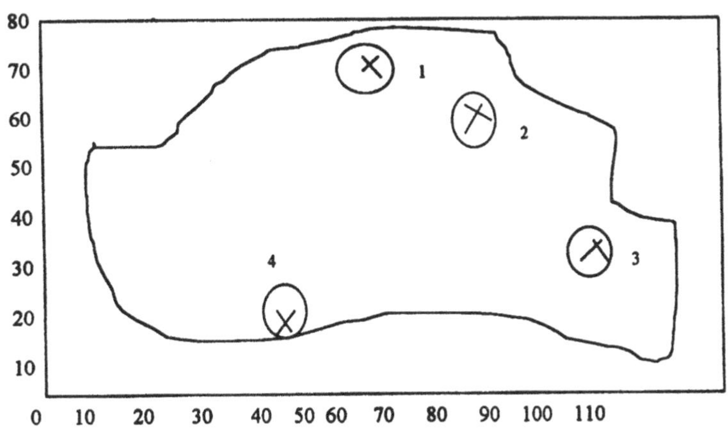

Rig	Coordinates	Trips/month
1	60, 70	7
2	80, 60	10
3	100, 30	8
4	40, 20	12

3.6 The island of Margarita is famous for its delicious lobster. They are caught at sea and brought to three main harbors around the island. A refrigerated warehouse is needed from which the lobster will be shipped by airplane from the harbors. Because freshness of the product is so important, the warehouse should be located to minimise the cost of losing market owing to lack of freshness. Determine the optimum location for the warehouse.

Harbors	Coordinates	Trips/day
1	4, 2	5
2	4, 7	7
3	10, 5	11

3.7 A manufacturing company is to be located along an east to west route in the southern section of the United States. The company will service the four states of Texas, Louisiana, Mississippi, and Alabama. The company produces outdoor lawn furniture. Listed are the coordinates and the number of trips to be made to each city. Determine the optimum location for the manufacturing facility.

City	Coordinates	Number of trips/monthly
1	400,225	6
2	350,100	8
3	60,90	9
4	150,50	11

3.8 After being cut at saws 1 and 2, wood is sent either to the sander at F1 or the lathe at F2. Determine the best location for the sander and the lathe to minimize the distance traveled. The number of trips per hour between each existing machine and the new machine along with the coordinates of the existing machine are given.

	Coordinates	Number of trips	
Machine	(x, y)	Facility 1	Facility 2
1	3, 5	5	2
2	10, 8	4	3
FI		—	0
F2		3	—

3.9 A corporation wishes to add on a view wing for its president and vice president. Secretaries are located at 1, engineering at 2, accounting at 3, and sales at 4. Locate these two offices to minimise distance traveled. The average number of trips per day that the president and vice president make to these offices is given.

	Coordinates	Number of trips	
Office	(x, y)	Facility 1	Facility 2
1	1, 1	10	15
2	1, 3	8	7
3	3, 1	12	10
4	5, 1	7	6
F1		—	6
F2		6	—

3.10 The resident society of an apartment complex decides to install a washer/dryer in one of its five-floor buildings. The weights associated with each floor depend on the number of people on each floor. Determine the place for locating the washer/dryer considering the location of the floors and their weights.

Floor	Location	Weight
1	(2, 3, 0)	7
2	(1, 1, 1)	10
3	(0, 1, 2)	6
4	(1, 2, 3)	9
5	(1, 0, 4)	8

3.11 In Ex. 3.6, cost is now measured as a p-power of the direct distance, where p for each harbor is as follows: $p_1 = .4$, $p_2 = .6$, and $p_3 = .8$. Determine the optimum warehouse location.

REFERENCE

Witzgall C. Optimal Location of Central Facility; Mathematical Models and Concepts. Gaithersberg, MD: National Bureau of Standards Report, 8388, 1964.

4

Alternative Objectives in Single-Facility Location

There are many instances for which minimization of total distance traveled (*minimax problem*) is not an appropriate criterion for finding the location for a new facility. For example, we might be interested in minimizing the maximum distance traveled by a customer: an emergency room should not be very far from the population it is required to serve. We may also be interested in placing a facility, such as a landfill site, as far away from the existing population, as possible. If a radio transmitter or a helicopter pad is to be placed within a group of population centers, it should be able to cover or respond within the shortest possible distance. In the following sections we will study a multitude of single-facility location problems with alternative objectives.

4.1 MINIMAX LOCATION PROBLEM

In this section we want to examine a problem for which maximum distance from a new facility to any of the existing facilities should be as small as possible. That is, we want to minimize the function:

$$f(x,y) = \max\{w_i[|x-a_i| + |y-b_i|]\} \qquad i = 1,\ldots,n$$

This problem is typical if the time required for travel is an important consideration. We do, for example, want to place a facility that serves a number of customers, such as a gas station or a grocery store, close to all the existing communities from which we expect the customers to come. Here the minimization of *total distance* traveled by all customers is not as effective a measuring criterion as that of minimization of maximum time a customer will take to come to the store. This, in turn, translates to the objective of minimization of the maximum distance from the store to a customer. The assumption here is that the travel is in a rectilinear fashion, not a bad assumption when most of the city roads are normally planned in north–south and east–west directions.

Initially, we will study a procedure called the *diamond covering procedure* in which the weight of each existing facility is assumed to be equal (weight of 1). Later we will introduce concepts of Tchebychev distances to simplify the procedure and expand on the objective function.

4.2 DIAMOND COVERING PROBLEM

Let us consider any two points, A and B in x–y coordinates. The rectilinear distance from A to B is the sum of the distance in the x direction and the distance in the y direction. Now, if we draw a 45-degree line through point A, not toward B, and consider any point on this line, within the limits derived by drawing parallel lines to the x and y axes from B intersecting the 45-degree line, then the rectilinear distance between that point B is the same as the original distance between A and B (Figure 4.1). This fact is used in developing a smallest diamond figure consisting of four 45-degree lines through existing facilities, such that the diamond will include all the facilities. Draw the diamond; make sure all the points are inside the

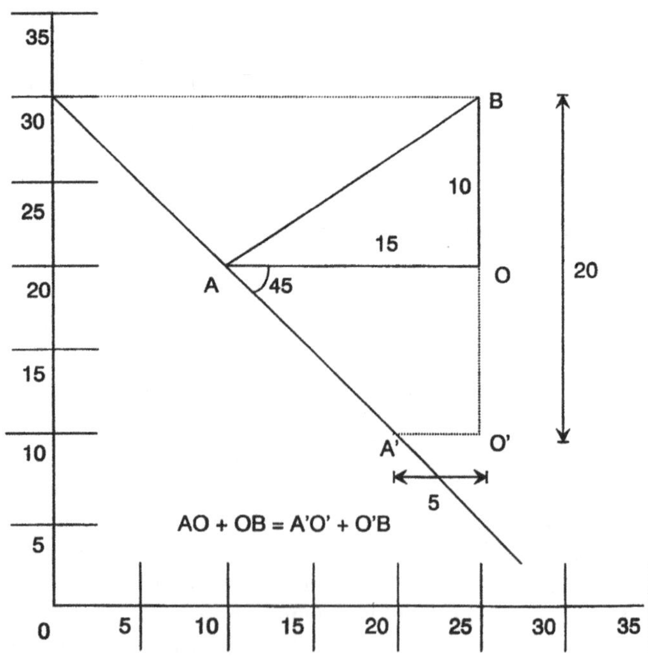

FIGURE 4.1 45-degree illustration.

diamond or on the line. The diamond thus created may not have equal sides, but the center of the diamond is the minimum distance away from the farthest existing facility. If now the sides of the diamond are made equal to the largest side (making it a square), then any point on any side of the square is equidistance away from the center. Thus, it is possible to generate a range of locations for the new facility that will satisfy our objective of minimizing the maximum distance. The locus of such points is obtained by first expanding only one short side at a time to form two squares. Then join the diagonal points across. This gives a small square in the center. Connect the points across the diagonal in the direction of expansion of this small square formed inside. This line provides the locus for facility location. The details are shown in Figure 4.2.

Consider the six existing facility, as shown in the following example.

Facility	1	2	3	4	5	6
Coordinate	(20, 46)	(15, 28)	(26, 35)	(50, 20)	(45, 15)	(1, 6)

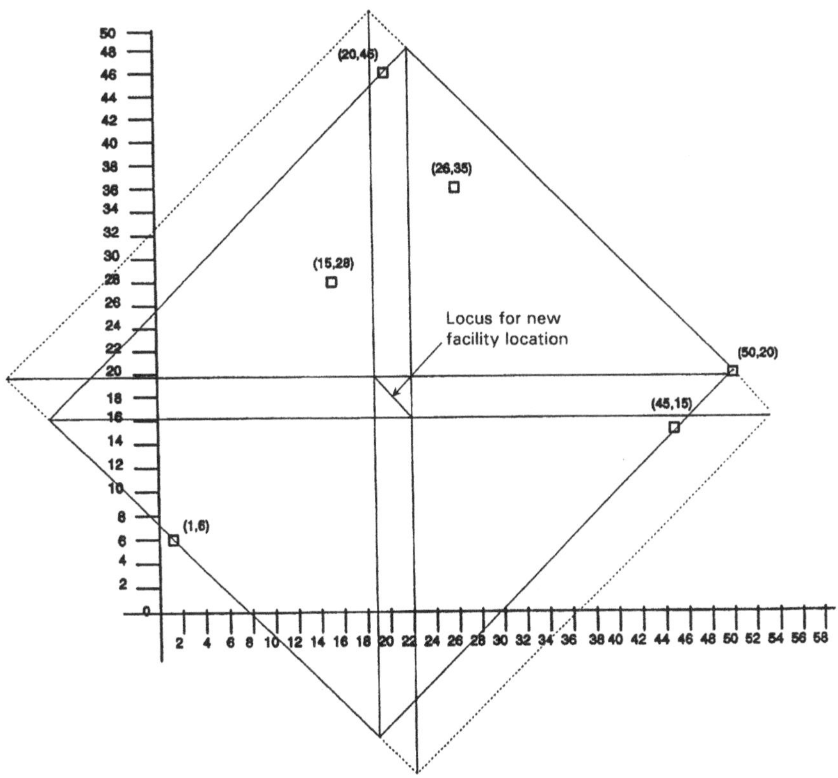

FIGURE 4.2 Diamond-covering problem.

Again, the objective is to find a location for a new facility so that the maximum rectilinear distance between it and any of the existing facilities is as small as possible. Alternatively, the objective is to find the center of a diamond of minimum radius that contains all the existing locations. For a given set of existing facility location we start by constructing the smallest rectangle, with each side making an angle of ±45 degrees with the axis. The outer lines pass through the points (50, 20), (45, 15), (1, 6), and (20, 46). The short side (i.e., the one through (20, 46) or (45, 15), is extended to form a square. The square obtained is the diamond of smallest radius, which is the minimum objective function value. The construction is shown in Figure 4.2, along with the locus for the new facility location

4.3 ALTERNATIVE TO DIAMOND COVERING PROBLEM USING TCHEBYCHEV DISTANCES

An alternative procedure that is easier and quicker to analyze, is to work with transformation known as Tchebychev transfer. Any point in x–y plane can be transformed on the Tchebychev plane. For a point $A(x, y)$ the Tchebychev coordinates are $A'(y+x, y-x)$.

For example, $A(20, 46)$ and $B(15, 28)$ can be transferred to Tchebychev coordinates as $A'(46+20, 46-20)$ or $A'(66, 26)$ and $B'(28+15, 28-15)$, or $B'(43, 13)$. The rectilinear distance between A and B is $(|20-15| +|46-28|) = 23$. This distance can also be found using Tchebychev coordinates. The distance between A and B is the maximum distance between x or y coordinates of $(A' - B')$. In our case, it is maximum of $(|66-43| = 23, |26-13| = 13) = 23$.

Because in Tchebychev coordinates, plotted on an x–y plane, the maximum distance between two facilities in the x and y directions is also rectilinear distance between these two facilities, we need only to create the smallest square that covers all existing facilities to know the maximum distance between facilities.

Returning to our example, we have, six existing facilities. Their x–y coordinates and transferred Tchebychev coordinates are as follows:

Facility	x–y Coordinate	Tchebychev coordinates
1	(20, 46)	(66, 26)
2	(15, 28)	(43, 13)
3	(26, 35)	(61, 9)
4	(50, 20)	(70, −30)
5	(45, 15)	(60, −30)
6	(1, 6)	(7, 5)

To find the location of the new facility that has a minimum maximum distance from the existing facilities, we want to enclose these transformed points into the smallest square and find the center of that square as the location point for the new facility. Start with constructing the smallest covering rectangle. The shorter side can then be extended on both sides to form a square, giving us some latitude on how the square is made.

In Figure 4.3, we can extend the lines forming the squares. This gives us two points (38.5, −5.5), and (38.5, 1.5), and the optimal locations lie on the line segment joining these two points in Tchebychev coordinates. If (X, Y) are Tchebychev coordinates, then the inverse transformation to planar form is $[(Y - X)/2, (Y + X)/2]$. Transforming these points into planar form by inverse transformation gives (22, 16.5) and (18.5, 20). Therefore, the line segment joining them is the set of optimal locations for the new facility.

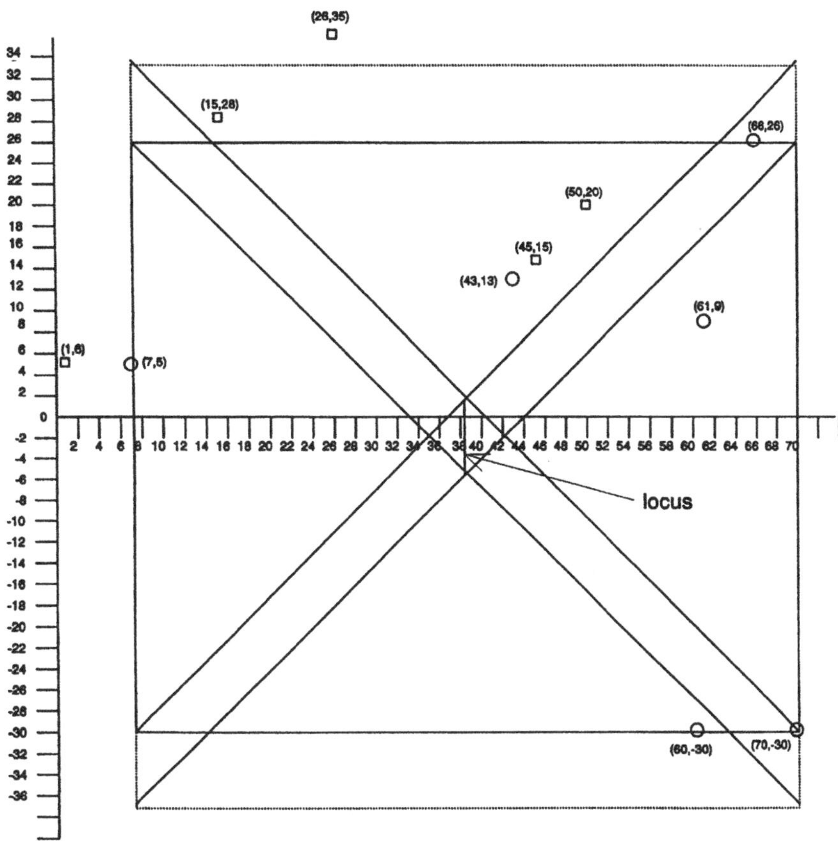

FIGURE 4.3 Tchebychev distance: ○, Tchebychev points; □, coordinates.

4.4 WEIGHTED CUSTOMER PROBLEM

In the previous problem, all customers were equally important and, therefore, the associated weight for each customer was 1. This is called an *unweighted problem.* Now, considering a problem in which some customers are more important than others. In developing a distribution center, for example, large and frequent customers may carry somewhat more weight than small or infrequent customers. This is called a *weighted problem.* We shall also include a new factor called *addendum* in this formulation.

We want to minimize the function $g(x,y)$ where,

$$g(x, y) = \max[\{w_i[|x - a_i| + |y - b_i|] + h_i: \quad i = 1, \cdots m\}]$$

The addendum has some significance. An ambulance may travel from its base location to a demand point in a community and then travel to the hospital. The hospital being already in existence, the distance between hospital and the center of the community is known and fixed and is considered as an addendum, h_i. If the hospital is very far from the community center (i.e., h_i is large), then the ambulance should respond as quickly as possible or minimize the distance from the base to the community center to minimize the total time (or distance) taken to bring a patient to the hospital.

The procedure is as follows: Let Tchebychev transformation of a point (x,y), where the new facility is to be located be (u,v), and that of a point (a_i, b_i) a location of an existing facility, be (α_i, β_i).

Define two functions $g_1(u)$ and $g_2(v)$ such that,

$$g_1(u) = \max\{w_i|u - \alpha_i| + h_i: i = 1, \cdots m\}$$
$$g_2(v) = \max\{w_i|v - \beta_i| + h_i: i = 1, \cdots m\}$$

and let

$$g(x, y) = \max\{g_1(u), g_2(v)\}$$

We obtain the optimum u^* by minimizing $g_1(u)$ and optimum v^* by minimizing $g_2(v)$. Then we apply the inverse transformation to (u^*, v^*) to obtain a point that minimized $g(x,y)$.

The procedure to minimize $g_1(u) = \max\{w_i|u - \alpha_i| + h_i: i = 1 \cdots m\}$ begins with arranging the facilities in ascending order of x-coordinates and then apply the following steps:

1. Calculate $\delta_{ij} = (w_i w_j|\alpha_i - \alpha_j| + w_i h_i + w_j h_j)/(w_i + w_j)$
2. Compute $\delta_{pq} = \max\{d_{ij}: \ 1 \le i \le j \le m\}$
3. Let $h_r = \max\{h_i: \quad 1 \le i \le m\}$
4. If $\delta_{pq} \le h_r$ then $u^* = \alpha_r$, $g_1(u^*) = h_r$ and stop the process. Otherwise determine the value of u^* by searching in the range of facilities from p through q such that $|u - \alpha_p| = \delta_{pq} - h_p/w_p$, such that at the optimum

value of $u = u^*$, the minimum value is associated with $g_1(u^*)$. Note that if any time the minimum $g_1(u)$ is equal to δ_{pq} then the associated $u = u^*$. Determine the associated x dimension by inverse transformation of u^*.

Similarly minimize $g_2(v)$ for the same sequence of facilities as with $g_1(\mathrm{u})$ and find v^*.

Example

Consider the same example that we addressed in the previous section except now each customer has a specific weight. Again, we wish to minimize the maximum weighted distance.

Customer	1	2	3	4	5	6
Coordinate (x, y)	(20, 46)	(15, 28)	(26, 35)	(50, 20)	(45, 15)	(1, 6)
Transformed coordinates (α, β)	(66, 26)	(43, 13)	(61, 9)	(70, −30)	(60, −30)	(7, 5)
Weights	20	15	30	18	20	15

Function $g_1(u)$ is associated with the transformed coordinates of each x, whereas $g_2(v)$ is associated with the transformed coordinates of each y. As the first step, function $g_1(u)$ is arranged in the ascending order of α_I. Addenda in this case are zeros.

The ascending order for α_I is 7, 43, 60, 61, 66, and 70 and the corresponding facilities are 6, 2, 5, 3, 1, and 4, respectively. The associated weights are 15, 15, 20, 30, 20, and 18. Therefore, we have

$$g_1(u) = \max\{15|u-7|+0, 15|u-43|+0, 20|u-60|+0, 30|u-61| \\ +0, 20|u-66|+0, 18|u-70|+0\}$$

Similar calculations lead to corresponding function for $g_2(v)$ as

$$g_2(v) = \max\{15|v-5|+0, 5|v-13|+0, 20|v-(-30)|+0, 30|v-9| \\ +0, 20|v-26|+0, 18|-(-30)+0\}$$

Now, according to step (1) of the procedure, we calculate δ_{ij} as:

$$\delta_{12} = (15^*15|7-43|+0+0)/(15+15) = 270$$
$$\delta_{13} = (15^*20|7-60|+0+0)/(15+20) = 454.28$$

Similarly the other values are

$$\delta_{14} = 540;\ \delta_{15} = 505.7;\ \delta_{16} = 515.45;\ \delta_{23} = 145.7;$$
$$\delta_{24} = 180;\ \delta_{25} = 197.14;\ \delta_{26} = 220.90;\ \delta_{34} = 12;$$
$$\delta_{35} = 60;\ \delta_{36} = 94.7;\ \delta_{45} = 60;\ \delta_{46} = 101.25;\ \delta_{56} = 37.89$$

According to step (2) we choose δ_{pq}, which is the max among the δ_{ij} values. In our example it is $\delta_{14} = 540$. Step (3) calculates the maximum of the addenda, which here is zero. Referring to step (4), because $\delta_{pq} = \delta_{14} = 540 > h_r = 0$ u^* is a point between α_1 and α_4, which are the first term and the fourth term, respectively, in the function $g_1(u)$. Thus u^* lies in between $\alpha_1 = 7$ and $\alpha_4 = 61$. For $\alpha_1 = 7$,

$$|u' - 7| = \delta_{14} - h_1/w_1 = (540 - 0)/15$$

which gives $u' = 43$. If we substitute the value of 43 in the foregoing function $g_1(u)$, we obtain $g_1(u) = \max\{15|43 - 7| + 0, 15|43 - 43| + 0, 20|43-60| + 0, 30|43 - 61| + 0, 20|43 - 66| + 0, 18|43 - 70| + 0\}$

$$g_1(u) = \max\{540, 0, 340, 540, 460, 486\}$$

The maximum is 540, which is the value for δ_{14}; therefore, we have found the optimum u^* and there is no need to evaluate for α_2, α_3, or α_4.

Similarly solving for $g_2(v)$, we obtain $v^* = -2$ and $g_2(v) = 560$.

Applying the inverse transformation to $(43, -2)$ to (43,-2) [i.e., $Q^{-1}(43, -2)$] we obtain, (22.5, 20.5), as the optimum location for the new facility.

4.5 CIRCLE-COVERING PROBLEM

Suppose we wish to place a radio station that will transmit its signals to known localities, at minimum cost. Here, the cost of transmitter is directly proportional to the wattage of the radio station which, in turn, also determines the distance the signal can travel. The greater the wattage more expensive the transmitter, but the more distance the signal can cover. Because the radio signal propagates directly, the cost of transmitter is also directly proportional to the maximum distance the signal must reach. The objective of this problem is then to reach all existing facilities with a circle. This circle will have the center (x, y), where the radio station will be located, and radius Z, such that Z is as small as possible. Thus, the objective is to minimize $G(x,y)$ such that

$$G(x, y) = \max\{[(x - a_i)^2 + (y - b_i)^2]^{0.5} 1 <= i <= m\}$$

This problem is known as the circle-covering problem.

The procedure for developing such a circle is as follows:

1. Select two extreme points among the given set (say A and B) and draw a circle with those points defining the diameter.
2. If the circle covers all the existing facilities, then it is the minimum covering circle, and the process stops; otherwise, select an outside point (say C) and move to step 3.
3. If the three points (say A,B,C) for a right angle or an obtuse angle,

exclude the point at which the corresponding angle is formed. Draw a circle, with the remaining two points defining the diameter of the circle; otherwise, if the three points form an acute angle-triangle, then draw a circle that passes through those three points. This is done by drawing perpendicular bisectors to any two sides, the intersection of the these lines is the center for the circle that will pass through all three points.

If the circle so formed, as defined by the three points, covers all the existing facilities stop; otherwise, move on to the next step.

4. Select an outside point *P*, that is farthest away from the circle, and determine the farthest point from P among the three defining points that were used in constructing the original circle. For illustration, let this point be A. Draw a line through the center of the circle and point A that divides the plane in to two halves. Note the points B and C that are in each half. Ignore the point that is in the same half as P, and draw a new circle using P, A, and the point in other half as the extreme points and follow the procedure of step 3. For example, say point B is on the same half-plane as P, and C is in the other half-plane. With points A, C, and P draw the new circle. Repeat step 4, until all the points are within or on the circle. The center of this circle is where the new facility should be located.

Example

Suppose there are six existing facilities with locations given by the coordinates as follows:

1. (7, 4)
2. (4, 6)
3. (3.5, 9)
4. (9, 10)
5. (12, 6)
6. (8, 14)

As shown in Figure 4.4, start by drawing a circle through two extreme points [i.e., in this case points 1 (7, 4) and 6 (8, 14)]. We find that except for point 5 (12, 6) the circle covers all the existing facilities. Taking outside point 5, define the three points 1, 5, and 6. Because these three points form an acute-angle triangle, draw a circle that passes through them. This is done by drawing bisects to the sides joining points 1, 5 and 5, 6. These bisects intersect at point (7.9, 9.0). With this point as the center draw the circle that passes through points 1, 5, and 6. This circle covers all the existing facilities. Suppose there is another point 7 at (2.5,

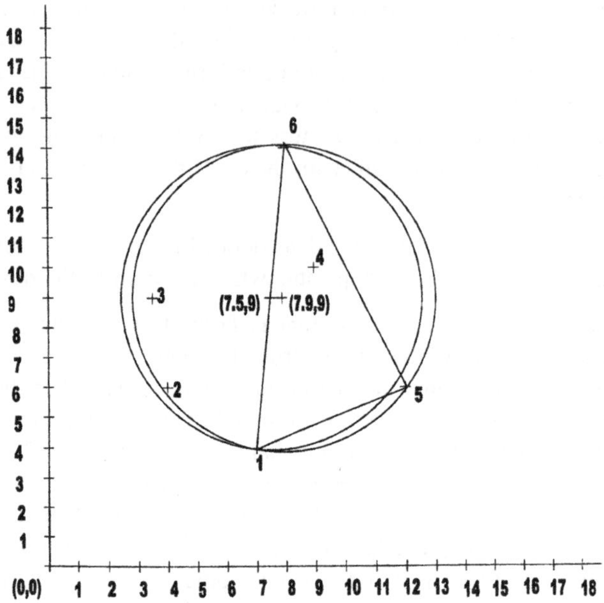

FIGURE 4.4 Circle-covering problem.

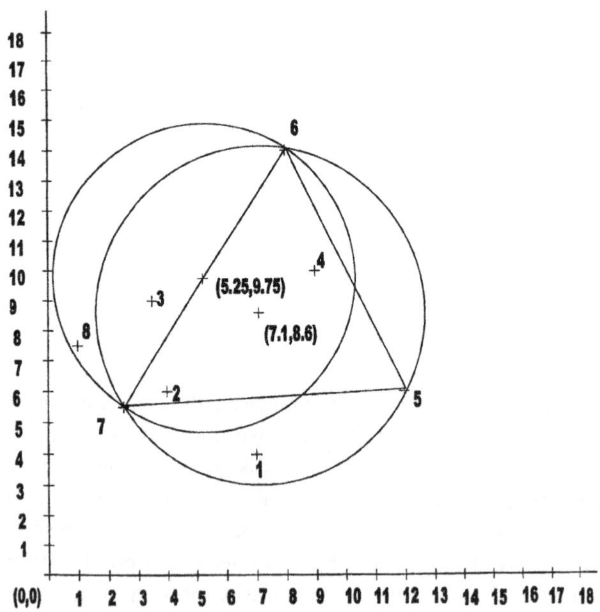

FIGURE 4.5 Circle-covering problem.

5.5) and 8 at (1, 7.5) in addition to the foregoing coordinates (Figure 4.5). Following the same procedure, we select two extreme points 6 and 7 and draw a circle with those two points defining the diameter. This circle does not cover points 1 and 5. Select the outside point 5, and with defining points 7, 6, and 5 develop a circle that covers all the existing facilities except 8.

Now, among the three points that define the circle (7, 6, and 5), point 8 is farthest away from point 5. By using step 4, draw a line passing through the center and point 5, which divides the plane into two. Points 7 and 8 fall into the same half; therefore, ignore point 7 and form the triangle with points 5, 6, and 8. As this forms an acute-angle triangle, draw a circle, defined by these points, that covers all the existing abilities, as shown in Figure 4.6.

The center for the circle is 6.7, 8.3, and it has a radius of 5.8.

Alternative Procedure

1. Draw the given points on the x and y coordinates.
2. Take the extreme points and create a rectangle, such that all the points fall inside the area covered by rectangle. To find the extreme points, draw two vertical lines through the points, which have minimum and maximum abscissa, and two horizontal lines through the points that have minimum and maximum ordinate.

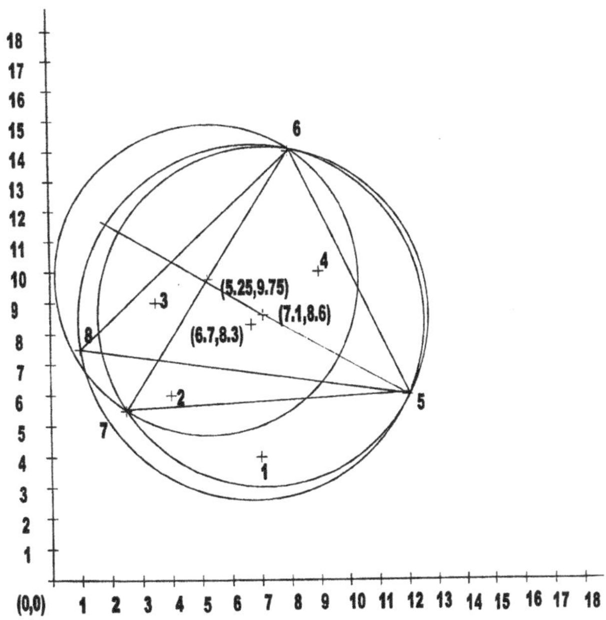

FIGURE 4.6 Circle-covering problem.

3. Join all the points that lie on the perimeter of the rectangle.
4. Creates as many triangles as possible, joining all the points that lie on the perimeter of the rectangle. In most cases, we obtain only two triangles, as we have just four points on the perimeter.
5. Choose a triangle with the largest area.
6. If the triangle obtained is an obtuse triangle, exclude the point where it forms a right angle or obtuse angle.
7. Join two extreme points of this triangle to define the diameter. Draw a circle using this diameter.
8. If the triangle is an acute one, draw the perpendicular bisectors on any two sides of the triangle, the intersection of which defines the center point for the circle. Draw a circle from this center that just cover three corner points of the triangle.
9. The circle thus obtained is the optimum location for the facility to be placed and covers all the points.

Example

Locations of the six existing facilities are given as follows:

A	B	C	D	E	F	G	H
(7, 4)	(4, 6)	(3, 5, 9)	(9, 10)	(12, 6)	(8, 14)	(2.5, 5.5)	(1, 7.5)

Draw the points on the *x-y* coordinates, as in Figure 4.7. Draw two vertical lines through the points with minimum and maximum abscissa (i.e., through E and H, respectively). Draw two horizontal lines through the points with minimum and maximum ordinates (i.e., through A and F, respectively). We now have a rectangle that covers all the points. Now join the points that lie on the perimeter of the rectangle (i.e., points A, E, F, and H). Draw as many triangles as possible by joining the points on the parameter with each other. Choose the triangle with the largest area, EHF. Draw the perpendicular bisectors of any two lines of the triangle to obtain an intersection. From the point of intersection draw a circle passing through the points E, F, and H. The obtained circle covers all the existing locations. The center of the circle is (6.7, 8.3) and the radius is 5.8.

4.6 WEIGHTED CIRCLE-COVERING PROBLEM

Now consider a problem for which population density or importance of a location varies from location to location. We want to consider this factor in the placement of the new facility. As an example, placement of a police call box on a campus. The telephone should serve all nearby buildings in a quadrangle. The weight from each building is decided by consideration of the population in each building. The

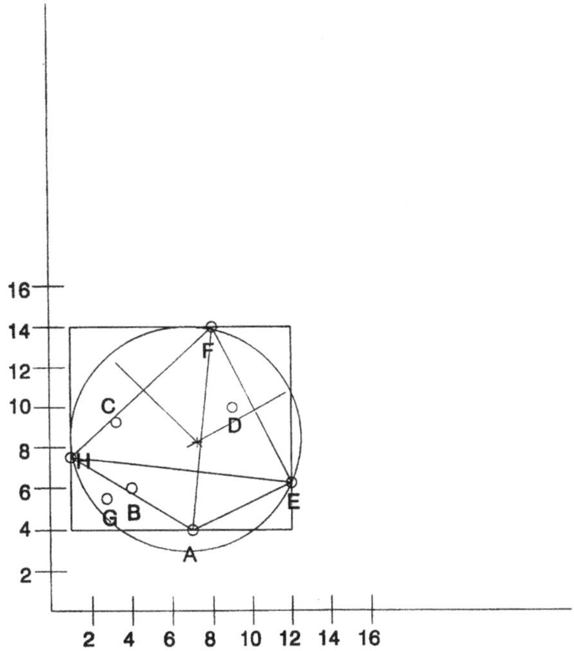

FIGURE 4.7 Alternative method for circle covering.

larger the student body in a building the more relative weight it carries. Perhaps, a building with a weigh of 2 is twice as significant as a building with a weight of 1.

The associated procedure is illustrated by applying it to a problem with six existing facilities, and their weights are given as follows:

	Coordinates	Weights
1.	(20, 46)	2.0
2.	(15, 28)	1.5
3.	(26, 35)	3.0
4.	(50, 20)	1.8
5.	(45, 15)	2.0
6.	(1, 6)	1.5

A sphere of influence from each demand point is measured by the circle with an appropriate radius. The point where all circles intersect is the weighted center point at which the new facility should be located. To achieve this, we start with each existing facility a center and draw a circle with radius equal to

$r = p/w$ where p is a constant.

Initially, the value of p is defined arbitrarily. This is a trial-and-error procedure, the value of p would be adjusted, up or down in successive drawings, based on our objective of finding an intersection with minimum value for p.

In this example, we took the initial guess of $p = 48$. The associated diagram is shown in Figure 4.8. Note that the radius of the circle from point 1 is 24, whereas that from point 2 is 32, and so on. This gives us a region A as the common points to all circles, or the region for locating the new facility. However, this region is fairly large; therefore, it may be possible to reduce the value of p from 48 down. So in the next iteration we try $p = 47$. Now the intersection is only a very small area A, as seen in Figure 4.9. A point within this area—point 25.9, 26.5—is chosen for the location for the police call box. Had we tried $p = 46$ we would not find a common area or even a point as illustrated in Figure 4.10. The

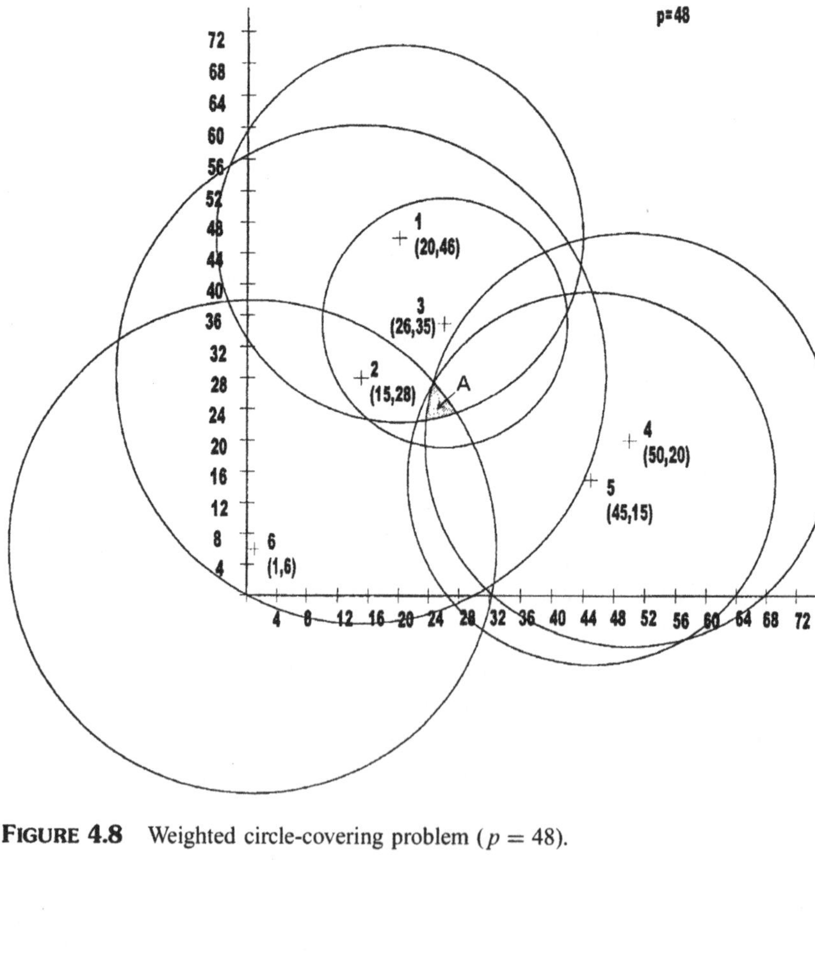

FIGURE 4.8 Weighted circle-covering problem ($p = 48$).

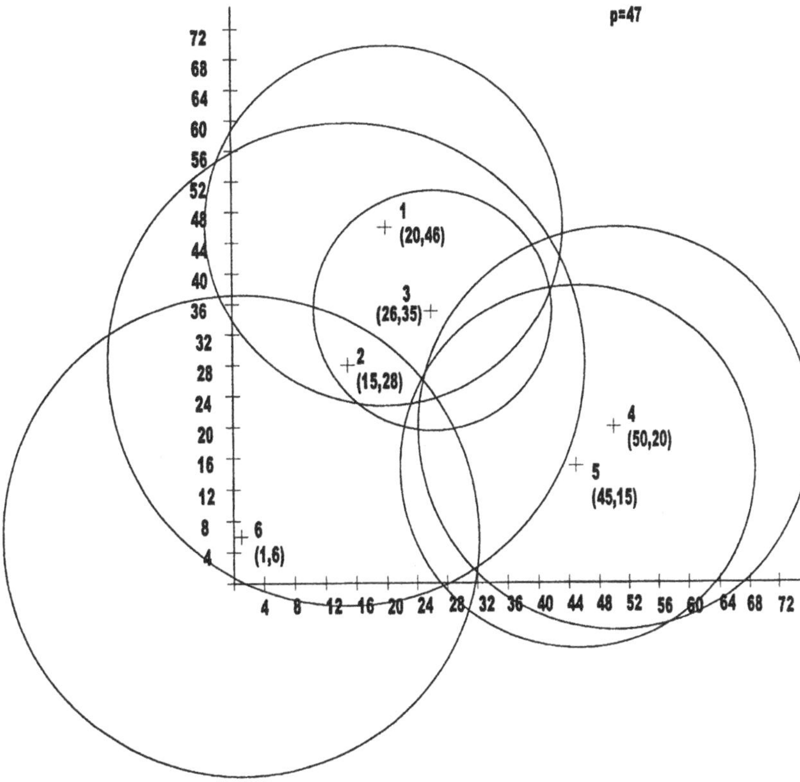

FIGURE 4.9 Weighted circle-covering problem ($p = 47$).

circles from point 1 do not have any simultaneous common area with the circles from point 5 and 6.

4.7 WEIGHTED CIRCLE PROBLEM WITH ADDENDUM

Now suppose we wish to place a dumpster to collect trash from all these buildings. It is obvious no one wants a trash collection point next to his or her building, but at the same time no one wants to walk very far. Adding the addendum serves as the minimum distance the dumpster should be away from each building. Then finding the optimum location that minimizes the travel will satisfy both goals.

Suppose in addition to the weights of the foregoing existing facilities, we add addendum h_I as follows:

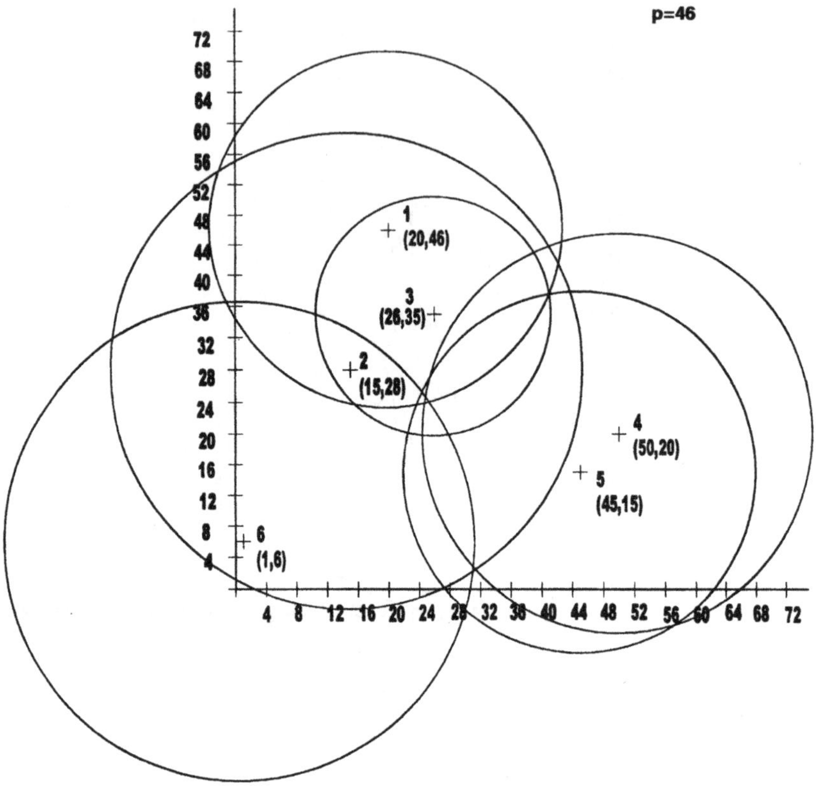

FIGURE 4.10 Weighted circle-covering problem ($p = 46$).

Coordinates	Weights	Addendum (h_I)
1. (20, 46)	2.0	4
2. (15, 28)	1.5	6
3. (26, 35)	3.0	5
4. (50, 20)	1.8	8
5. (45, 15)	2.0	6
6. (1, 6)	1.5	7

The procedure requires only slight modification. We draw circles with centers at each exiting facility and radius:

$$r = (p - h_I)/w$$

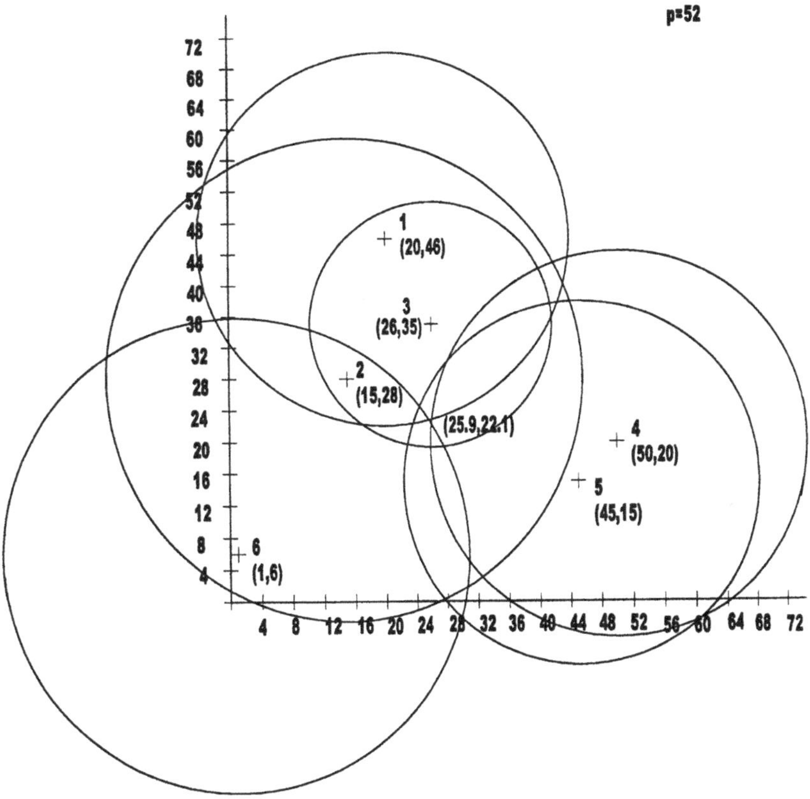

FIGURE 4.11 Weighted circle-covering problem with addendum ($p = 52$).

and follow the same procedure as in the weighted circle problem.

After applying the iterative procedure, at $p = 52$, intersection is at one point for circles 1, 4, and 6, and the point is within reach of circles 2, 3, and 5. The point is at (25.9, 22.1). The details are shown in Figure 4.11.

4.8 UNDESIRABLE FACILITY LOCATION

Contemplate the location of an "undesirable" facility. An undesirable facility is one that needs to be located as far away as possible from the existing facilities and yet should also be within a defined perimeter to minimize the cost of servicing these existing locations. This situation is very often encountered in the modern society. Not very many persons want a facility, such as nuclear power plant, chemical factory, or waste disposal site, in their backyard. But they are necessary (in most societies) and provide jobs for the same population. In the following development for determination of the desired location for an undesirable facility, we consider the distance is measured in the rectilinear manner, as most often is the travel on the connecting roads.

The objective function is to maximize the minimum distance between any existing facility and the new facility; that is,

$$\phi(x, y) = \max\{\min(|(x - a_i| + |(y - b_i)|)\} \qquad \text{for } i = 1 \cdots n$$

subject to the constraint that the new facility must be within a boundary specified. The procedure is best illustrated by simultaneously applying it to an example.

Suppose there are seven existing facilities (or demand points), with locations (20, 34), (37, 36), (35, 26), (8, 20), (16, 16), (27, 10), and (18, 4), respectively. The boundary region given is $0 < x < 40$ and $0 < y < 40$ (Figure 4.12).

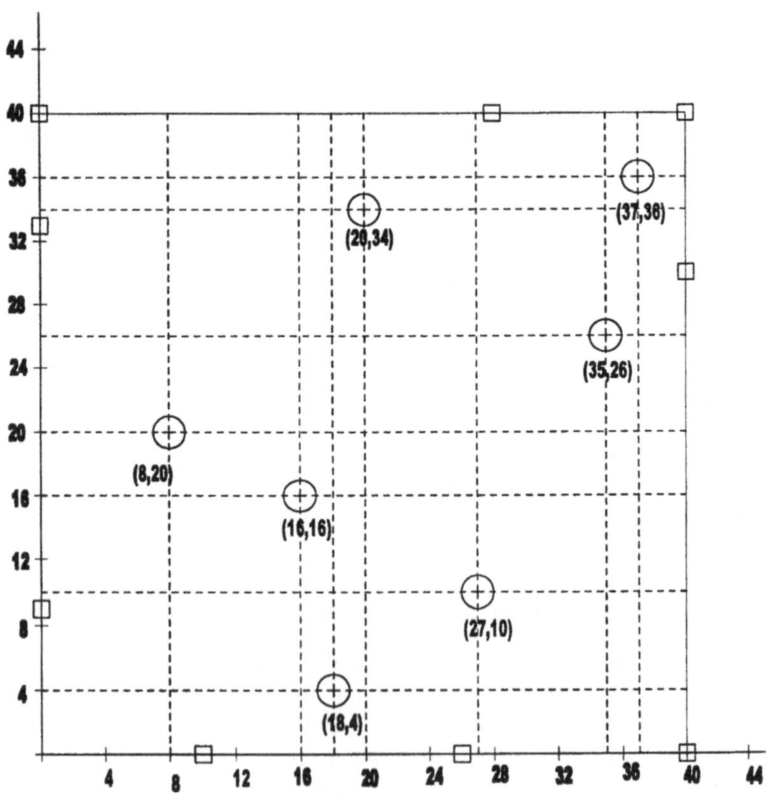

FIGURE 4.12 Locations and coordinates: ○, existing locations; □, new locations.

The new facility should be as far away as possible but within the desired boundary.

The first step of the solution procedure is to draw horizontal and vertical lines through every demand point. This should given $(n + 1)^2$ rectangles, where n is the number of the existing facilities. Henceforth, we refer to these rectangles as cells. Figure 4.12 shows this construction.

Consider a largest rectangle with one of the existing location on the corner. Find the farthest point in that rectangle from this existing facility. This point is the opposite corner point of that triangle. Calculate the rectilinear distance to this from a facility that is closest to it. Repeat the process for each existing facility.

Next, consider existing locations in pairs and find the point that is farthest away from them. This point is also equidistant from these two facilities. It is achieved by constructing the largest possible rectangle (cell), with the pair of demand points under consideration. For this cell, we need to find the locus of all the points that are equidistant from these two demand points. Such a locus is a line that is constructed geometrically using the following constructive procedure. Draw a line at 45 degrees from the perimeter of the cell to the midpoint (bisector) of the straight-line joining the existing demand points. The 45-degree line should be taken from the side of the cell that is the longest. Because this is an equidistant line, it should be extended to the maximum possible, and the rectilinear distance should be measured from the farthest point found, unless there is another existing facility, distance between that and the new point is smaller. The procedure is shown in Figure 4.13.

Perform similar calculations for all the points and determine the shortest rectilinear distance between each point and the facility that is closest to it. Choose from among these distances the largest value, and the point associated with it is the location for the obnoxious facility.

Consider the problem again, observing dominating (largest) rectangle with a single facility on the corner, for example, point (20, 34), and the dominating rectangle (0, 40), (0, 34), (20, 34), and (20, 40). This rectangle dominates smaller rectangles within it, such as (8, 40), (8, 34), (20, 34), and (20, 40). The farthest point from the existing facility is on the opposite corner [i.e., (0, 40)]. This point is noted and the rectilinear distance from the facility closest to it, in this case the facility on (20, 34) is determined. It is not necessary to consider rectangles (0, 26), (0, 34), (20, 34) or (20, 26), because just by observation we can see that the point opposite to (20, 34)in this rectangle, namely (0, 26), is quite closer to point (8, 20) than the distance we observed before. Thus, judicial selections of points will reduce the necessary evaluations considerably. Table 4.1 shows some of the points that are evaluated.

Points of interest considering two points at a time are listed in Table 4.2. These points are selected by looking ahead so that any one of them may be the point farthest away from an existing facility. If a point is clearly not going to be a

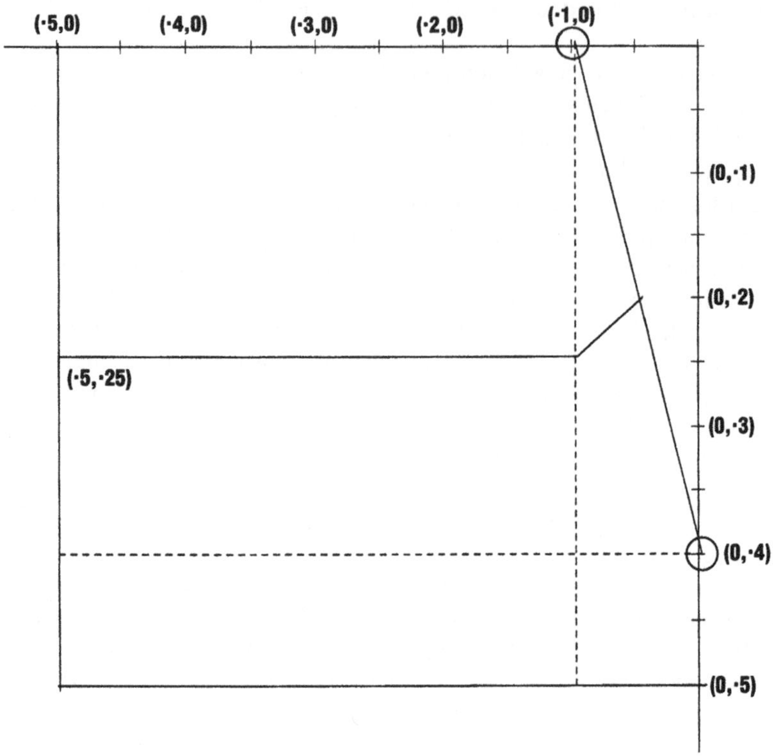

FIGURE 4.13 Drawing of equidistance line.

TABLE 4.1 Evaluated Points and Associated Distances

Point	Farthest point	Rectilinear distance
(20, 34)	(0, 40)	26
(37, 36)	(40, 40)	7
(27, 10)	(40, 0)	23
(18, 4)	(0, 0)	22
(8, 20)	(0, 40)	26 from point (20, 34)
(8, 20)	(0, 0)	22 from (18, 4)

point of contention, there is no need to evaluate that point. Drawing the diagram to equal scale on both x and y axis is very helpful in visualizing these points.

For example, point (20, 34) and (37, 36) are on the opposite corners of a rectangle, as shown in Figure 4.12. To obtain the locus of equidistant points,

TABLE 4.2 Evaluated Points from a Pair of Existing Locations

Points	Farthest point	Rectilinear distance
(8, 20)(20, 34)	(0, 33)	21
(20, 34)(37, 36)	(27.5, 40)	13.5
(35, 26)(27, 10)	(40, 30)	9
(18, 4)(27, 10)	(25.5, 0.0)	11.5
(18, 4)(16, 16)	(0.0,9.0)	23
(8, 20)(16, 16)	(10.0, 0.0)	22

connect these two points and find the midpoint of the diagonal. Draw a line with a 45-degree angle to the length of the rectangle passing the line through the midpoint. This equidistant locus can be extended toward the boundary by drawing a line perpendicular to the side of the rectangle from the point of intersection. Observe that as we go farther out, the rectangular distance between a point on the line and existing facility increases.

If we observe Table 4.1 and 4.2, we see the maximum distance is 26 associated with point (0, 40) and, therefore, the good location for an obnoxious facility is (0, 40).

4.9 LOCATION OF A LINEAR PATH FACILITY

Designing a transportation access path, such as a road or a conveyer lane, that serves all customers in the optimum manner is another application of facility location problems. The facility to be located is now a road or a conveyor that serves numerous destinations or customers. And the best manner of placement of such facility is to minimize the sum of perpendicular distance from each customer on to the facility, representing the shortest distance traveled by the customer to reach the facility. The facility or road, therefore, can be considered as the main road, and the roads connecting customers to the main road are the feeder roads; we want to minimize the total length of the feeder roads, assuming each customer has to have one feeder road.

The problem becomes somewhat complex as each customer has a weight representing perhaps the number of trips made by that customer to the main road, or the value or weight associated with demand from that customer as compared with other customers. If we plot these customers on an x–y graph, call them points, there are numerous lines, representing the main road that can be drawn among these points. However, two observations (proved by the number of applications) serves to reduce this number considerably. First the road must pass through at least two existing points, and second, it serves as a median; that is, there are some points above the line and there are some points below the line. We will utilize these properties in solution procedures.

The first procedure is to merely to try to draw a line through each of two-point combinations. Determine the perpendicular distance from all remaining points on this line and obtain the sum. The line with the minimum cost is the optimum path We can draw the line joining points 1 and 2 by first calculating the slope m as $(y2 - y1)/(x2 - x1)$, and then extend the line to intersect in y axis at c. The perpendicular distance d from any other point to the line is given by the formula:

$$\text{Perpendicular distance } d_i = |c - (y_i - m^* x_i)|/(m^2 + 1)^{1/2}$$

Figure 4.14 shows the pictorial view of a typical linear facility layout. For example consider two points (2, 6) and (5, 7). For the line joining these two points, the slope is $m = (7 - 6)/(5 - 2) = 1/3$. The intercept can be obtained by substituting coordinates for one of these two points say point 1. $y = mx + c$ or $6 = 1/3^*2 + c$ or $c = 16/3$.

The distance from a point (3,1) is given as

$$d = |16/3 - (1 - 1/3^*3)|/\sqrt{(1/9 + 1)} = 5.06$$

With the help of computers we can easily program the procedure and obtain the optimum path by examining all possible pair combinations. The number of combinations equals ${}^nC_2$, which are not excessive. For example, even for 100 customers the combinations to examine would be $100 \times 99/2 = 4950$. With a modern computer this does not take very long.

For those who want a quick and good approximate solution, which is often very close to optimum if not optimum, the following procedure is suggested. This procedure reduces the number of combinations that must be examined to a maximum of 8.

Rank the abscissa (and ordinates) of points in increasing order of weights. Determine the points that are on 33.33 and 66.66 percentile. Lines joining combinations associated with x and y coordinates are the prospective lines. Evaluate each set and select the best line.

Consider following example of 8 customers. The data is given in Table 4.3.

With $8 \times 7/2 = 28$ different pairs through which lines can be drawn, following the first method we examine them all. The results are shown in Table 4.4.

The optimum line passes through points 4–5, 4–8, and 5–8 with total perpendicular distances from other points being 22.50.

Application of the approximate method requires ranking of the points as based on their weights results in Table 4.5.

The 33.33 percentile is $15^*0.333 = 5$, with customer 4 based on the abscissa and customer 8 based on the ordinate. Similarly, the 66.66 percentile

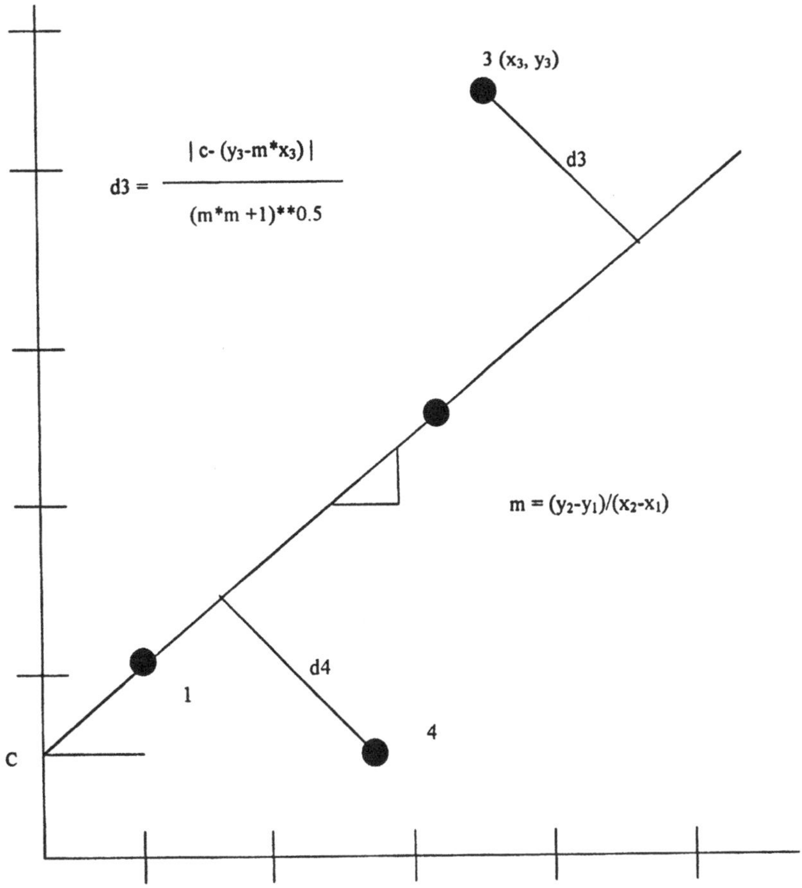

FIGURE 4.14 Linear facility: perpendicular distances.

TABLE 4.3 Data for Linear Facility Location

Customer number	Coordinates	Weights
1	(2, 6)	1
2	(5, 7)	3
3	(3, 1)	1
4	(4, 4)	2
5	(1, 5)	2
6	(8, 8)	1
7	(6, 2)	3
8	(7, 3)	2

TABLE 4.4 Pair of Customers on the Linear Path and Total of Perpendicular Distances from All Other Customers

Pair	Distance	Pair	Distance	Pair	Distance	Pair	Distance
1–2	35.39	2–3	23.92	3–5	28.17	4–8	22.50
1–3	39.15	2–4	23.92	3–6	28.75	5–6	30.74
1–4	22.62	2–5	32.62	3–7	35.40	5–7	31.24
1–5	45.25	2–6	35.40	3–8	37.11	5–8	22.50
1–6	35.40	2–7	23.31	4–5	22.50	6–7	31.24
1–7	22.62	2–8	28.17	4–6	28.28	6–8	38.13
1–8	25.52	3–4	23.92	4–7	22.62	7–8	39.59

TABLE 4.5 Ranking of Customers Based on Weights

x-Axis				*y*-Axis			
Customer	Abscissa	Weight	Cumulative weight	Customer	Ordinate	Weight	Cumulative weight
5	1	2	2	3	1	1	1
1	2	1	3	7	2	3	4
3	3	1	4	8	3	2	6
4	4	2	6	4	4	2	8
2	5	3	9	5	5	2	10
7	6	3	12	1	6	1	11
8	7	2	14	2	7	3	14
6	8	1	15	6	8	1	15

is 15*0.55 = 9.9. From the abscissa we select customers 7 and from the ordinates we select customer 5. The customer pairs we can form and associated costs are as follows:

Pairs	Cost
4, 7	22.62
8, 7	39.59
4, 5	22.50
8, 5	22.50

The line passing through customer pairs 4 and 5 or 8 and 5 gives the best solution of 22.50.

Although in this problem the procedure did select the optimum pair, it does not guarantee the optimum solution every time. It does give a very good solution.

4.10 FACILITY LOCATION WITH LIMITED DISTANCE

On occasion, the service provided by the facility, after it has exceeded a critical value, is insensitive to a measuring criteria, such as distance or time. For example, when renting a machine from a rental shop, the charges may be by hour or day. After a certain number of hours, say about 5 h, renting by day (8 h) is cheaper than the hourly rate. Therefore, 5 h becomes the critical value. Before the critical value, charges are proportional to the time, and after that value, the charges are constant (assuming 8 h is the maximum time we need the machine).

In some other problem of placing a new facility to serve the customers with different demands we may have an alternative. There might be a stand-in service facility, or an outside agency that can provide service to any customer with the cost equal to its critical value if a customer cannot be served by the new facility within the critical cost. Here, there might be two objectives:

1. To minimize the maximum total cost for serving all the customers at the same time
2. To minimize the maximum cost of serving any customer

We can formulate this as a set of mixed integer programming problems (Drezner et al. *Transportation Science* 25(3) 1991). However, a simple procedure that gives close to the optimum solution for the first objective is as follows:

1. Apply the 50th percentile method to the existing points (call it the list) and determine the coordinates of the 50th percentile point.
2. Find the distance of all the existing points to the 50th percentile point. If this distance, for any point, exceeds the maximum set distance, reduce the value of that distance to the maximum set distance.
3. Calculate the cost of travel from the new location to all the existing points. The cost is calculated as:

$$\text{Cost} = (\text{distance})^{*}(\text{weight of that point})$$

4. Find the total cost for serving all customers; also note the maximum cost.
5. For the next iteration, drop the customer with the least weight among all the existing customers from the list for future calculations.
6. Using the remaining customers, apply the 50th percentile method and determine the coordinates of the 50th percentile point for these points. Repeat the steps 2, 3, 4, 5, and 6 until no customer is left in the list.
7. Go on repeating this procedure, every time dropping the point with the next minimum weight until we reach only one customer.

To find the point on the existing network that serves to minimize the maximum total cost, compare the total costs in each iteration and select the value that is minimum. The coordinates of the 50th percentile point that corresponds to the associated iteration ware the coordinates of the ideal location to place new facility to meet the first objective.

To find the point on the network that serves to minimize the maximum cost to serve one customer, we can similarly develop a procedure using weighted customer algorithm described in Section 4.4. We will, however, illustrate only the first procedure here.

Example

Consider the following problem (data in Table 4.6) with 8 existing customers to be served by a new facility. The maximum distance, or critical value for the distance, is 10 units. The travel is rectilinear.

Iteration 1

Step 1. Apply the 50th percentile method to all existing points

For the x coordinates

x-Coordinate	Weight	Cumulative
2	2	2
3	3	5
3	9	14
4	15	29
6	11	40
8	19	59
8	9	68
13	4	72
20	13	85

50th percentile: $85/2 = 42.5$.
Hence, x-coordinate: 8

TABLE 4.6 Data for Limited Distance Problem

Existing facilities	1	2	3	4	5	6	7	8	9
Coordinates	(2, 6)	(3, 9)	(8, 15)	(13, 6)	(4, 19)	(20, 2)	(6, 15)	(8, 27)	(3, 19)
Weights	2	3	9	4	15	13	11	19	9

For y-coordinate

y-Coordinate	Weight	Cumulative
2	13	13
6	2	15
6	4	19
9	3	22
15	11	33
15	9	42
19	15	57
19	9	66
27	19	85

50th percentile: $85/2 = 42.5$
Hence, y-coordinates: 19
Coordinates of the 50th percentile point are (8, 19).

Steps 2, 3, 4 and 5. Distances of the various points for the 50th percentile point with the third column representing the converted values of the distances that exceed the maximum set distance of 10 units are given in Table 4.7. The table also shows the cost.

Iteration 2

For iteration 2, drop the point with the minimum weight among all the existing points; that is, point 1 with the coordinates of (2, 6) and with a weight of 2.

TABLE 4.7 Cost Calculations: Iteration 1

No. of customer	Distance from 50th percentile point	Converted distance	Weight	Weight* distance	Weight* distance (considering converted distance)
1	19	10	2	38	20
2	15	10	3	45	30
3	4		9	36	36
4	18	10	4	72	40
5	4		15	60	60
6	29	10	13	377	130
7	6		11	66	66
8	8		19	152	152
9	5		9	45	45

Step 1. Apply the 50th percentile to all the remaining points.

For x-coordinate

x-Coordinate	Weight	Cumulative
3	3	3
3	9	12
4	15	27
6	11	38
8	19	57
8	9	66
13	4	70
20	13	83

50th percentile: $83/2 = 41.5$.
Hence, x-coordinate: 8.
For y-coordinate

y-Coordinate	Weight	Cumulative
2	13	13
6	4	17
9	3	20
15	11	31
15	9	40
19	15	55
19	9	64
27	19	83

50th percentile: $83/2 = 41.5$.
Hence, y-coordinate: 19.
Coordinates of the 50th percentile point are (8, 19).

As there is no change in the location of the 50th percentile point, there is no need to check the costs of transportation for this iteration.

Subsequent iteration results are tabulated in Table 4.8. The minimum cost of 564 is associated with point (4,15); hence, that is the location for the new facility. As a reference the table also shows the maximum cost associated with any one customer when the associated 50th percentile point is used as a facility location.

TABLE 4.8 Summary of Results

Iteration number	Customers considered for 50th percentile	50th percentile point for iteration	Total cost in serving all customers	Maximum cost in iteration
1	1, 2, 3, 4, 5, 6, 7, 8, 9	(8, 19)	579	152
2	2, 3, 4, 5, 6, 7, 8, 9	(8, 19)	579	152
3	3, 4, 5, 6, 7, 8, 9	(8, 19)	579	152
4	3, 5, 6, 7, 8, 9	(8, 19)	579	152
5	5, 6, 7, 8, 9	(4, 15)	564	190
6	5, 6, 7, 8	(6, 19)	565	190
7	5, 6, 8	(8, 19)	579	152
8	5, 8	(8, 27)	660	150
9	8	(8, 27)	660	150

Exercises

4.1 A new community hospital is to be located in Shreveport. The hospital has identified 5 communities that will be benefited by it. They all have the same need for hospital. The 5 communities exist on the following coordinates:

1. (8, 10)
2. (30, 15)
3. (2, 4)
4. (25, 18)
5. (14, 20)

Assume that the travel is measured in rectilinear distance and the measuring criteria to locate the hospital is minimization of the maximum time any patient from these 5 communities will take to reach the hospital. Determine the optimum location for the hospital.

4.2 What is the Tchebychev distance. Solve problem 1, by using Tchebychev distances. Do you obtain the same optimum location as in problem 4.1?

4.3 In problem 4.1, depending on the population, each community is assigned a weight as shown. Determine the optimum location of the hospital assuming the number of calls to the hospital is directly proportional to the population of the community.

	Coordinate	Weight
1.	(8, 10)	3
2.	(30, 15)	3
3.	(2, 4)	4

4.	(25, 18)	2
5.	(14, 20)	1

4.4 A new FM radio station is to be set up in the town. It has identified six core regions to which the station will broadcast. To do so, it needs to determine the location of the transmitter. Because the cost is directly proportional to the distance the signal needs to travel, the location should be such that it minimizes the cost and serves al the identified six regions. The distances considered are direct distances. Identify the location of the transmitter. The coordinates of the six regions are

1. (5, 9)
2. (12, 8)
3. (10, 2)
4. (11, 18)
5. (6, 13)
6. (1, 2)

4.5 A towing company wants to open their service for six cities in the region. The location of the cities along with the estimated market for towing for each city are as follows:

	Coordinates	Market
1.	(3, 8)	8
2.	(10, 15)	10
3.	(9, 2)	6
4.	(5, 12)	9
5.	(12, 6)	12
6.	(6, 9)	5

The towing company charges the customer based on the distance it has to travel from its office location. If any towing truck is required to travel more than 12 units in a day, the distance consideration is ignored because the truck will not be available for other customers on that day. Determine the location of its office to minimize the total cost and serve all the cities.

4.6 Four different industries, namely foundry, lumber plant, cement factory, and coal processing plant, are located in a region. A rail line is to be passed near to these plants to minimize the transportation cost. Determine the linear path along which the rail line should e placed to achieve the objective. Depending on the amount shipped, each plant has its weight. The location of the plants along with their weights are

	Coordinates	Weight
Foundry	(3, 5)	5

Lumber	(8, 12)	3
Cement	(6, 10)	2
Coal	(10, 4)	2

4.7 A dump site is to be located to serve communities in the area shown in the following figure. Locate the dump site such that it is as far as possible from all the communities, but located within the area shown in the figure.
The communities are located as follows:

1. (1, 5)
2. (3, 2)
3. (6, 5)
4. (8, 1)

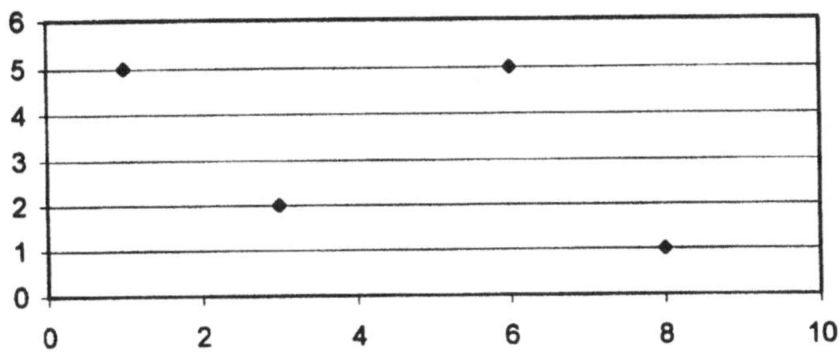

4.8 The isolation ward in a hospital needs to be far away from the cafeteria, operating room, and waiting room. The location of these facilities is as shown in the following figure. Determine the location of the isolation ward within the given area.

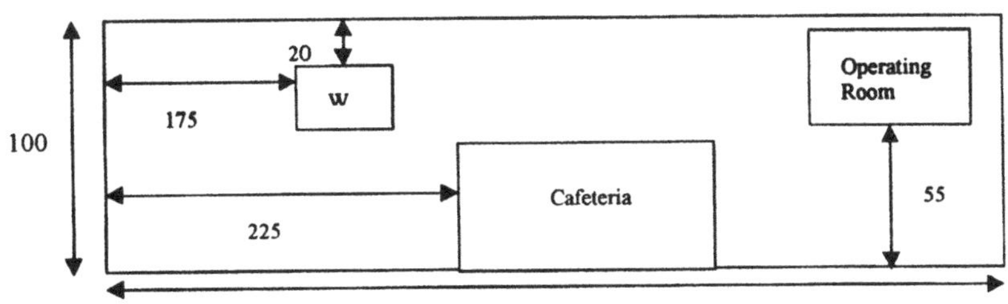

4.9 A major road is to be constructed that will serve five localities. The state government pays a major portion of the major road, but not for the feeder roads. At the same time the state government does expect to minimize its cost by constructing only a straight path. Determine the path that these localities should suggest to minimize the sum of feeder road cost (which they agree to share equally). The localities are situated at following locations:

1. (5, 9)
2. (12, 5)
3. (7, 15)
4. (6, 12)
5. (3, 6)

4.10 A central light is to be installed that will illuminate six parking areas. The cost of the light depends on the intensity of light required. Determine the location of the light to minimize its cost. The parking areas are located as follows:

P1(3, 8)
P2(8, 13)
P3(4, 10)
P4(6, 15)
P5(2, 7)
P6(13, 4)

4.11 In an hurricane-prone area, hurricane response teams are formed to provide quick support in cases of emergencies. It is believed that these teams should not be more than 2-h flying time from any of the areas they are assigned to serve. Assume that a unit of distance on the *x*–*y* plot is equal to 1 h of flying time. How many teams need to be formed and where they should be located to cover following areas?

Area	Frequency of hurricanes
(4, 8)	5
(10, 14)	3
(6, 10)	4
(1, 4)	3
(8, 12)	6
(2, 7)	2

5

Multiple Facility Location

So far we had been interested in optimally locating a single facility that is designed to serve all the existing customers. In Chapters 3 and 4, we outline numerous different objectives that can be associated with such placement and observed that each objective may lead to a different solution procedure and a different optimum location.

In this chapter, we extend the analysis further to encompass multiple facilities. These varied facilities can serve the customers in the same or in different ways, and the objective is to find the optimum site for each facility. The problem of this type is designated multifacility location problem.

There are two prominent variations in multiple facility problems. They are

1. Multiple facilities of different types: Each type is independently demanded by the existing customers. Furthermore, there may also be a demand in between the new facilities.
2. Multiple facilities of the same type: All facilities serve the same purpose. Customers are divided among them such that any one customer is served by only one new facility.

5.1 MULTIPLE FACILITIES OF DIFFERENT TYPES

Cost once more becomes the major factor in the location decisions. When the cost of travel between the new facilities and the customers and also among the new facilities, is proportional to the rectilinear distance, then the solution procedure presented next is applied. The quadratic and euclidean costs also follow a similar procedure and are discussed in the subsequent section.

5.1.1 Rectilinear Cost

Consider a manufacturing plant within which four machines (customers) will require the services of two new facilities that are to be installed in the plant. The present location of each machine and the expected number of trips between them and each of the new facilities are given in the Table 5.1. There are also nine trips expected between the two new facilities. The problem is to determine the optimum locations for the new facilities.

5.1.2 Linear-Programming Formulation

The problem can be formulated as a linear-programming problem. (For a similar procedure, go to 5.1.3.) As usual the coordinates of the ith existing facility are denoted by (a_i, b_i) and that of the new facilities denoted by (x_i, y_i). Let us develop the model for two new facilities p and q. Let the flow between the new facility p and the existing facility i be denoted by f_{pi} and that the cost per unit distance of travel by m_{pi}. Also let the flow between new facilities p and q be denoted by t_{pq} and the cost per unit distance of travel by n_{pq}.

The objective function then is to minimize

$$t_{pq}n_{pq}[|x_p - x_q| + |y_p - y_q|] + \sum_{i=1}^{n} f_{pi}m_{pi}[|x_p - a_i| + |y_p - b_i|] + \sum_{i=1}^{n} f_{qi}m_{qi}[|x_q - a_i| + |y_q - b_i|]$$

The model is linear, but cannot be easily optimized because of absolute value terms. Fortunately, the following substitutions can convert it to a linear-programming problem. Let,

$$x_{pq}^{+} = \{x_p - x_q\} \quad \text{if} \quad (x_p - x_q) >= 0 \quad \text{otherwise } 0$$
$$x_{pq}^{-} = \{x_q - x_p\} \quad \text{if} \quad (x_p - x_q) <= 0 \quad \text{otherwise } 0$$

TABLE 5.1 Multiple Facilities of Different Types

		Number of trips	
Machine	Coordinates (a, b)	Facility 1	Facility 2
1	10, 15	10	8
2	8, 20	4	6
3	4, 7	6	12
4	15, 9	0	30
F1	$x1, y1$	—	9
F2	$x2, y2$	9	—

We can therefore say that when

$$x_p < x_q$$

$$|x_p - x_q| - x_{pq}^+ + x_{pq}^-$$

otherwise

$$x_p - x_q = x_{pq}^+ - x_{pq}^-$$

Similarly,

$$|y_p - y_q| = y_{pq}^+ + y_{pq}^-$$

$$y_p - y_q = y_{pq}^+ - y_{pq}^-$$

$$|x_p - a_i| = xa_{pi}^+ + xa_{pi}^-$$

$$x_p - a_i = xa_{pi}^+ - xa_{pi}^-$$

$$|y_p - b_i| = yb_{pi}^+ + yb_{pi}^-$$

$$y_p - b_i = yb_{pi}^+ - yb_{pi}^-$$

Therefore, the objective function is to minimize

$$t_{pq}n_{pq}(x_{pq}^+ + x_{pq}^- + y_{pq}^+ + y_{pq}^-) + \sum_i^n f_{pi}m_{pi}(xa_{pi}^+ + xa_{pi}^- + yb_{pi}^+ yb_{pi}^-)$$
$$+ \sum_{i=1}^n f_{qi}m_{qi}(xa_{qi}^+ + xa_{qi}^- + yb_{qi}^+ yb_{qi}^-)$$

Subject to

$$x_{pq}^+, x_{pq}^-, y_{pq}^+, y_{pq}^- >= 0$$

$$xa_{pi}^+, xa_{pi}^-, yb_{pi}^+, yb_{pi}^- >= 0$$

x_p, y_q are unrestricted in sign.

Now, in our problem, if (x_5, y_5) and (x_6, y_6) denote the location of the two new facilities then the objective function will be

$$9x_{56}^+ + +9x_{56}^- + 9y_{56}^+ + 9y_{56}^- + 9x_{65}^+ - 9x_{65}^- + 9y_{65}^+ + 9y_{65}^- + 10xa_{51}^+$$
$$+ 10xa_{51}^- + 10yb_{51}^+ + 10yb_{51}^- + 4xa_{52}^+ + 4xa_{52}^- + 4yb_{52}^+ + 4yb_{52}^-$$
$$+ 6xa_{53}^+ + 6xa_{53}^- + 6yb_{53}^+ + 6yb_{53}^- + 8xa_{61}^+ + 8xa_{61}^- + 8yb_{61}^+ + 8yb_{61}^-$$
$$+ 6xa_{62}^+ + 6xa_{62}^- + 6yb_{62}^+ + 6yb_{62}^- + 12xa_{63}^+ + 12xa_{63}^- + 12yb_{63}^+$$
$$+ 12yb_{63}^- + 30xa_{64}^+ + 30xa_{64}^- + 30yb_{64}^+ + 30yb_{64}^-.$$

subject to

$$-x_{56}^{+} + x_{56}^{-} + x_5 - x_6 = 0 \quad (1)$$
$$-x_{65}^{+} + x_{65}^{-} - x_5 + x_6 = 0 \quad (2)$$
$$-y_{56}^{+} + y_{56}^{-} + y_5 - y_6 = 0 \quad (3)$$
$$-y_{65}^{+} + y_{65}^{-} - y_5 + y_6 = 0 \quad (4)$$
$$-xa_{51}^{+} + xa_{51}^{-} + x_5 = 10 \quad (5)$$
$$-xa_{52}^{+} + xa_{52}^{-} + x_5 = 8 \quad (6)$$
$$-xa_{53}^{+} + xa_{53}^{-} + x_5 = 4 \quad (7)$$
$$-xa_{54}^{+} + xa_{54}^{-} + x_5 = 15 \quad (8)$$
$$-xa_{61}^{+} + xa_{61}^{-} + x_6 = 10 \quad (9)$$
$$-xa_{62}^{+} + xa_{62}^{-} + x_6 = 8 \quad (10)$$
$$-xa_{63}^{+} + xa_{63}^{-} + x_6 = 4 \quad (11)$$
$$-xa_{64}^{+} + xa_{64}^{-} + x_6 = 15 \quad (12)$$
$$-yb_{51}^{+} + yb_{51}^{-} + y_5 = 15 \quad (13)$$
$$-yb_{52}^{+} + yb_{52}^{-} + y_5 = 20 \quad (14)$$
$$-yb_{53}^{+} + yb_{53}^{-} + y_5 = 7 \quad (15)$$
$$-yb_{54}^{+} + yb_{54}^{-} + y_5 = 9 \quad (16)$$
$$-yb_{61}^{+} + yb_{61}^{-} + y_6 = 15 \quad (17)$$
$$-yb_{62}^{+} + yb_{62}^{-} + y_6 = 20 \quad (18)$$
$$-yb_{63}^{+} + yb_{63}^{-} + y_6 = 7 \quad (19)$$
$$-yb_{64}^{+} + yb_{64}^{-} + y_6 = 9 \quad (20)$$

when

$$x_{pq}^{+}, x_{pq}^{-}, y_{pq}^{+}, y_{pq}^{-} >= 0 \qquad p, q = 1, 2, 3, \ldots, n$$
$$xa_{pq}^{+}, xa_{pq}^{-}, yb_{pq}^{+}, yb_{pq}^{-} >= 0$$

Constraints set (1) through (4) define the variables such as $x_p - x_q - x_{pq}^{+} - x_{pq}^{-}$ whereas the constraints set (5) through (20) define the geometry using equations such as $x_p - a_i = xa_{pi}^{+} - xa_{pi}^{-}$.

The solution to the linear programming is that both new facilities should be located at the same location (10, 9) and with the associated minimum cost of 532.

5.1.3 Iterative Procedure

Even though linear-programming formulation gives the optimum solution, it requires a careful formulation. For large number of locations the problem dimension (i.e., the number of variables and constraints), can become fairly unwieldy. An alternative and simpler procedure is that of the iterative technique in which the new facilities are located one at a time.

We will begin the process, called "one-at-a-time," by determining the abscissa for the new facilities. As a first step, ignore the trips between the new facilities and determine the x-axis value for facility 1 (F1) relative to the present machines (customers). Recall from Section 3.1 that the procedure calls for listing the present machines in order of increasing abscissa as shown in Table 5.2, and then, selecting the x value associated with the 50th percentile of the cumulative number of trips.

The 50th percentile is ten trips (20/2), and the corresponding abscissa is 8. Therefore, as a first try the abscissa for the first facility is fixed at 8.

Repeat the process for the second facility (F2), except that we will also consider the trips between the facilities because the first facility is at least temporarily located on the x-axis. Table 5.3 lists the data in proper sequence to make the determination.

The 50th percentile is 32.5, and the associated x value, which must be the same as that of one of the existing customers, is 10. Therefore, the temporary abscissa for the second facility is 10.

Now, go back and determine the abscissa for facility 1 knowing that the temporary x location for facility 2 is 10. Table 5.4 shows the necessary update of Table 5.2 to find the new F1 abscissa.

The 50th percentile, 14.5, is associated with $x = 10$, and the new temporary abscissa for F1 is different from the previous value for the same facility. The process must be repeated with $x = 10$ as the new temporary abscissa for facility 1. The details are shown in Table 5.5.

TABLE 5.2 x-Axis (Abscissa) Data, Facility 1 (1st Iteration)

Machine	Abscissa	Number of trips	Cumulative trips
3	4	6	6
2	8	4	10
1	10	10	20
4	15	0	20

TABLE 5.3 x-Axis (Abscissa) Data, Facility 2 (1st Iteration)

Machine	Abscissa	Number of trips	Cumulative trips
3	4	12	12
2	8	6	18
F1	8	9	27
1	10	8	35
4	15	30	65

TABLE 5.4 x-Axis (Abscissa) Data, Facility 1 (2nd Iteration)

Machine	Abscissa	Number of trips	Cumulative trips
3	4	6	6
2	8	4	10
1	10	10	20
F2	10	9	29
4	15	0	29

TABLE 5.5 x-Axis (Abscissa) Data, Facility 2 (2nd Iteration)

Machine	Abscissa	Number of trips	Cumulative trips
3	4	12	12
2	8	6	18
1	10	8	26
F1	10	9	35
4	15	30	65

The abscissa associated with the 50th percentile of 32.5 trips is 10 (i.e., $x = 10$), which is the same as before. Therefore, the search for the abscissa is complete and the latest temporary value becomes permanent.

Similarly, the search process for the F1 and F2 ordinates leads to the following tables. Keep in mind that the nine trips between the two new facilities are ignored during the first iteration (Table 5.6).

The median cumulative trip value is 20/2; therefore, the temporary ordinate is 15 (machine 1). Table 5.7 presents the data in the appropriate order to determine the first temporary ordinate for F2.

The median cumulative trip value is 32.5, and this corresponds with an ordinate of 9, determined by machine 4. Table 5.8 continues the procedure providing the 2nd iteration data for F1.

The median cumulative trip value is 14.5, which equates to an ordinate of 9, determined by facility 2. Because this is a value different from that found during

TABLE 5.6 y-Axis (Ordinate) Data, Facility 1 (1st Iteration)

Machine	Ordinate	Number of trips	Cumulative trips
3	7	6	6
4	9	0	6
1	15	10	16
2	20	4	20

TABLE 5.7 *y*-Axis (Ordinate) Data, Facility 2 (1st Iteration)

Machine	Ordinate	Number of trips	Cumulative trips
3	7	12	12
4	9	30	42
1	9	8	50
F1	15	9	59
2	20	6	65

TABLE 5.8 *y*-Axis (Ordinate) Data, Facility 1 (2nd Iteration)

Machine	Ordinate	Number of trips	Cumulative trips
3	7	6	6
4	9	0	6
F2	9	9	15
1	15	10	25
2	20	4	29

the first iteration (see Table 5.6), we must perform at least one more iteration, as shown in Table 5.9.

The median is 32.5, and this leads to an ordinate of 9, determined by machine 4. This y value remains unchanged from that of the first iteration; hence, the procedure terminates. The optimum locations for the new facilities are

Facility 1 (10, 9)
Facility 2 (10, 9)

This is the same solution that we obtained earlier with linear programming. The solution yields the same location for both facilities, which could present a problem. In practice, the facilities would be set side by side within existing practical limitations, such as the shape and size of the facilities, space required for

TABLE 5.9 *y*-Axis (Ordinate) Data, Facility 2 (2nd Iteration)

Machine	Ordinate	Number of trips	Cumulative trips
3	7	12	12
4	9	30	42
F1	9	9	51
1	15	8	59
2	20	6	65

operating and maintenance personnel, and the area necessary for material flow and inventory storage.

5.2 QUADRATIC AND EUCLIDEAN DISTANCE COST

The one-at-a-time method is also applicable when the cost function is either quadratic or euclidean. The only modification needed is using the appropriate placement method as described in Sections 3.3 and 3.4.

5.3 MULTIPLE FACILITIES OF THE SAME TYPE

5.3.1 Rectilinear Cost

With multiple facilities that are of the same type, the problem of facility placement becomes that of determining the demand points (customers) that should be assigned to each facility and then deciding each facility's location so that it will minimize the total cost of travel from the customers.

The linear-programming formulation that worked so well in the previous section cannot be applied directly to this problem. Here, the customer needs service from only one new facility, and in formulating the problem by linear programming, difficulty arises in determining which customers should be served by which new facility because we do not even know the location of the new facility. In fact, these two are the decision variables of the problem.

5.3.2 One-Dimensional Problem

For one-dimensional problems, that is, where all the existing facilities are located on either the x or y direction, a dynamic-programming approach proves to be very efficient. One-dimensional problems are not rare. Placement of stores, warehouses, repair facilities, distribution centers, manufacturing facilities along a transportation flow artery, such as road, railroad, or a river, can be analyzed as a one-dimensional problem.

This type of problem is best handled by dynamic programming and, therefore, before addressing the application of dynamic programming in our problem, we first review the basic procedure of dynamic programming in the next section.

5.3.3 Dynamic Programming

Dynamic programming is a mathematical technique for optimizing a multistage decision process. A problem requiring "n" decisions simultaneously is broken down into n sequential problems requiring only one decision at a time. It is much easier to work with one variable at a time than to manipulate n variables simultaneously.

The procedure was first developed by Richard Bellman (1957) in the 1950s. It has been applied to many areas, including production and maintenance planning, reliability, and financial analysis. The following allocation example illustrates the basics of the procedures, which is then applied to the location analysis.

Allocation Problem

Suppose a chain department store has four retail locations within a single city. The purchasing agent has just bought six television (TV) sets from a major manufacturer, and he wants to distribute these in the four stores. Previous experience has indicated that the number of television sets that can be sold within a specific time varies from store to store. It depends on, for example, the location and appearance of the store, the clientele, and the sales personnel within a store. This also makes the profit from the sale of a certain number of television sets vary, because the expense is different at each store. The expected amount of profit if certain numbers of televisions are placed in a store is given in Table 5.10. The question is: How should the TV sets be distributed among the stores?

There are many ways of distributing the sets, which may realize different profits. For example, store A may receive two sets, store B two, and stores C and D one each, for a total profit of $95 + 90 + 60 + 50 = $295; or all six sets can be assigned to store A for the total profit of $240. There are many other combinations, and it is possible to examine each, but clearly it would be a very tedious exercise in the view of the fact that there can be hundreds of such combinations.

In examining each combination, we are asked to form the combination first. This is equivalent to asking for an assignment in each of the four stores, or

TABLE 5.10 Profit Data

	Stores			
TV	A	B	C	D
0	0	0	0	0
1	50	45	60	50
2	95	90	110	100
3	135	135	160	145
4	175	180	200	190
5	210	220	230	235
6	240	250	260	250

making four assignment decisions simultaneously. This has created, in many other examples, almost unmanageable combinations.

The process can be simplified by applying dynamic programming. A four-decision problem (how much to allocate in each of the four stores) is now broken down into four problems in series. Graphically the process can be represented in the following sketch (Figure 5.1) in which each box represents a store. There are six sets requiring distribution in four stores.

The procedure called the "backward pass" starts at store D and proceeds backward toward store A. Suppose we are at store D with an unspecified number of sets to distribute. How many should we place in store D? This depends on how many are available, which we do not know. However, we can determine what to do if we assume different values of available sets. The minimum is when no set is available; the maximum is when all six sets are available. Table 5.11 shows what our decisions would be for different numbers of TV sets at our disposal. Obviously, we will assign whatever is available to the store to make the maximum profit. This is shown by the decision column in Table 5.11.

Now suppose we are in front of store C with our available TV sets. Once more the number available can vary anywhere from zero to six, and we are asked to distribute the sets between stores C and D such that the total profit is maximized. How can we achieve this?

The first possibility is the trivial case. Suppose we have no available sets; then no profit can be realized, as shown in Table 5.12, because no set can be placed in any store.

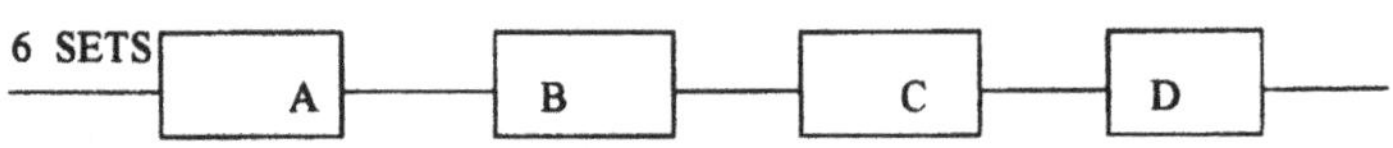

FIGURE 5.1 Graphical representation of the stages.

TABLE 5.11 Store D

Available TV sets	How many to assign	Profit	Decision for store D
0	0	0	0
1	1	50	1
2	2	100	2
3	3	145	3
4	4	190	4
5	5	235	5
6	6	250	6

TABLE 5.12 Store C

	To be assigned to store C								
Available sets	0	1	2	3	4	5	6	Maximum profit	Decision for store C
0	0							0	0
1	50	60						60	1
2	100	110	110					110	1, 2
3	145	160	160	160				160	1, 2, 3
4	190	205	210	210	200			210	2, 3
5	235	250	255	260	250	230		260	3
6	280	295	300	305	300	280	260	305	3

If we have one set available, then there are two alternatives. First, we may assign no set to store C, in which case the set can be assigned in the next store—namely, store D—resulting in a profit of 0 from store C and 50 from store D for a total of $50. The second alternative is to assign the set in store C, which will leave no set for the subsequent store, resulting in a $60 profit from store C and none from store D, for a total profit of $60. These numbers are shown in the Table 5.12 for one available set.

Similarly, if there are two sets available, then store C can get either 0, 1, or 2 sets. If no set is assigned to store C, it will result in no profit from store C, but the two sets then can be assigned in the next stor, resulting in $100 profit, and giving a total profit of $0 + 100 =$ $100.

If store C is given one set, it will generate $60 profit for the store. It will also leave a set for the following store, generating $50 profit there, for a total profit of $60 + 50 =$ $110. If both sets are assigned to store C, they will produce $110 profit from that store. There being no set left for the next store, no profit is generated there, resulting in total profit of $110 + 0 = 110$.

Similar analyses are performed for all possible values of available input, the maximum profit is found, and the corresponding value or decision variable is indicated. For example, if there are four available sets, the maximum profit is $210 and is associated with the decision of assigning either two or three sets to store C, as shown in the table.

Now, let us proceed one step farther backward, in front of store B. Again we will analyze the problem for a different number of available sets. For example, if we have two sets, we can assign either 0, 1, or 2 sets to store B, with the remainder being passed on to the following stores. If no units are assigned to store B, two units are available for the following stores. To obtain the profit from the succeeding stores, stores C and D, we do not need to recalculate the profits for different distributions in each store, but simply to refer to Table 5.12. As we have just seen, it shows the maximum cumulative profit for different numbers of available sets in front of store C, with the optimum distribution between stores C and D. With two units we can make a maximum of $110 profit, for a total profit of $0 + 110 = \$110$.

Similarly, if one set is assigned to store B, then the total profit is $90 + 60 = \$150$. The $90 is the profit from store B and the $60 is the maximum we can obtain from the remaining stores, with $2 - 1 = 1$ set going into the remaining stores.

The results of the calculations are shown in Table 5.13.

Let us now proceed one more step back to the front of store A. We know we have six sets here, and we are going to make the decision on how many should be assigned to store A and how many should be passed onto the remaining stores. Table 5.14 shows the result.

The procedure followed in developing Table 5.14 is the same as the one we used previously. For example, if out of six, two units are assigned to store A, we will obtain $95 from store A (see Table 5.10), and with the remaining $6 - 2 = 4$ units (see Table 5.13), we could have $210 from the other stores, resulting in a total profit of $95 + 210 = \$305$.

The maximum profit shown in Table 5.14 is $310. Let us retrace the steps to see what allotment gives us this profit. We proceed backward from Table 5.14

TABLE 5.13 Store B

	To be assigned to store B								
Available sets	0	1	2	3	4	5	6	Maximum profit	Decision for store B
0	0							0	0
1	60	45						60	0
2	110	105	90					110	0
3	160	155	150	135				160	0
4	210	205	200	195	180			210	0
5	260	255	250	245	220			260	0
6	305	305	300	295	290	280	250	305	0, 1

TABLE 5.14 Store A

Available sets	To be assigned to store A							Maximum profit	Decision for store A
	0	1	2	3	4	5	6		
6	305	310	305	295	285	270	240	310	1

to Table 5.11 to find this information. Table 5.14 indicates we should assign 1 unit to store A. With $6 - 1 = 5$ units remaining in front of store B, Table 5.13 lists the optimum assignment for that location as zero, leaving 5 units in front of store C. Going on to Table 5.12, we observe the assignment of 3 units to store C. That leaves $5 - 3 = 2$ units in front of store D. Table 5.11 advises that we should assign however much (in this case, two sets) is available to store D. The result is summarized as

Store	A	B	C	D
Sets assigned	1	0	3	2

This principle, dynamic programming, is based on what is called the principle of optimality. Simply stated it says: It does not matter what state we are presently in (i.e., how many TV sets we now have), or what stages we are in (i.e., in front of which store); from here on we must perform in an optimal manner. Each of the tables 5.10 through 5.14 were developed with this principle in mind.

5.3.4 Dynamic Programming Application to a One-Dimensional Problem

In dynamic programming we must first define the stages. In our problem, the number of stages is the number of new facilities that are to be located. We have seen in Chapter 3 that to serve a given set of customers in one dimension, the optimum location of a new facility coincides with one of the existing facilities. We use this result to examine only a select number of locations, associated with the existing facilities, as possible sites for the new facilities. Let us consider the following example to illustrate this approach.

There are six existing communities on an Interstate 20 highway, A through F. A gas company is interested in providing gas for each community. It believes it can build three storage tanks and supply gas to each community. The amount of gas each community needs is coded in terms of the weight (wt). The cost of transportation is directly proportional to the distance between the storage tank and the city (for convenience we will take the proportionality constant as 1); the coded

```
0      2   3      5      7   8     10
|------|---|------|------|---|-----|-------------------
0      A   B      C      D   E     F
```

TABLE 5.15 Data for Location of Storage Tanks

Existing facilities	A	B	C	D	E	F
Location	2	3	5	7	8	10
Weight	5	1	2	2.5	4	3

location of each community, in increasing order of x, is also noted; The data is given in Table 5.15.

Because we have three facilities to place, we have three stages in dynamic programming. In each stage we evaluate the total cost of locating a storage tank. As usual in dynamic programming, we start from the last stage, placement of the third facility. Let $f1(\ )$ denote the point from which all further locations are to be served by i facilities (storage tanks). For example, $f1(5)$ means all locations from and including 5 (locations at 5, 7, 8, and 10) are to be served by one facility, which may be located on any one of the existing locations, 5, 7, 8, or 10. Table 5.16 shows the cost calculation for stage 1. In calculating the cost, we must place the facility in each of the existing community locations and determine the resulting cost. For example, for $f1(8)$, we have two alternatives. Place the new facility at location 10 or place it on location 8. Both options are evaluated and the minimum cost 6 is chosen as the optimum cost; the associated location, location 8, is selected as the optimum location.

We have not shown in Table 5.16 the values for $f1(3)$ and $f1(2)$. In this problem with three facilities to place, there was indeed no need to check $f1(3)$ or

TABLE 5.16 Stage 1 Solution

F1(*)	Possible facility location	Associated cost	Minimum cost	Optimum location policy p1(*)
10	10	3*0 = 0	0	10
8	8	3*2 + 4*0 = 6	6	8
	10	3*0 + 4*2 = 8		
7	7	3*3 + 4*1 = 13	8.5	8
	8	3*2 + 2.5*1 = 8.5		
	10	3*0 + 4*2 + 2.5*3 = 15.5		
5	5	2*0 + 2.5*2 + 4*3 + 3*5 = 32	14.5	8
	7	2*2 + 2.5*0 + 4*1 + 3*3 = 17		
	8	2*3 + 2.5*1 + 4*0 + 3*2 = 14.5		
	10	2*5 + 2.5*3 + 4*2 + 3*0 = 25.5		

$f1(2)$. This is because, at this stage, we are placing the third facility. The third facility can never be placed in the location 2 or 3 because if optimally required we can always place the first facility in location 2 and the second facility in location 3.

Stage 2 is associated with placing the second facility so now we have two facilities to serve. The details of the calculations are shown in Table 5.17. There is no need to check location 10 because it can be served by one facility. In the table, under "Combinations," the first bold entry denotes the location of the first facility and its range shows the range of locations it is serving. The second facility is placed in the optimum location determined in the first stage and is indicated by $f(*)$. For example, **3**–5 $f1(7)$ means that the first facility is located at 3, serving location 3 and 5, and the second facility is serving optimally from location 7 on; the actual location of the facility was determined in stage 1 by policy $p1(7)$—in our case the facility is located at 8. The total cost is obtained by calculating the cost for policy **3**–5 and adding to it the optimum cost of $f1(7)$ from stage 1.

Again, following the same reasoning as mentioned in the first-stage calculations, $f2(2)$ is not checked.

TABLE 5.17 Stage 2 Solution

f2(*)	Combinations	Cost	Minimum cost	Optimum location policy p2(*)
8	**8** f1(**10**)	0	0	8, p1(10)
7	**7**–8 f1(**10**)	0 + 4 + 0 = 4	2.5	7–**8**, p1(10)
	7–**8** f1(**10**)	2.5 + 0 + 0 = 2.5		
	7 f1 (**8**)	0 + 6 = 6		
5	**5**–7–8 f1(**10**)	0 + 5 + 12 + 0 = 17	8	5–7–8, p1(10)
	5–**7**–8 f1(**10**)	4 + 0 + 4 + 0 = 8		
	5–7–**8** f1(**10**)	6 + 2.5 + 0 + 0 = 8.5		
	5–7 f1(**8**)	0 + 5 + 6 = 11		
	5–**7** f1(**8**)	4 + 0 + 6 = 10		
	5 f1(**7**)	0 + 8.5 = 8.5		
3	**3** f1(**5**)	0 + 14.5 = 14.5	10.5	3–**5**, p1(7)
	3–5 f1(**7**)	0 + 4 + 8.5 = 12.5		
	3–**5** f1(**7**)	2 + 0 + 8.5 = 10.5		
	3–5–7 f1(**8**)	0 + 4 + 10 + 6 = 20		
	3–**5**–7 f1(**8**)	2 + 0 + 5 + 6 = 13		
	3–5–**7** f1(**8**)	4 + 4 + 0 + 6 = 14		
	3–5–7–8 f1(**10**)	0 + 4 + 10 + 20 + 0 = 34		
	3–**5**–7–8 f1(**10**)	2 + 0 + 5 + 12 + 0 = 19		
	3–5–**7**–8 f1(**10**)	4 + 4 + 0 + 4 + 0 = 12		
	3–5–7–**8** f1(**10**)	5 + 6 + 2.5 + 0 + 0 = 13.5		

TABLE 5.18 Stage 3 Solution

f3(*)	Combinations	Cost	Minimum cost	Optimum location policy
2	2 f2(3)	0 + 10.5 = 10.5	9	2–3, p2(5)
	2–3 f2(5)	0 + 1 + 8 = 9		
	2–3 f2(5)	5 + 0 + 8 = 13		
	2–3–5 f2(7)	0 + 1 + 6 + 2.5 = 9.5		
	2–3–5 f2(7)	5 + 0 + 4 + 2.5 = 11.5		
	2–3–5 f2(7)	15 + 2 + 0 + 2.5 = 19.5		
	2–3–5–7 f2(8)	0 + 1 + 6 + 12.5 + 0 = 19.5		
	2–3–5–7 f2(8)	5 + 0 + 4 + 10 + 0 = 19.0		
	2–3–5–7 f2(8)	15 + 2 + 0 + 5 + 0 = 22.0		
	2–3–5–7 f2(8)	25 + 4 + 4 + 0 + 0 = 33		

In stage 3 (Table 5.18), one additional facility is placed making total of three facilities. Now we must cover the range of all facilities; that is, from location 2 through 10. Again, because we have three facilities to place, the last condition we want to check is when one facility covers from location 2 through 7, and the second and third facilities are placed on locations 8 and 10, respectively, that is, $p2(8)$. The calculations are similar to stage 2, except for the optimum policy for placement of next two storage facilities we refer back to stage 2 Table 5.17.

Thus the optimum locations for three facilities are 2–7–10 with the expected cost of 9 units.

5.3.5 Multiple Facility Placement in Two Dimensions

For the placement of multiple new facilities when the existing customer locations can be displayed in x–y coordinates is not an easy task. Again the problem is twofold, to decide where to locate the new facility and which customers should be served by which facility. Rather than working on these two decision problems simultaneously, we will divide it into two one-dimensional problems. First we will group the existing customers into clusters and then determine for each cluster the optimal location of the new facility. Finally, we will assign all the customers in each cluster to the new facility in that cluster.

The procedure to optimally locate one facility was described in Chapter 3, so here we will study the procedure to form the clusters. The number of clusters we want to form is equal to the number of new facilities we want to place. The procedure has the following steps:

1. Begin with the customer with the largest weight. Assume that the first facility is located at the coordinates of this customer.

2. Find the "cost" of assigning other customers to this facility. This cost for a customer is equal to the weight of the customer times the distance (either rectilinear or euclidean) between the customer and the new facility location. Choose the coordinates of the customer with the largest cost as the location for the second facility. If we have more than two new facilities to place, go to step 3; otherwise, go to step 5.
3. Calculate the cost associated with assigning a customer to each of the designated new facility locations. Choose the minimum of these as the associated cost for the customer. Repeat the procedure for all customers. For the location of the next new facility, choose the customer with maximum cost and locate the new facility at the coordinates of that customer.
4. Repeat step 3 until we obtain as many new facility locations as the number of new facilities we want to locate.
5. Once the locations of all new facilities are chosen, assign each customer to a location that is closest to it based on the criteria used to measure the distance. After all the customers are assigned we will obtain the necessary number of clusters.
6. To serve each cluster of customers find the optimum location of a facility by applying techniques from Chapter 3.

Example

Rectilinear Distance. Sixteen consumers are to be serviced by three distribution centers. The locations of the existing customers and expected number of trips from a distribution center to the customer (i.e., weight) are as given in Table 5.19. Group the localities and find the servicing location in each group. Figure 5.2 shows the initial distribution of the customers.

TABLE 5.19 Data on Existing Customers

No.	Consumer	X	Y	Weight	No.	Consumer	X	Y	Weight
1	A	5	5	50	9	I	15	10	5
2	B	5	15	80	10	J	17	15	88
3	C	4	31	90	11	K	20	25	72
4	D	8	10	30	12	L	23	6	12
5	E	7	27	60	13	M	22	15	80
6	F	11	7	20	14	N	25	18	85
7	G	10	23	40	15	O	30	15	64
8	H	12	31	20	16	P	31	30	60

It is always good to visualize the problem by plotting the existing customer location on graph paper. Keeping the scale on the x and y axes the same makes it easier to calculate the distances (we can measure just the actual scale distance, rather than calculating the exact distance between facilities). The grouping with rectilinear distances is shown in Table 5.20 and Figures 5.2–5.5.

TABLE 5.20 Rectilinear Distance Location Selection

No.	Customer	X	Y	Wt		RD from "C"	RD* Wt "C"		RD from "N"	Minimum RD "C"&"N"	Minimum RD*WT		RD from "B"
1	A	5	5	50	S	27	1350	S	33	27	1350	S	10
2	B	5	15	80		17	1360	E	23	17	1360	E	—
3	C	4	31	90	L	—	—	L	—	—	—	L	—
4	D	8	10	30		25	750	E	25	25	750	E	8
5	E	7	27	60		7	420	C	27	7	420	C	14
6	F	11	7	20		31	620	T	25	25	500	T	14
7	G	10	23	40		14	560		20	14	560		13
8	H	12	31	20		8	160		26	8	160		23
9	I	15	10	5		32	160		18	18	90		15
10	J	17	15	88		29	2552		11	11	968		12
11	K	20	25	72		22	1584	N	12		72	B	25
12	L	23	6	12		44	528		14	14	168		27
13	M	22	15	80		34	2720		6	6	480		17
14	N	25	18	85		34	2890		—	—	—		—
15	O	30	15	64		42	—		8	8	512		25
16	P	31	30	60		28	1680		18	18	1080		41

Service center	Customer
B	A, B, D, F, G, I
C	C, E, H
N	J, K, L, M, N, O, P

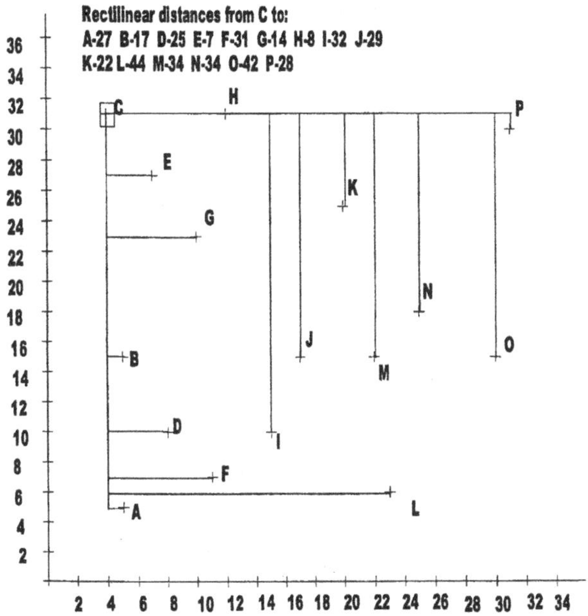

FIGURE 5.2 Rectilinear distances from C.

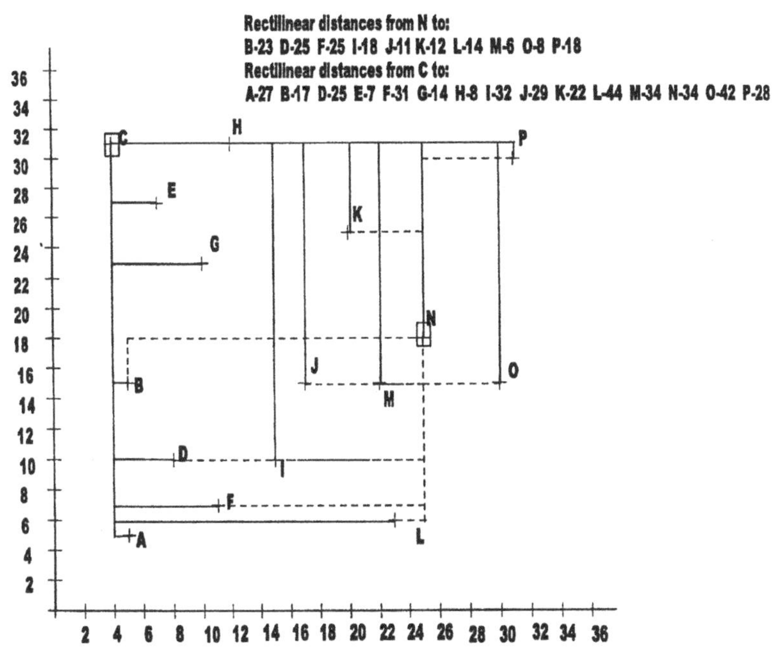

FIGURE 5.3 Rectilinear distances from C and N.

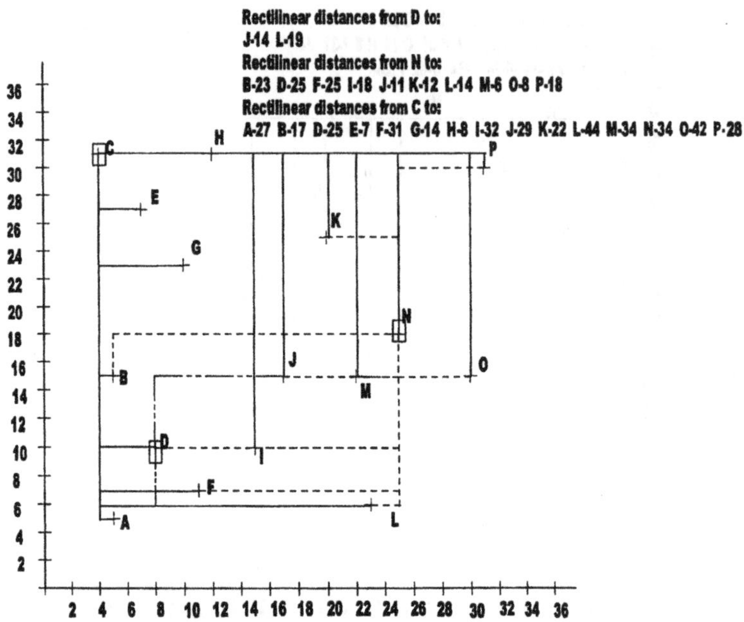

FIGURE 5.4 Rectilinear distances from C, N, and D.

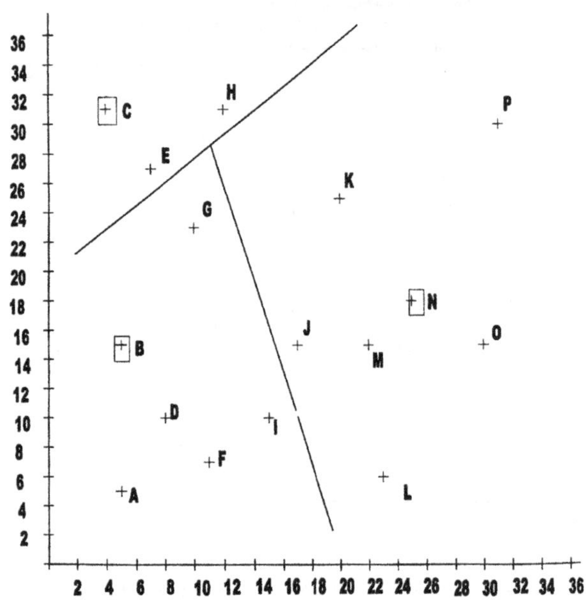

FIGURE 5.5 Grouping with rectilinear distance.

Because there are three facilities to place, we need to form three groups. First we will find the centers for the groups. The first center is located in location C because it has the largest weight of 90 among all the demand points. Next, rectilinear distance (RD) from each demand point to C and the product of distance times weight (RD*WT) is determined. The point with the largest product (of 2890), point N, becomes the next center. To determine the third center point, from each nonselected demand points determine the closest selected demand point. This is done by evaluating the distances from a demand point to point C and point N and choosing the smallest distance. Calculate the product of weight times the smallest distance and the point with the largest value for this product, location B, is the site of the next center. Because we have three centers, we now need to distribute other demand points into groups.

From each demand point the shortest distance among the demand centers, is determined, and the demand point is attached to the center to which it has the shortest distance. For example, point A is attached to B because it is closer to B than to N or C. The final grouping is shown in the Table 5.20 and Figure 5.5.

As each group has only one facility to place, it can be placed independently of other groups. As such, for each group, optimum facility location is developed by applying single-facility location algorithm from Chapter 4. The optimum locations are is played in Table 5.21.

Euclidean Distance. The same procedure is repeated if we have euclidean or direct distance as the measuring criteria. The solution to the data set from Table 5.19 is illustrated in Table 5.22. As before the first location selected is C. Figure 5.6 shows the direct measurement from C to other customers. This distance, measured in millimetres, is used in calculating the euclidean cost as the product of weight times the distance. The next point selected is customer N, and Figure 5.7 shows measurement from both C and N to each customer. Minimum distance is used to calculate associated cost. The final three groupings (Figure 5.8), in this case, are identical with those obtained in rectilinear distance analysis. However, optimum location of the new facilities are different, as shown in Table 5.23.

TABLE 5.21 Optimum Grouping of the Customers and Associated Optimum Facility Location

Group	Customers	Optimum location for new facility
1	A, B, D, F, G, I	(5, 15) location B
2	C, E, H	(4, 31) location C
3	J, K, L, M, N, O, P	(25, 18) location N

TABLE 5.22 Euclidean Distance Location Selection

No.	Customer	X	Y	Wt	D from	D* Wt	D From	Minimum D	Minimum	D from
					"C"	"C"	"N"	"C"&"N"	D*WT	"B"
1	A	5	5	50	26.02	1301	23.85	23.85	1192.5	10
2	B	5	15	80	16.03	1282.4	20.22	16.03	1282.4	—
3	C	4	31	90	—	—	—	—	—	—
4	D	8	10	30	21.38	641.4	18.8	18.8	564	5.83
5	E	7	27	60	5	300	20.12	5	300	12.16
6	F	11	7	20	25	500	17.8	17.8	356	10
7	G	10	23	40	10	400	15.81	10	400	9.43
8	H	12	31	20	8	160	18.38	8	160	17.46
9	I	15	10	5	23.71	118.55	12.81	12.81	64.05	11.18
10	J	17	15	88	20.62	1814.56	8.54	8.54	751.52	12
11	K	20	25	72	17.08	1229.76	8.6	8.6	619.2	18.03
12	L	23	6	12	31.4	376.8	12.17	12.17	146.04	20.12
13	M	22	15	80	24.08	1926.4	4.24	4.24	339.2	17
14	N	25	18	85	24.7	2099.5	—	—	—	—
15	O	30	15	64	30.53	—	5.83	5.83	373.12	25
16	P	31	30	60	27.02	1621.2	13.42	13.42	805.2	30.02

Service center	Customer
B	A, B, D, F, G, I
C	C, E, H
N	J, K, L, M, N, O, P

TABLE 5.23 Optimum Location of New Facilities (Euclidean Distance)

Group	Customers	Optimum location for new facility
1	A, B, D, F, G, I	(5.0, 14.9)
2	C, E, H	(4, 31)
3	J, K, L, M, N, O, P	(24.2, 17.8)

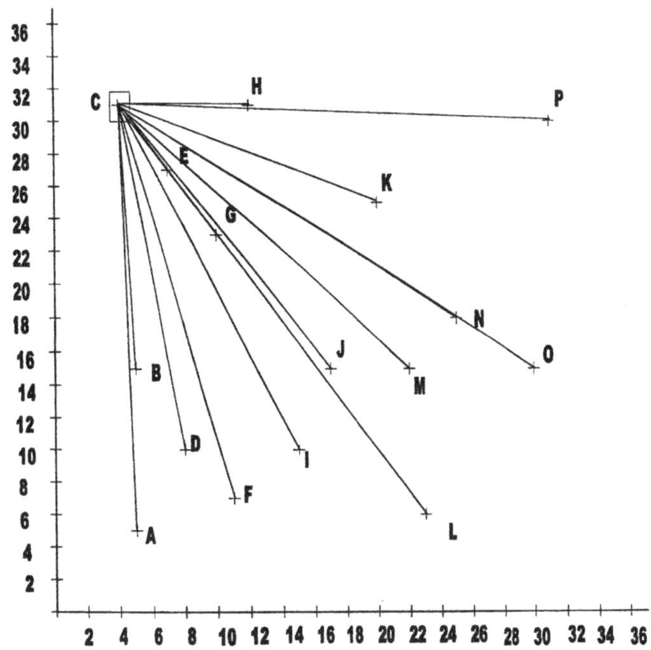

FIGURE 5.6 Distances from C.

5.4 EQUITY CRITERIA

In some business and social problems it is not the minimization of maximum distance from a customer to the facility that is important, but the thought that the facilities should be serving "equitably" to all customers may be more desirable. For example, fire trucks should take about the same time to respond to an emergency, irrespective of the income level of the people living in different districts. A post office should be about the same maximum distance from all the customers it is supposed to serve. A backup facility for a group should take about the same time to respond to an emergency in the group. Thus, in a multifacilities location problem a measure such as the maximum response time or maximum

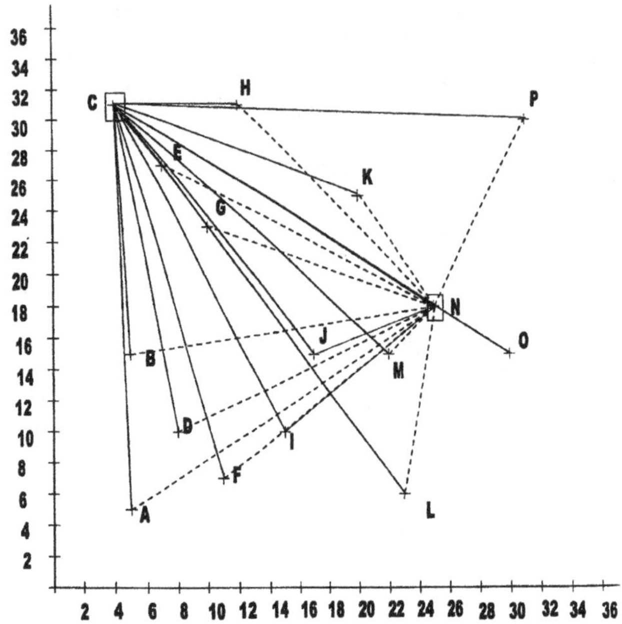

FIGURE 5.7 Distances from C and N.

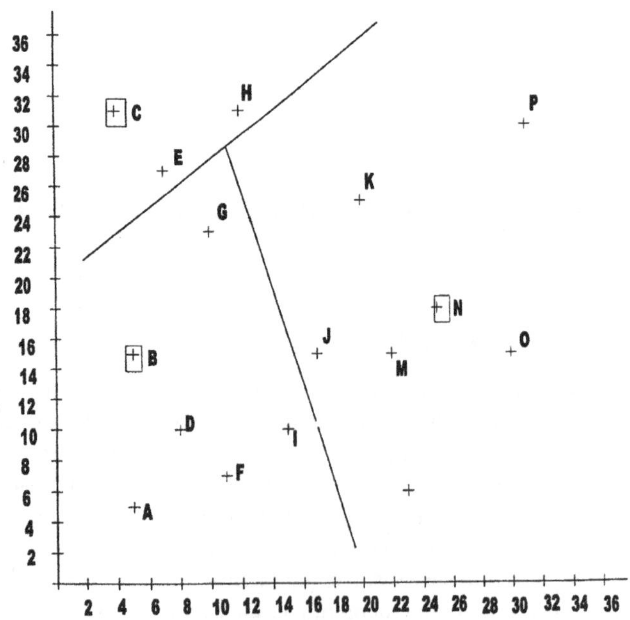

FIGURE 5.8 Grouping with euclidean distance.

distance between a facility and a customer, should have approximately the same value in different groups.

One way to resolve this problem is to start by placing a facility in a feasible location and then determining the grouping and associated cost. Repeat this process by starting on every feasible location. The best solution is then selected from among all the solutions thus generated.

We illustrate the procedure by an example. The data is given in Table 5.24. By considering customer 1 as the location of the first facility results in the groupings indicated in the first trial. Sixteen trials are needed because we have 16 customers. The minimum of the maximum difference in the response time is associated with trial 14. Table 5.25 shows the details.

For expanded discussion on this problem, the interested reader may refer to an excellent review by Marsh and Schilling (1994).

5.5 MACHINE LAYOUT MODELS

Thus far we have been dealing with "facilities," such as plants, stores, hospitals, police stations, and so on, each providing services to a set of customers. There is, however, another set of problems for which customers provide each other services, thus in effect act as facilities. For example, consider a machine shop.

TABLE 5.24 Initial Data for Equity Criteria

Point	*Xi*	*Yi*
1	5	5
2	5	15
3	4	31
4	8	10
5	7	27
6	11	7
7	10	23
8	12	31
9	15	10
10	17	15
11	20	25
12	23	6
13	22	15
14	25	18
15	30	15
16	31	30

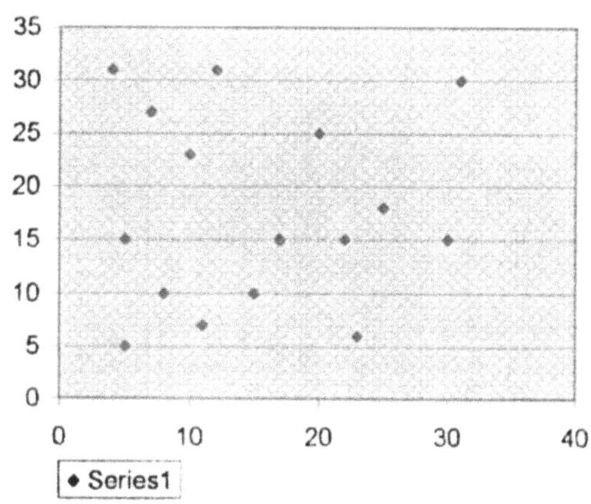

TABLE 5.25 Solution to Equity Criteria

Trial	First facility point	Location of service centers and customer assignment			Maximum distance from service center			Maximum difference
1	**1**	**1**	**3**	**16**	**1**	**3**	**16**	
		1,2,4,6,9,10,12	3,5,7,8	11,13,14,15,16	15.62	10	15.03	5.62
2	**2**	**2**	**12**	**16**	**2**	**12**	**16**	
		1,2,3,4,5,6,7,8	9,10,12,13,14,15	11,16	16.03	12.16	12.08	3.95
3	**3**	**3**	**12**	**16**	**3**	**12**	**16**	
		2,3,5,7,8	1,4,6,9,10,12,13,14,15	11,16	16.03	18.02	12.08	5.94
4	**4**	**3**	**4**	**16**	**3**	**4**	**16**	
		3,5,7,8	1,2,4,6,9,10,12,13	11,14,15,16	10	15.52	15.03	5.52
5	**5**	**5**	**12**	**16**	**5**	**12**	**16**	
		2,3,5,7,8	1,4,6,9,10,12,13,14,15	11,16	12.16	18.02	12.08	5.94
6	**6**	**3**	**6**	**16**	**3**	**6**	**16**	
		3,5,7,8	1,2,4,6,9,10,12,13	11,14,15,16	10	13.6	15.03	5.03
7	**7**	**7**	**12**	**16**	**7**	**12**	**16**	
		2,3,4,5,7,8,10,11	1,6,9,12,13,14,15	16	21.54	18.02	0	21.54
8	**8**	**1**	**8**	**15**	**1**	**8**	**15**	
		1,2,4,6,9	3,5,7,8,11	10,12,13,14,15,16	11.18	10	15.03	5.03

9	**9**	**3**	**9**	**16**	**3**	**9**	**16**	
		3,5,7,8	1,2,4,6,9,10,12,13,14	11,15,16	10	12.8	15.03	5.03
10	**10**	**3**	**10**	**16**	**3**	**10**	**16**	
		3,5,7,8	1,2,4,6,9,10,11,12,13,14,15	16	10	15.62	0	15.62
11	**11**	**1**	**3**	**11**	**1**	**3**	**11**	
		1,2,4,6,9,12	3,5,7,8	10,11,13,14,15,16	18.03	10	14.14	8.03
12	**12**	**3**	**12**	**16**	**3**	**12**	**16**	
		2,3,5,7,8	1,4,6,9,10,11,12,13,14,15	11,16	16.03	19.23	12.08	7.15
13	**13**	**1**	**3**	**13**	**1**	**3**	**13**	
		1,2,4,6	3,5,7,8	9,10,11,12,13,14,15,16	10	10	17.49	7.49
14	**14**	**1**	**3**	**14**	**1**	**3**	**14**	
		1,2,4,6,9	3,5,7,8	10,11,12,13,14,15,16	11.18	10	13.41	***3.41***
15	**15**	**1**	**3**	**15**	**1**	**3**	**15**	
		1,2,4,6,9	3,5,7,8	10,11,12,13,14,15,16	11.18	10	15.03	5.03
16	**16**	**1**	**3**	**16**	**1**	**3**	**16**	
		1,2,4,6,9,10,12	3,5,7,8	11,13,14,15,16	18.02	10	17.49	8.02

As parts are produced, the material flows from machines to machines for successive machining operation. The flow, therefore, is between machines, and it is a cause of a major expense. Location problem results in determining the optimum place for each machine to minimize the total material movement cost. The quadratic assignment problem (QAP), a method by which a job shop layout is developed, is one such special case that we will study in detail in Chapter 10. Here, we describe another layout problem, simply called machine layout model, in which the optimum locations of the machines are determined if they must be placed in one of the following arrangements.

1. Linear or in a single row
2. Loop or circular flow
3. Double or multiple row serpentine flow

Figure 5.9 shows these flow arrangements. We can develop a mathematical formulation of the problem first by defining the following terms:

Let

X_i x-coordinate for the center of machine i
f_{ij} Material flow between machine i and j
d_{ij} Distance between machine i and j
C_{ij} Cost of transporting a unit of material for unit distance between machine i and j
l_i Length of machine i
S Minimum space required between two machines

Then we have the following

$$\text{Minimize } \sum\sum C_{ij} f_{ij} |X_i - X_j|$$

$$\text{Subject to } |X_i - X_j| >= \frac{1}{2}(l_i + l_j) + S$$

The constrain set assures us that machine work area has the necessary clearance between two machines, whereas the objective function minimizes the cost of transportation or material handling. The problem can be solved by a commercially available optimization software package, such as GINO. There is, however, a simpler heuristic method that can be applied and is presented next.

1. Compute a matrix that shows the cost $C_{ij} f_{ij}$ from each machine i to each machine j.
2. Select the largest value element from the matrix and place associated machine i^* and j^* adjacent to each other with $1/2(l_i^* + l_j^*) + S$ distance apart. The sequence is $i^* - j^*$.
3. Follow the remaining steps for each machine that is not yet placed. Select a machine, say k, and place it alternately on each side of the

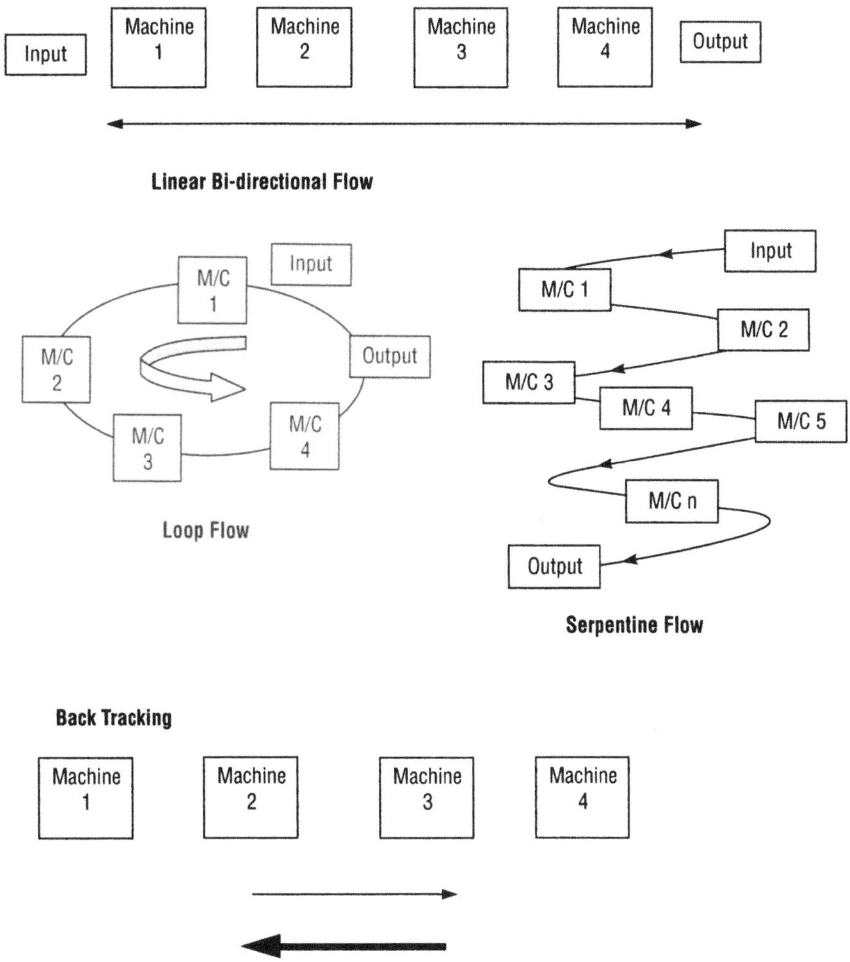

FIGURE 5.9 Significant flow patterns in machine layout.

sequence developed so far (Initially, the sequence is $i^*–j^*$ and, therefore, we need to try, $k–i^*–j^*$ and $i^*–j^*–k$). In each case, determine the flow cost by maintaining the minimum necessary separation between the machines. From among all the sequences thus tested, select the sequence that gives the minimum cost. The associated machine k^* is then placed in the position that gave the minimum cost sequence, With newly developed machine sequence, repeat step 3 again.

4. Continue the process until all machines are placed.

5.5.1 Single-Row Layout Example

Suppose we have four machines to be placed in a single row where travel in either direction is permitted. The dimensions of the machines are given as follows and the minimum clearance required between the machines is 2 units.

Machine	1	2	3	4
Dimensions	2×2	4×4	6×6	8×8

The loaded material handling trips between the machines are

(To)		1	2	3	4
	1	—	10	15	12
Machine	2	7	—	20	18
(From)	3	5	15	—	25
	4	6	3	15	—

Assume $C_{ij} = 1$ for all i and j.

Solution. Because the flow in both directions is permitted, the first step is to construct a table showing the total flow between machines. The matrix is symmetrical about the diagonal and, therefore, only the top or the bottom half needs to be displayed.

(To)		1	2	3	4
	1	—	17	20	18
Machine	2		—	35	21
(From)	3			—	40
	4				—

The maximum flow is 40, between 4 and 3, so these two machines enter the layout first. The order of the placement, 4–3 or 3–4 is not important at this point, because other machines will be placed optimally based on the order we select. Let

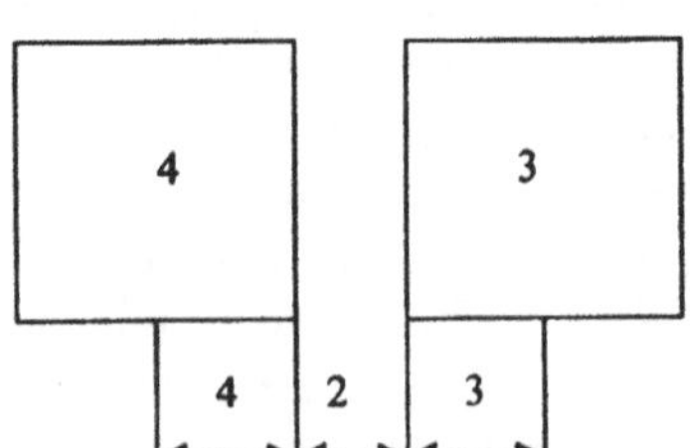

us start by selecting the order 4–3. Their centers are placed $(8+6)/2+2=9$ units apart.

For the next machine's placement, evaluate the effectiveness of each machine that has not yet been placed. Because there are two machines, 1 and 2, that need to be examined, the following placement orders are examined.

$$2-4-3, 4-3-2, 1-4-3, 4-3-1$$

In each case, the appropriate distances from center to center of the machines are determined, with the assumption that the machines are placed as close together as permissible, allowing for the minimum clearance. The sum of the flows between the test machine and existing layout multiplied by the appropriate distances gives material-handling cost. This is shown, for example, for test machine 2 placed before 4–3 as, f2–4 + f2–3 (i.e., flow between 2 and 4 times the distance plus the flow between 2 and 3 times the distance between 2 and 3). An arrangement that gives the minimum cost is selected as the optimum for this stage. The costs are:

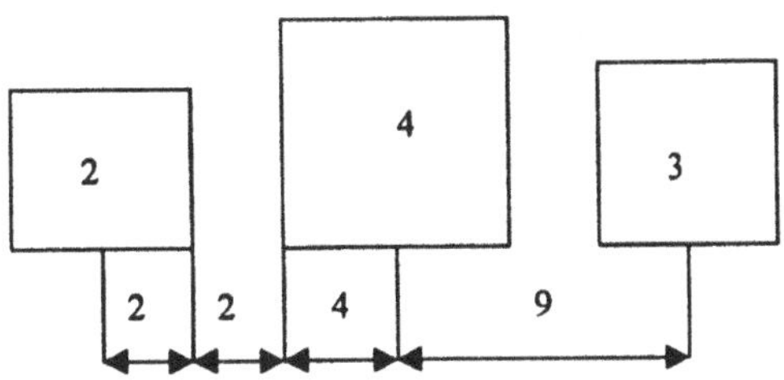

1. For order 2–4–3 → f2–4 + f2–3

$8(21)+17(35)=763$

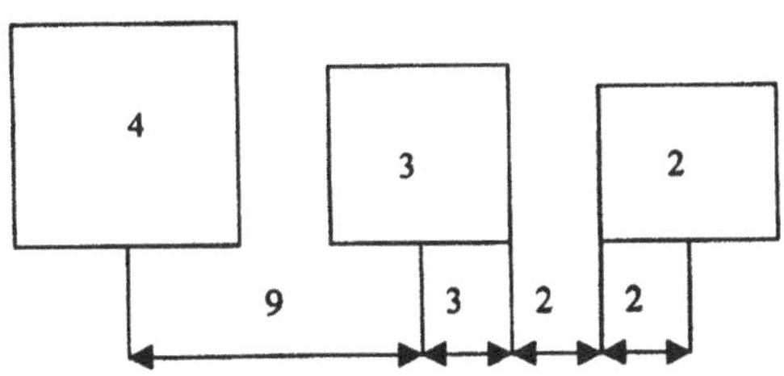

2. For order 4–3–2 → f4–2 + f3–2

$16(21)+7(35)=581$

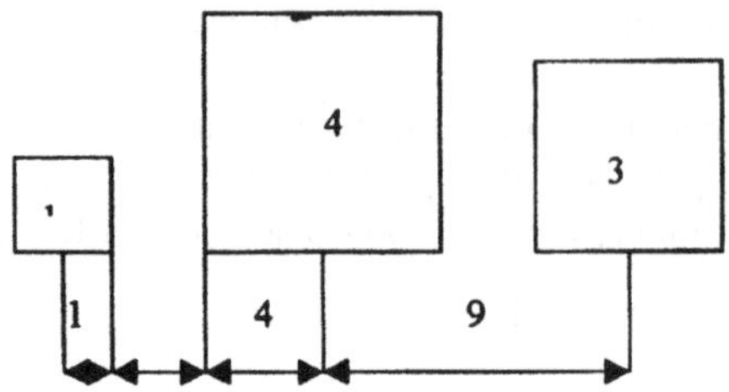

3. For order 1–4–3 → f1–4 + f1–3

7(18) + 16(20) = **446**

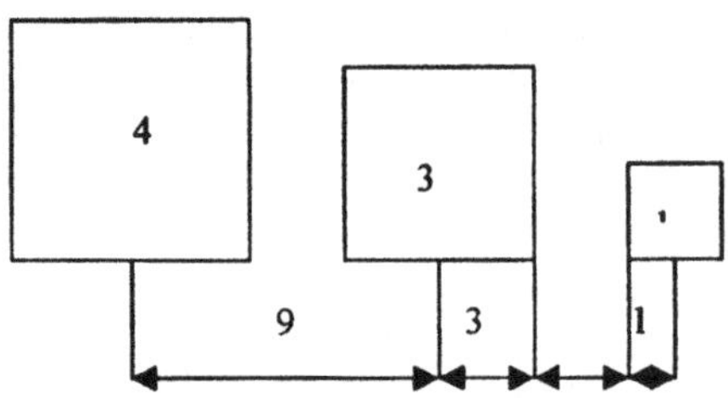

4. For order 4–3–1 → f4–1 + f3–1

15(40) + 6(20) = 720

The minimum placement cost is 446, so select the associated placement order **1–4–3**.

Repeat the process for other unplaced machines relative to the presently selected placement order. Only machine 2 is unassigned at this stage, therefore, the remaining placement orders that need evaluations are 2–1–4–3 and 1–4–3–2.

1. For order 2–1–4–3 → f2–1 + f2–4 + f2–3

5(17) + 12(21) + 17(35) = **932**

2. For order 1–4–3–2 → f1–2 + f4–2 + f3–2

23(17) + 16(21) + 7(35) = 972

Thus, as the minimum cost is associated with the sequence **2–1–4–3**, it is selected as the final sequence.

5.5.2 Loop Layout

Consider the same problem of four-machine placement, but this time the machines are placed in a loop so that the material flow is in only one direction.

This results in variation in travel distance between two machines based on the direction of flow, travel from machine *i* to *j* may not have the same distance as from *j* to *i*. In fact, if we know the loop length, *l*, and if travel distance in *i*–*j* direction is $d(ij)$ then the travel distance in *j*–*i* direction $d(ji) = l - d(ij)$. The flows from *ij* and *ji* must be kept independent.

For convenience we repeat the data on the dimensions of the machines. Let us assume that minimum clearance required between two machines is still 2 units.

Machine	1	2	3	4
Dimensions	2×2	4×4	6×6	8×8

The material flow between the machines is

(To)		1	2	3	4
	1	—	10	15	12
Machine	2	7	—	20	18
(From)	3	5	15	—	25
	4	6	3	15	—

Solution. The loop length is the length of all machines, assuming the machines are placed lengthwise on the loop, plus the number of machines times the clearance required between two successive machines. In our case, the total length of the loop is $2 + 4 + 6 + 8 + 2(4) = 28$ units.

Start with the flow table again. The maximum flow is 25 units; therefore, the associated machines 3 and 4 are selected and enter into layout first. Following

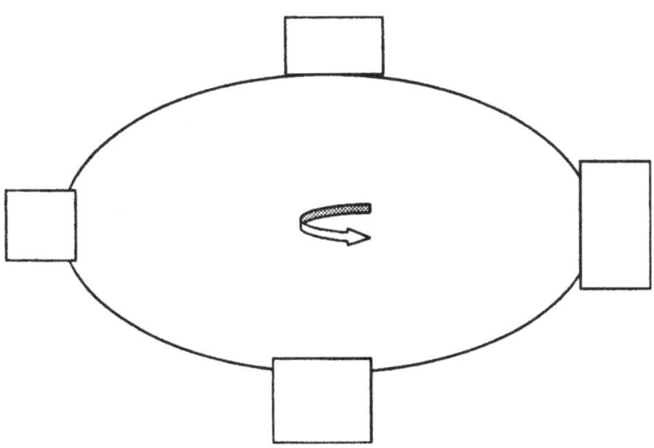

the same procedure as earlier, for placing the next machine we evaluate the following orders:

$$2-3-4, 3-4-2, 1-3-4, 3-4-1$$

For order 2–3–4 → f2–3 + f2–4 + f3–2 + f4–2

$$7(20) + 16(18) + [(28-7)(15)] + [(28-16)(3)] = 779$$

For order 3–4–2 → f3–2 + f4–2 + f2–3 + f2–4

$$17(15) + 8(3) + [(28-17)(20)] + [(28-8)(18)] = 859$$

For order 1–3–4 → f1–3 + f1–4 + f3–1 + f4–1

$$15(6) + 12(15) + [(28-6)(5)] + [(28-15)(6)] = \mathbf{458}$$

For order 3–4–1 → f3–1 + f4–1 + f1–3 + f1–4

$$5(16) + 6(7) + [(28-16)(15)] + [(28-7)(12)] = 554$$

The minimum relative placement cost is 458, so associated placement order **1–3–4** is selected.

There is only one more unassigned machine and associated placement orders are

$$2-1-3-4 \quad \text{and} \quad 1-3-4-2.$$

The costs are

For order 2–1–3–4 → f2–1 + f2–3 + f2–4 + f1–2 + f3–2 + f4–2

$$5(7) + 11(20) + 20(18) + [(28-5)(10)] + [(28-11)(15)] + [(28-20)(3)] = 1124$$

For order 1–3–4–2 → f1–2 + f3–2 + f4–2 + f2–1 + f4–1 + f3–1

$$23(10) + 17(15) + 8(3) + [(28-23)(7)] + [(28-17)(20)] + [(28-8)(18)] = 1124$$

The minimum cost is associated with the sequence **2–1–3–4** or **1–3–4–2**. We should have expected this, as both sequences form an identical sequence in a loop arrangement. In fact, the last calculations were not even necessary, because the last machine, machine 2, could have been placed either at the beginning or at the end (same physical position) of the sequence obtained when all other machines, namely 1, 3, and 4, are placed optimally.

5.5.3 Backtrack

In some instances backtracking cost is considerably larger than forward flow cost. For example, the flow in the forward direction may be achieved by a conveyer, whereas bringing units back in the reverse direction of flow may require special handling. In such instance, the objective may be to minimize backtracking, rather than to minimize the total flow cost.

The procedure described before also achieves this objective, with minor modification. Suppose, for example, we wish to minimize the backtracking for the data given in the previous example, again the data is reproduced here for convenience.

Machine	1	2	3	4
Dimensions	2 × 2	4 × 4	6 × 6	8 × 8

(To)		1	2	3	4
	1	—	10	15	12
Machine	2	7	—	20	18
(From)	3	5	15	—	25
	4	6	3	15	—

Consider the "from–to" chart (not the total flowchart) and determine the two machines with maximum of minimum backtracking flow. For example machines 1 and 2, the flow from 1 to 2 is 10 and from 2 to 1 is 7. Thus the minimum backtracking flow between these two machines is the minimum of 10 and 7 = 7. Similarly calculate minimum backtracking in all other pairs of machines. Select the pair with maximum value. If there is a tie select all tied pairs.

In our case, it is between machines 2 and 3 and 2 and 4 have the same backtrack of 15, the maximum of minimum backtrack value for all pairs. Select 2–3. These machines can be placed either as 2–3 or 3–2. In the case of 2–3 the backtrack cost is 105 and the flow from 3 to 2, whereas the backtrack cost for 3–2 is 140. Because the objective is to minimize the backtrack, choice 2–3 is taken as the layout.

Steps 2 and 3 require similar calculations, placing the candidate machine in the front of the fixed sequence and then in the back of the fixed sequence to determine the backtracking costs. A sequence is selected at each step, as shown in following table, which gives the minimum cost.

Step	Sequence	Backtrack cases	Backtrack cost	Optimum sequence
1	2–3	3–2	15*7 = 105	**2–3**
	3–2	2–3	20*7 = 140	
2	1–2–3	3–1, 2–1	5*12 + 7*5 = 95	**1–2–3**
	2–3–1	1–2, 1–3	10*13 + 15*6 = 220	
	4–2–3	3–4, 2–4	25*15 + 18*8 = 519	
	2–3–4	4–2, 4–3	3*16 + 15*9 = 183	
3	4–1–2–3	3–4, 2–4, 1–4	25*19 + 12*18 + 7*12 = 775	
	1–2–3–4	4–1, 4–2, 4–3	6*21 + 3*16 + 15*9 = 309	**1–2–3–4**

The machine sequence is 1–2–3–4, with backtracking cost equal to sum of the optimum cost at each step or,

Total backtracking cost = 105 + 95 + 309 = **509**

Similarly, starting with other tied pair 4–3 leads us to following results:

Step	Sequence	Backtrack cases	Backtrack cost	Optimum sequence
1	3–4	4–3	15*9 = 135	**3–4**
	4–3	3–4	25*9 = 225	
2	1–3–4	4–1, 4–3	6*15 + 12*7 = 120	**1–3–4**
	3–4–1	1–3, 1–4	15*16 + 12*7 = 324	
	2–3–4	4–2, 4–3	3*16 + 15*7 = 153	
	3–4–2	2–3, 2–4	20*17 + 18*8 = 484	
3	2–1–3–4	4–2, 4–1, 4–3	3*20 + 15*11 + 10*5=275	**2–1–3–4**
	1–3–4–2	2–1, 2–3, 2–4	7*23 + 20*17 + 18*8 = 645	

The best sequence is 2–1–3–4 and associated backtracking cost is, 135 + 120 + 275 = **530.**

Of the two sequences select the optimum which is **1–2–3–4.**

REFERENCES

Marsh MT, Schilling DA. Equity measurement in facility location analysis: a review and framework. Eur J Operational Res 74 (Apr 17) 1994.

Bellman R. Dynamic Programming. Princeton, NJ: Princeton University Press, 1957.

6

Basic Location–Allocation Model

In the previous two chapters our objective was to find the optimum location for one or more facilities. This location, although optimum in terms of cost or some other goal, may or may not be feasible for locating the facility. In this and in the next few chapters we study a class of problems for which we install the new facilities in one of the known locations. These locations are said to be available because they may have also satisfied other intangible objectives, some of which were discussed in Chapters 1 and 2. The objective here is not only to select the best locations from among those available, but also to assign the customers to these locations in an optimum manner. Such problems are called *location–allocation problems*.

We start with the basic model for choosing locations. The only cost considered here is the cost that is directly proportional to the unit of demand. For example, the total cost of transportation may be the product of unit cost of transportation from a customer to a location times the number of units to be transported, or the demand from the customer. In our approach, we will assume that each facility is sufficiently large that it can serve all the customers if necessary; that is, there is no limitation on the capacity of a facility. Sample problems are presented in the following discussion and are divided into two parts: (1) The single-facility placement problem, and (2) The multifacility placement problem.

6.1 SINGLE-FACILITY PROBLEM

Six product lines originating from a plant need the services of a paint booth. The paint booth can be set up in one of five possible locations, to which these six product lines will be assigned. The demand from each product line (i.e., the expected number of trips made daily) and the cost (measured in terms of the time required to transport the products to and from each location) are given in Table 6.1.

This table shows that it takes 4 min for customer A to go to and return from location 1. We wish to determine where the paint booth should be installed to minimize the total cost; that is, the total time spent moving to and from the paint booth.

6.1.1 Solution Procedure

When there is only one facility to place, the solution procedure consists of three very basic steps. They are as follows:

1. Calculate the total "transportation cost" matrix (also called the demand cost matrix). An element of the matrix is obtained by multiplying the cost of assigning a unit of demand, from a customer to a location, by the total demand from customer. For example, the cost of assigning customer A to location 1 is $4*30 = 120$ units of time.
2. Take the sum of each column. These values represent the total cost of assigning all customers to each location.
3. Select the location with the minimum total cost. This is the location in which the machine should be placed.

TABLE 6.1 Time and Demand

Customer	Location					Demand
(Product line)	1	2	3	4	5	
A	4	8	3	1	7	30
B	5	5	3	4	3	40
C	4	9	2	4	6	20
D	6	6	7	5	7	24
E	4	12	4	2	9	20
F	10	13	6	9	3	15

6.1.2 Application to the Sample Problem

The total transportation cost matrix, generated in step 1, is given in Table 6.2. The last row is the total sum of each column. Because the total cost for location 4 is the least, this site is selected, and all the customers are assigned to this machine. The procedure is quick and requires a minimum of effort.

6.2 MULTIFACILITY PROBLEM

Best Transmission Repair Shop (BTS) has purchased two identical transmission diagnostic machines. He presently has five service stations in operation within the city, and he has identified five areas from which demand will occur. The transportation time from the center of gravity of each area cluster to each location is given in Table 6.3. Also shown is the demand from each customer (area) for such a machine. The requirement is to place the two new machines in locations that will minimize the total transportation time because management considers this to be the measure of efficiency.

TABLE 6.2 Demand Cost

Customer	Location				
	1	2	3	4	5
A	120	240	90	30	210
B	200	200	120	160	120
C	80	180	40	80	120
D	144	144	168	120	168
E	80	240	80	40	180
F	150	195	90	135	45
Total	774	1199	588	565	843

TABLE 6.3 Transportation Time and Demand Data

Customer	Time for location					Demand
	1	2	3	4	5	
A	4	8	3	1	7	30
B	5	5	3	4	3	40
C	4	9	2	4	6	20
D	6	6	7	5	7	24
E	4	12	4	2	9	20
F	10	13	6	9	3	15

A unique feature of this type of problem is that because the capacity of each facility is unlimited, there is no need to divide the demand from a customer and assign parts of it to the two different facilities. Indeed, a customer should be assigned to the facility for which the cost of transportation is at a minimum. The problem, then, is twofold: first, choose from the available locations those at which the facilities should be placed and, then, assign the customers to these facilities such that the cost of these assignments is a minimum.

6.2.1 Brute Force Approach

Because we wish to place K facilities in M available locations it can be done in $({}^{m}C_{K}) = m![k!(m-k)!]$ different ways. In our problem, as there are five locations and two facilities to place, we have $({}^{m}C_{K}) = 5!/2!\,3! = 10$ possible ways of placing these facilities. If all combinations are examined and each demand is assigned so that we obtain the minimum cost for the combination, then the overall minimum can be easily found. The following example illustrates the brute force approach.

First calculate the total cost of assigning all the demand from a customer to a location. Table 6.4 gives the data for our problem. Consider one possible combination; say, locations 1 and 2. This means that one of the two available machines is placed in location 1 and the other in location 2. If that is so, then each customer should be assigned to the location for which the cost of transportation is minimum. For example, customer A is assigned to location 1; customer B to location 1 or 2; customer C to location 1; customer D to location 1 or 2; customer E to location 1; and customer F to location 1. This solution has the cost of $120 + 200 + 80 + 144 + 80 + 150 = 774$. Similarly analyses for the other ten combinations lead to the data listed in Table 6.5.

The minimum cost is that associated with the combination of location 4 and 5; hence, these locations are selected as the best in which to place the facilities, and the customers are assigned in the manner that led to the minimum cost.

TABLE 6.4 Demand Cost

Customer	Location				
	1	2	3	4	5
A	120	240	90	30	210
B	200	200	120	160	120
C	80	180	40	80	120
D	144	144	168	120	168
E	80	240	80	40	180
F	150	195	90	135	45

TABLE 6.5 Minimum Cost

Customer	Location pairs (combinations)									
	1 and 2	1 and 3	1 and 4	1 and 5	2 and 3	2 and 4	2 and 5	3 and 4	3 and 5	4 and 5
A	120	90	30	120	90	30	210	30	90	30
B	200	120	160	120	120	160	120	120	120	120
C	80	40	80	80	40	80	120	40	40	80
D	144	144	120	144	144	120	144	120	168	120
E	150	80	135	45	90	135	45	90	45	45
F	150	90	135	45	90	135	45	90	45	45
Total	774	564	565	565	589	564	565	819	543	435

Computationally, the method is fast and simple, and relatively small problems can be quickly worked out by hand. But the calculations increase very rapidly, almost exponentially, with the number of locations that need examination; therefore, this method is not recommended for problems involving many combinations.

6.2.2 Heuristic Methods

Most heuristic methods follow certain logical steps that generally lead to optimum or near-optimum results. For the preceding problem, the heuristic method suggested in this section works well. The procedure is based on the following logic.

If we are to place the facilities one at a time, then we should place each in the location where the benefit derived from it is maximum. To begin, if we have only one facility, it should be located in a position where the cost of assigning all customers to that location is the minimum. If additional facilities are available, they should be placed one at a time in positions that contribute to the maximum savings. Thus, each subsequent facility that is placed in an additional location should successively improve on each previous solution. If the solution does not improve, then we do not need the additional facility. Therefore, the method not only indicates where the facilities should be located and how the customers should be assigned, it also points out the optimum number of facilities we should have. The procedure continues until either all the facilities are located, or there is no further improvement in the solution.

The steps involved in the procedure are as follows:

1. Formulate the total cost table. An entry *ij* in the total cost table represents the cost of allocating all demand from customer *i* to location *j*.
2. Total each column. The sum represents the total cost if demands from all customers are assigned to that location.

3. Assign the first facility to the location with the minimum total cost. We will denote a location with a facility as an assigned location and one without a facility as an unassigned location.
4. If no additional facility is available for assignment, go to step 8, otherwise continue.
5. Determine the savings to be derived by moving each customer from the assigned location(s) to a nonassigned location. The savings may occur because of a change in transportation cost, but if there are no savings, mark a dash (—) in the appropriate column.
6. Total each unassigned column. The sum represents savings that could be achieved if an assignment is made in that location.
7. Make an assignment in the location that indicates the maximum savings. Transfer the customers that contributed to these savings to the new location, which now becomes an assigned location. Return to step 4.
8. All the assignments are made; calculate the minimum cost and schedule.

Application

Consider the BTS problem. The cost data, in terms of time, are the same as in Table 6.3, but suppose, however, that the shop has three machines to place. Table 6.4, a demand cost table, is reproduced for convenience as Table 6.6 and is expanded to include the customers' demands and the totals of the columns.

Because column 4 has the minimum total cost of 565, the first machine could be assigned to location 4. Now, to decide where to place the second machine, we must calculate the savings (if any) to be gained by moving customers that are presently assigned to location 4 to any other unassigned location *j*. For example, consider customer B. If he or she moved from location 4 to location 3, it

TABLE 6.6 Demand Cost (Time)

Customer	Location				
	1	2	3	4	5
A	120	240	90	30	210
B	200	200	120	160	120
C	80	180	40	80	120
D	144	144	168	120	168
E	80	240	80	40	180
F	150	195	90	135	45
Total	774	1199	588	565	843

will save 160 – 120 = 40. Similarly, for location 5 the saving is 160 – 120 = 40. Other savings are calculated in the same manner, and Table 6.7, "Savings Table 1", provides the results. Note that if moving to a different location results in negative savings, then that location is marked by a dash. For example, there is a negative saving when customer B is moved from location 4 to location 2 (160 – 200 = –40), and accordingly, the dash notation is entered in the appropriate position. As shown in Table 6.7, the savings are calculated for all the customers and for each of the unassigned locations.

The total savings for each location is obtained by adding the column entries. The second machine should be placed in a location that has the maximum savings; here, location 5. The customers that contributed to these savings, customers B and F are now reassigned to location 5. The others, customers A, C, D, and E, are retained at location 4. The assignments at this stage are shown by asterisks in Table 6.8.

TABLE 6.7 Savings Table 1 for Moving from Location 4

Customer	Location				
	1	2	3	4	5
A	—	—	—	0	—
B	—	—	40	0	40
C	0	—	40	0	—
D	—	—	—	0	—
E	—	—	—	0	—
F	—	—	45		90
Total	0	—	125		130

TABLE 6.8 Assignments After Two Iterations

Customer	Location				
	1	2	3	4	5
A	—	—	—	*	—
B	—	—	40		40*
C	0	—	40	*	—
D	—	—	—	*	—
E	—	—	—	*	—
F	—	—	45		90*
Total	0	—	125		130

To decide where to place the third machine, we must recalculate the savings to be realized by moving the customers from their presently assigned locations to the unassigned locations. Again if the amount saved is negative, we will mark the unassigned location with a dash. Customer A is presently attached to location 4, and if we move him to location 1, the saving is 30 − 120 = −90. This −90 is a cost and, therefore, is marked with the dash. Similarly, for location 2 it is 30 − 240 = −210. For customer B, who is presently assigned to location 5, the corresponding calculations are as follows: for location 1, the amount saved is 120 − 200 = −80; for location 2, the figure is 120 − 200 = −80; and for location 3, the saving is 120 − 120 = −0. All of the results are shown in Table 6.9, savings table 2.

Because location 3 has the only savings, a total of $40, we will place the third machine in this location. The customers contributing to this saving, customers B and C are now assigned to location 3. Thus, our final assignments are as follows:

Customer	A	B	C	D	E	F
Location	4	3	3	4	4	5
Cost	30	120	40	120	40	45

The total cost for this solution is 395.

Now, suppose a fourth machine is available. Could we profitably use it? We decide by determining whether there are any additional savings possible; the results of our calculations are displayed in Table 6.10.

The only unassigned locations are 1 and 2. We find no savings in either of these two locations, thus indicating no need for the fourth machine. Our heuristic approach has identified the optimum number of machines as three.

TABLE 6.9 Savings Table 2 for Moving from Location 4 or 5

Customer	Location				
	1	2	3	4	5
A	—	—	—		—
B	—	—	0		
C	0	—	40		—
D	—	—			—
E	—	—	—		—
F	—	—	—		
Total			40		

TABLE 6.10 Savings Table 3 for Moving from Locations 3, 4, and 5.

Customer	Location				
	1	2	3	4	5
A	—	—		0	
B	—	—	0		
C	—	—	0		
D	—	—		0	
E	—	—		0	
F	—	—			0
Total	—	—			

Customer	A	B	C	D	E	F
Location	4	3	3	4	4	5
Cost	30	120	40	120	40	45

How good is the method? We have found by experimentation that this procedure provides the exact result about 90% of the time. In the remaining 10% of the time, when it deviated from the minimum, it differed by an average of 2% of the optimum solution. In most applications, data is estimated to start with, and the 2% variation is negligible. This is especially important when one observes the simplicity of the procedure, and the speed with which it can be applied.

6.2.3 Linear-Programming Formulation

We will briefly review the mathematical formulation of the model. Let us define

X_{ij} = 1 if the demand from source i is assigned to location j.
X_{ij} = 0 otherwise.
C_{ij} = cost of assigning a unit of demand from source i to location j where $i = 1, 2, \ldots, n$, and $j = 1, 2, \ldots, m$.
d_i = demand from source i.
K = number of facilities available for placement.

Then, we wish to minimize total cost (TC) where,

$$TC = \sum_{j=1}^{m} \sum_{i=1}^{n} C_{ij} {*} d_i {*} X_{ij}$$

subject to the constraints that are developed as follows:

Because the demand for each source must be assigned to at least one location, we have

$$\sum_{j=1}^{m} X_{ij} >= 1 \text{ for; each } i$$

Similarly, because only K locations are to be selected and only X_{ij} variables assigning demands to these locations can take positive values, we have:

$$\sum_{i=1}^{n} X_{ij} <= nI_j \text{ for each } j$$

$$\sum_{j=1}^{m} I_j = K$$

$I_j = 1$ if the facility is assigned to location j.
$I_j = 0$ if the facility is not assigned to location j.

The problem requires an application of an integer-programming software, such as LINDO.

6.3 LOCATION ANALYSIS WITH FIXED COSTS

Under most circumstances, the site at which a facility is to be located is not in a suitable condition to accept the facility on an "as is" basis, and site preparation or modification is required. This involves an additional expenditure; however, it is a one-time, initial cost, known as a fixed cost. The numeric values of the fixed costs for installing a particular facility may vary from location to location. For example, in choosing a site for a plant, locations may be found to differ in available conveniences. In some locations, land may have to be purchased and a building constructed; whereas, in others, suitable existing buildings may be available that can be converted to the desired plant. In yet other locations, modifications to transportation facilities may be needed to use the existing buildings as production facilities. Thus, the one-time costs to build the warehouses (or other facilities) in each location are generally different.

In some other instances, location (or relocation) of a piece of equipment may be planned. Examples include relocation of a lathe in a machine shop, placement of a new x-ray machine in a hospital, and installation of a new copy machine in an office building. In each case there may be a number of suitable locations available; however, the cost of placing a machine may be different in each location. It could depend on factors such as the present condition of the site or the modifications required to the foundation as well as the surroundings. In some areas, preparation of the site would demand no additional expenditure, but in other areas, it might require extensive work, such as the installation of air-conditioning systems.

Environmental regulations might also dictate an additional fixed cost for a site. For example, in locating a chemical or power plant, regulations governing exhaust emissions may be important. Such regulations within a city might require a special piece of equipment (e.g., a scrubber for a coal-burning power plant), whereas in other outlying areas these restrictions might not exist.

The objective is to select the locations and assign the demands in a manner that minimizes the total cost (i.e., the transportation and fixed costs). Before proceeding, however, it is important to understand the time units for which these costs are defined. The fixed cost is a one-time cost and, therefore, is to be recovered throughout the entire life of the facility. On the other hand, a transportation cost resulting from assigning a customer's demand(s) to a location is dependent on the amount of time for which the demands are defined. A demand may be expressed for any convenient unit of time (e.g., a week, a month, or a year), provided all the fixed costs are for the same time period. If this is not already true, they must be converted to a convenient unit. For fixed costs, this can be easily done by estimating the life of each facility and then using an annuity factor to spread the fixed cost (e.g., depreciation and interest cost) for each facility over a unit time period.

6.3.1 Procedure for Solving Problems with Fixed Costs

The steps in the procedure are the following:

1. Construct the demand cost table as before. If the number of facilities available for placement is two or more, there are two ways to obtain the initial solution. Take the sum of each column, which will represent the total variable (transportation) cost if the demands from all sources are assigned to that location. Then add the total variable cost for each column to the corresponding fixed cost. Select the location with the minimum overall cost and assign all customers to that location. This is also the optimum solution if there is only one available facility. The second alternative is described in step 2. Each initial solution may lead to different final solution; then, we can select the best alternative. Although most often the initial solution of step 1 leads to the least-cost solution, some times the initial solution obtained by applying step 2 leads to a better alternative.
2. A good initial solution when there are more than one facility is obtained by constructing a "minimum-savings table" as follows: For each customer, calculate the difference between the cost of assigning the customer to the minimum-cost location and the next higher-cost location. Let us call this difference the *minimum increment*. This represents the minimum amount saved if the customer is assigned to

the location with the minimum cost. All other locations to which this customer could be assigned would only increase the cost of assignment. In constructing the minimum-savings table, assign each customer's minimum increment to its minimum-cost location. Total the savings for each column in the minimum-savings table and subtract the fixed cost for each location for the corresponding column total. This gives the net savings for each location. Because we must have at least one facility to satisfy the demands, the first facility should be located where the cost is the least. Assign all of the customers to this facility. This location now becomes an assigned location.

The remaining steps 3–6 are applied to the initial solution.

3. Determine the savings in moving each customer from the assigned location(s) to the nonassigned location(s). If there are no savings for a particular location, mark it by a dash in the appropriate column.
4. Total each unassigned column. Each sum represents the savings that could be achieved if the customers are moved to that column. From each sum, subtract the fixed cost associated with that location that has the maximum net savings. This is because we would use the additional facility only if it is going to reduce the total cost. The location just chosen now becomes an assigned location and a part of the "assigned group." Move the customers that contributed to the savings to this location. If all the facilities have been assigned, or if no column with savings can be found, proceed to step 5. Otherwise return to step 3.
5. This step is denoted as the final check analysis. For each assigned location, determine the cost that would be incurred if each of the presently assigned customers for that location were moved to the assigned location with the next higher transportation cost. Total these costs. If this sum is less than the fixed cost for the assigned location under consideration, then additional savings could be achieved by not selecting the present location for placing a facility (i.e., by making this location an unassigned location). Perform this check on all the assigned locations, and select the assigned location with the maximum savings, if it exists. Make this location an unassigned location by moving each of the customers presently assigned to this location to the assigned location with the next higher transportation cost. Repeat step 5 until no further savings can be obtained; we then have the optimum solution.

Application

Consider the same BTS problem analyzed previously. Recall that we had six customers and five possible locations in which the facilities would be placed.

Previously, we assumed that the costs were in units of time; here we will measure them in terms of dollars.

Suppose now that the fixed cost associated with each location, if a facility is placed there, is as follows:

Location	1	2	3	4	5
Fixed cost	350	0	250	300	100

We wish to place the facilities and assign the customers in a manner that will minimize the total cost of operation.

Because Table 6.6 gives the total cost table data. We illustrate here the application of step 2 to obtain the initial solution. For customer A, location 4 has the minimum transportation cost of $30, and the next larger amount is $90 for location 3. The difference of $60 is the minimum savings if this customer were serviced by location 4. This amount for customer A is entered in location 4 of the "minimum savings table or initial table" (Table 6.11). Similar analyses are performed for all other customers and the results are entered in the table. Finally, the fixed cost for each location is entered as a negative saving.

The net savings for each location is determined by taking the sum of each column. Here, each one has a net negative saving, which actually indicates additional cost, rather than a saving. Because the equipment must be placed and customers must be assigned, the first machine is placed where the cost is minimum (i.e., location 2). All of the customers are assigned to this location, as shown in Table 6.12.

TABLE 6.11 Initial Assignment

Customer	Location				
	1	2	3	4	5
A				60	
B			0		
C			40		
D				24	
E				40	
F					45
Fixed cost	−350	0	−250	−300	−100
Net savings	−350	0	−210	−176	−55

TABLE 6.12 Demand Assignments 1

Customer	Location				
	1	2	3	4	5
A		30			
B		40			
C		20			
D		24			
E		20			
F		15			

The next step is to determine whether a second machine is needed, and if so, where it should be placed. To do so, we need to calculate the saving resulting from moving a customer from its presently assigned location to an unassigned location. Table 6.13 shows the results.

Again, a negative saving is indicated by a dash. Taking the total of the savings and the fixed cost for each location makes it clear that the minimum net savings are at location 3. The facility, therefore, is placed in location 3, and the customers that contributed to these savings, A, B, C, E, and F are assigned there. The present demand assignments are shown in Table 6.14.

Table 6.15 details the process of selecting a location for the third machine (facility). After including the fixed cost, the net savings are found to be negative for all unassigned locations; therefore, the third machine is not profitable, and the process is stopped here. Thus, the previous solution is still the best at this stage, and Table 6.16 presents the costs for those assignments.

TABLE 6.13 Savings Table 1 (for Moving from Location 2)

Customer	Location				
	1	2	3	4	5
A	120	—	150	210	30
B	—	—	80	40	80
C	100	—	140	100	60
D	—	—	—	24	—
E	160	—	160	200	60
F	45	—	105	60	150
Total savings	425		635	634	380
Fixed cost	−350		−250	−300	−100
Net savings	75		385	334	280

TABLE 6.14 Demand Assignment 2

Customer	Location				
	1	2	3	4	5
A			30		
B			40		
C			20		
D		24			
E			20		
F			15		

TABLE 6.15 Savings Table (for Moving from Locations 2 and 3)

Customer	Location				
	1	2	3	4	5
A	—			60	—
B	—			—	—
C	—			—	—
D	—			24	—
E	—			40	—
F	—			—	45
Total	0			124	45
Fixed cost	−300	0	−250	−300	−100
Net savings	−300	0	−250	−176	−55

TABLE 6.16 Total Cost for the Final Solution

	1	2	3	4	5	Total
A			90			90
B			120			120
C			40			40
D		144				144
E			80			80
F			90			90
Demand cost		144	420			564
Fixed cost		0	250			250
Total cost		144	670			814

The next step is to perform a final-check analysis. Because there are only two assigned locations, the final check is rather simple. Customers, A, B, C, E, and F are presently assigned to location 3. If all of them are moved to location 2, the transportation cost is increased by 150 + 80 + 140 + 160 + 105 = $635. The fixed cost saving associated with making location 3 unassigned is $250. Because the saving is smaller than the cost, this move is not made.

Similarly, if customer D is moved from location 2 to location 3, the transportation cost is increased by $24. The fixed cost saving for location 2 is $0. Consequently, this move is also rejected. Thus, the present solution is indeed the best solution.

6.3.2 Unassignable Location

Quite often we come across a problem in which one or more customers cannot be assigned to certain locations. For example, consider the placement of a fire station in a community. Fire codes may dictate the maximum allowable response time of the fire department to an alarm; obviously, travel time is a major component of that time. This restriction on the maximum value for response time may not permit the selection of certain locations as suitable sites for placement of fire stations to serve the outlying areas. To solve this type of problem, we modify the fixed cost procedure by substituting an asterisk (*) wherever it appears in the demand cost table (which indicates the designated customers cannot be assigned to that location) by some large cost. However, assigning an infinite cost will make it difficult to determine the net savings (they will be indistinguishable from one location to another) as we move the demands from assigned locations to unassigned locations. As a general rule, this cost should be more than the largest fixed cost plus the two largest transportation costs. With this modification of the demand cost table, we can then apply the method described in Section 6.2 to solve the resulting problem.

6.4 UNASSIGNABLE FACILITY LOCATION

Consider the case of eight customers who are to be assigned to facilities placed in five locations. The demand cost data (i.e., demand multiplied by cost per unit demand and the fixed cost data) are given in Table 6.17.

Remember, an asterisk (*) for a cost entry means that the indicated customer cannot be assigned to the associated location. The first task is to determine the appropriate value for the asterisk based on the data in Table 6.17, the largest fixed cost is 400; the largest transportation cost is 240, and the next largest is 210. With this data, the asterisk could be replaced by 400 + 240 + 210 = 850 or, say, $900. This is shown in Table 6.18.

TABLE 6.17 Demand Cost and Fixed Cost

Customer	Location				
	1	2	3	4	5
A	120	210	180	210	170
B	180	*	190	190	150
C	100	150	110	150	110
D	*	240	195	180	150
E	60	55	50	65	70
F	*	210	*	120	195
G	180	110	*	160	200
H	*	165	195	120	*
Fixed cost	200	200	200	400	300

TABLE 6.18 Modified Demand Cost and Fixed Cost

Customer	Location				
	1	2	3	4	5
A	120	210	180	210	170
B	180	900	190	190	150
C	100	150	110	150	110
D	900	240	195	180	150
E	60	55	50	65	70
F	900	210	900	120	195
G	180	110	900	160	200
H	900	165	195	120	900
Fixed cost	200	200	200	400	300

Application of the steps of the procedure now leads to the final solution, shown in Table 6.19.

It may seem fortuitous that we have customers assigned only to locations where the assignments are possible; however, this is not a matter of luck. As long as we are free to choose the number of facilities, and as long as there is a location to which a customer can be assigned, then a valid assignment for each customer is always possible.

The problem may become infeasible if there is only a limited number of facilities. For instance, if we had only one facility in the previous example, all of the customers would have been assigned to location 1 because of its lowest total cost. But customers D, E, and H cannot be assigned to this location, thus making

TABLE 6.19 Total Cost for the Final Solution

Customer	Location 1	2	3	4	5	Total
A	120					120
B	180	*				180
C	100					100
D	*	240				240
E		55				55
F		210				210
G	*	110				110
H	*	165	*		*	165
Demand cost	400	780				1180
Fixed cost	200	200				400
Total cost	600	980				1580

the solution infeasible. One way to resolve this is to drop from the data the location where a customer with an artificially large cost is assigned in the final solution, and solve the problem again with the remaining data. If a solution exists, continued application of this modification process will obtain the minimum-cost solution. If there is no solution, the process will indicate so.

Let us return to the example. It is obvious that if there is only one facility to locate, it will not be in location 1. By using the modification process, we find that it will be placed in location 4 because this is the only location at which it is possible to assign all customers.

Exercises

6.1 A warehouse is to be located in one of two possible locations to store merchandise. The warehouse will supply the merchandise to three stores. The individual demand from each store per week and the distances from stores to each possible location is shown in the table.

Store	Location A	B	Demand
1	12	30	45
2	15	22	55
3	18	4	40

Determine the best location for the warehouse.

6.2 If fixed cost associated with each of the locations is as shown in the table, determine the location of the warehouse

Locations	A	B
Fixed cost	100	150

6.3 Three locations are possible for a copier to serve four departments. It is desired to minimize the time required to travel from the department to the copier. The distances from the departments to the locations and number of employees in each department are shown in the table.

Department	Location			Employees
	1	2	3	
1	20	30	18	3
2	42	21	12	5
3	55	70	44	4
4	32	70	80	2

1. Use the brute force method to determine the best location.
2. Determine the location by heuristic methods and compare the result with those from 1.
3. Determine the best locations if two copiers are to be located.

6.4 Four locations in a plant are being considered for the installation of a new subassembly station. The subassembly will be used in five departments. A forklift will transport the subassembly to the individual departments. The

Department	Location				Demand
	A	B	C	D	
1	4.50	3.00	3.20	0.60	6
2	2.40	0.70	2.50	1.75	12
3	0.80	2.60	1.00	1.90	15
4	1.20	0.50	1.00	2.60	20
5	1.80	0.70	0.75	0.40	7

time required for the forklift to travel to each location and the number of trips to be made by each are shown in the table.

1. Determine the best location for the subassembly station.
2. If two such subassembly stations can be placed, determine their locations.

6.5 There is a fixed cost associated with placing the subassembly station at each location in problem 6.4.

Locations	A	B	C	D
Fixed cost	6.50	0	5.40	2.80

Determine the location of the subassembly station.

6.6 A CNC machine is purchased for use by six departments in a manufacturing firm. The transportation time from the department to each of the possible locations for the CNC is shown along with the demand from each department to use the machine. Owing to the nature of the layout, some departments cannot access some of the possible locations shown here by an asterisk (*).

Department	Location						Demand
	1	2	3	4	5	6	
A	5.0	6.0	2.5	4.0	4.5	1.0	10
B	*	3.5	*	0.0	6.0	3.0	14
C	3.0	2.0	1.5	5.0	3.0	2.5	20
D	2.0	4.0	4.0	*	2.5	0.0	5
E	*	6.0	3.0	6.5	4.5	6.0	7
F	2.0	0.0	10.0	3.0	3.0	*	20

1. Determine the location of the CNC machine.
2. If two CNC machines are to be placed, determine their locations.
3. Determine the optimum number of CNC machines.

7

Network Facility Location

Network representation provides a powerful setting to display a variety of situations. In the past, we may have seen network illustrations for problems in transportation, production–distribution, scheduling, and other areas of resource utilization. In this chapter, we study the problems in location analysis that can be represented and analyzed using networks. Typically, nodes in a network represent either demand points, customers, or possible location site, whereas the branches connecting these node symbolize the available paths. New facilities may be located anywhere on the network, either on a branch or on a node. Such problems are referred to as *network location models*. Nodes, perhaps representing cities or a large concentration of populations, may in fact present more plausible sites for locating the new service facilities. If the facilities can be located only on the nodes, then the problems are called a *vertex problem*. On the other hand, if a facility can also be located on a branch, that is on a path joining two nodes, then it may be possible to reduce the value of the solution in a minimax (minimize the maximum value) objective. These problems are referred to as absolute *p-center problems*. In the next section we study the vertex problem, and p-center problems are discussed later in the chapter.

7.1 A NETWORK PROBLEM

One example to which a network analysis can be applied is a problem finding an optimum number of facilities to minimize cost and at the same time serve all the

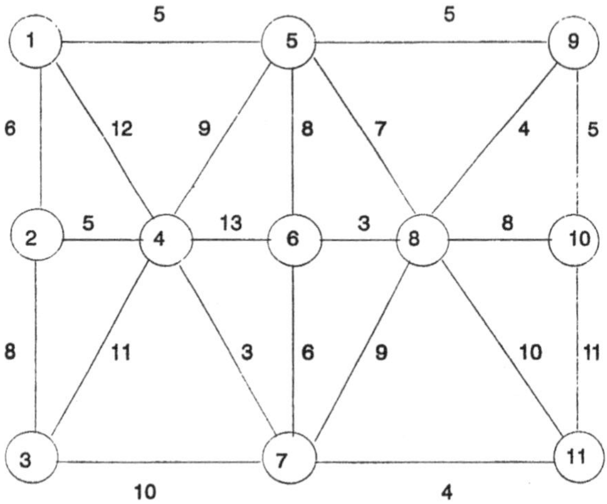

FIGURE 7.1 Network representing dealerships and connection paths.

customers satisfying some constraints. For example, a large automobile-manufacturing plant has its dealerships in 11 cities and wants the spare part distribution policies. Each distribution center means additional cost, but then prompt service is also a major consideration. For this reason the plant wishes to set a minimum number of distribution centers in the existing cities, such that each center is no more than 80 miles from any existing city. The road map for this example is displayed by the network in Figure 7.1. The distance between cities (nodes or vertices) in ($\times 10$) miles is shown on each arc.

The problem can be redefined as a discrete optimization problem of zero–one variables. We identify what are called covering sets. Each covering set displays the vertices that can be reached from itself within the constraint distance specified. For example, from node 1, if a facility is placed in node 1, we can reach node 1 itself, and nodes 2 and 5 which are within 8 units of node 1. The set is identified as set S1. Similarly other sets are displayed.

$$\begin{aligned} S1 &= \{1, 2, 5\},\ S2 = \{1, 2, 3, 4, 7\},\ S3 = \{2, 3\}, \\ S4 &= \{2, 4, 7, 11\},\ S5 = \{1, 5, 6, 8, 9\},\ S6 = \{5, 6, 7, 8, 9\}, \\ S7 &= \{2, 4, 6, 7, 11\},\ S8 = \{5, 6, 8, 9, 10\},\ S9 = \{5, 6, 8, 9, 10\}, \\ S10 &= \{8, 9, 10\},\ S11 = \{4, 7, 11\} \end{aligned}$$

By observation we might select nodes 2, 8, and 11 as the centers that will reach all other nodes within 8 units of distance, but such findings may not be always so obvious.

7.1.1 Row–Column Dominance Method

A better method might be to display the data as a 0–1 matrix, if a node (vertex) can be reached from a center located in one of the nodes (facility being placed in a center), it is marked as 1, otherwise it has a value 0. In our example, this is represented by the matrix in Table 7.1, in which rows represent centers (j) and columns vertices (i).

The problem has 11 variables (each column) and 11 constraints (each row), and needs some reduction. One way to reduce the matrix is by identifying relations in rows that are automatically satisfied by the relations in other rows. Let C_i be the set of all centers that can be reached from vertex i. Let R_j be all the vertices that can be reached from center j, then the following rules are used to reduce the dimension of the matrix without losing the optimum solution. From the reduced matrix it may be (and generally is) possible to observe the optimum solution.

Reduction Rules

1. Row infeasibility rule: If a vertex cannot be reached from any center the problem is infeasible.
2. Column dominance rule: If there are two columns, r and q such that $C_r \subseteq C_q$ (the numbers of centers from which vertex r can be reached are fewer than the numbers of centers from which vertex q can be reached, and the centers from which vertex r can be reached are also the vertex q can be reached), then column q may be dropped.
3. Row dominance rule: If there are two distinct rows s and t, such that $R_s \subseteq R_t$, then remove row s. The condition states that all the vertices

TABLE 7.1 Representation of Network in 0–1 Matrix

Centers	Vertices										
	1	2	3	4	5	6	7	8	9	10	11
1	1	1	0	0	1	0	0	0	0	0	0
2	1	1	1	1	0	0	1	0	0	0	0
3	0	1	1	0	0	0	0	0	0	0	0
4	0	1	0	1	0	0	1	0	0	0	1
5	1	0	0	0	1	1	0	1	1	0	0
6	0	0	0	0	1	1	1	1	1	0	0
7	0	1	0	1	0	1	1	0	0	0	1
8	0	0	0	0	1	1	0	1	1	1	0
9	0	0	0	0	1	1	0	1	1	1	0
10	0	0	0	0	0	0	0	1	1	1	0
11	0	0	0	1	0	0	1	0	0	0	1

that can be reached from center t also covers vertices that can be reached from center s and, furthermore the number of vertices that can be reached from s are fewer than those that can be reached from t.

By applying the column dominance rule to foregoing problem, we see that vertex at column 2 can be reached by 1, 2, 3, 4, and 7, whereas vertex at column 3 can be reached only by 2, and 3. Therefore, $C_3 \subseteq C_2$, which means any center that covers vertex 3 will also cover vertex 2; hence, we remove column 2. Similarly, $C_4 \subseteq C_7$; hence, we remove column 7 from the matrix, and also $C_{11} \subseteq C_4$; hence, column 4 is also removed. Proceeding in this manner we obtain the reduced matrix shown in Table 7.2.

Now further reducing the matrix by row dominance rule. For example, $R_2 \subseteq R_2$, which means that the potential center location at 2 covers all the vertices covered by potential center located at 3; hence, row 2 dominates row 3 and row 3 can be removed. Continuing in the similar manner we obtain Table 7.3.

TABLE 7.2 Reduced Matrix by Column Dominance Rule

Center	Vertex					
	1	3	5	6	10	11
1	1	0	1	0	0	0
2	1	1	0	0	0	0
3	0	1	0	0	0	0
4	0	0	0	0	0	1
5	1	0	1	1	0	0
6	0	0	1	1	0	0
7	0	0	0	1	0	1
8	0	0	1	1	1	0
9	0	0	1	1	1	0
10	0	0	0	0	1	0
11	0	0	0	0	0	1

TABLE 7.3 Reduced Matrix by Row Dominance Rule

Center	Vertex					
	1	3	5	6	10	11
2	1	1	0	0	0	0
5	1	0	1	1	0	0
7	0	0	0	1	0	1
8	0	0	1	1	1	0

In the foregoing matrix we can make the following observations. Vertex 10 and 11 are covered only by centers at 8 and 7, respectively, so we must have centers in these locations. The vertices not covered by these centers, vertices 1 and 3, can be covered by a center at 2. So three centers at 2, 7, and 8 are the minimum number of distribution centers required to cover the network within 8 time units from any center.

7.1.2 Alternative Approach for No Cost Problem

Another method that may frequently be quicker and easier to apply is as follows: For each row and column calculate the total in the matrix. Determine the least total among the columns. For that column, for each active cell—that is, a cell with 1—compare the totals for the associated rows. From these rows, choose the row with the maximum total as the new center. Remove the vertices that are covered from this center, as well as the center itself and form the new matrix and repeat the process.

The reasoning is as follows: The column total gives the number of centers that can reach this vertex. By choosing the smallest value from the column total we are selecting a vertex that is most difficult to reach. Observing the active cells in this column, we select a center that can reach the largest number of vertices. Once a vertex becomes accessible from a center in which a facility is established, it is dropped from any further considerations. Thus, in the next iterations only the vertices that are yet to be attached to a center are present. Successive application of this procedure selects the centers that can additionally contribute to vertex connection in the most optimum manner.

TABLE 7.4 Row–Column Totals

Center	Vertex										
	1	2	3	4	5	6	7	8	9	10	11	Total
1	1	1	0	0	1	0	0	0	0	0	0	**3**
2	1	1	1	1	0	0	1	0	0	0	0	**5**
3	0	1	1	0	0	0	0	0	0	0	0	**2**
4	0	1	0	1	0	0	1	0	0	0	1	**4**
5	1	0	0	0	1	1	0	1	1	0	0	**5**
6	0	0	0	0	1	1	1	1	1	0	0	**5**
7	0	1	0	1	0	1	1	0	0	0	1	**4**
8	0	0	0	0	1	1	0	1	1	1	0	**5**
9	0	0	0	0	1	1	0	1	1	1	0	**5**
10	0	0	0	0	0	0	0	1	1	1	0	**3**
11	0	0	0	1	0	0	1	0	0	0	1	**3**
Total	**3**	**5**	**2**	**4**	**4**	**5**	**5**	**5**	**5**	**3**	**3**	

Application of the first step of the procedure to Table 7.1 data requires adding the rows and columns. In our problem the results we shown in Table 7.4.

The least among the column total is 2, associated with column 3. The active cells in that column are in rows 2 and 3. Among these, row 2 gas the largest total value. Therefore, we place one center at 2, which covers vertices 1, 2, 3, 4, and 7. These vertexes and center 2 are dropped from further considerations. The reduced matrix is given in Table 7.5.

TABLE 7.5 Reduced Matrix After First Assignment

Center	Vertex						
	5	6	8	9	10	11	Total
1	1	0	0	0	0	0	**1**
3	0	0	0	0	0	0	**0**
4	0	0	0	0	0	1	**1**
5	1	1	1	1	0	0	**4**
6	1	1	1	1	0	0	**4**
7	0	1	0	0	0	1	**2**
8	1	1	1	1	1	0	**5**
9	1	1	1	1	1	0	**5**
10	0	0	1	1	1	0	**3**
11	0	0	0	0	0	1	**1**
Total	**5**	**5**	**5**	**5**	**3**	**3**	

TABLE 7.6 Reduced Matrix After Second Assignment

Center	Vertex	
	11	Total
1	0	0
3	0	0
4	1	1
5	0	0
6	0	0
7	1	1
9	0	0
10	0	0
11	1	**1**
	3	

TABLE 7.7 Fixed Cost for Locating New Centers

Potential centers	Cost in (×1000) dollars
1	80
2	70
3	110
4	60
5	120
6	110
7	90
8	70
9	105
10	115
11	65

By repeating the foregoing process, the least total among the columns is 3 associated with columns 10 and 11. The largest total among the active rows for these columns is 5 associated with row 8 or 9. So place a new facility in either one, say 8. The new center at 8 serves vertices 5, 6, 8, 9, and 10. Reducing the matrix further, we obtain Table 7.6.

Because column vertex 11 is the only vertex yet to be covered, we place the new facility at vertex 11. Thus, the centers are 2, 8, and 11 reach all the vertices within an 8-unit distance away. Also an alternative solution would be centers 2, 9, and 11.

The method is much simpler to follow and eliminates the need of evaluating row and column dominance.

7.2 COST CONSIDERATION

Suppose there is a fixed cost associated with locating a center at a potential node, This cost may be due to factors such as cost of land, building construction, and other expenses. The problem now is to find the number of centers that will cover the entire network within specific constraint time with minimum cost.

Continuing with the previous example, suppose the cost associated with the 11 centers is given in Table 7.7.

7.2.1 Linear-Programming Approach

The problem can be formulated as a linear-programming (LP) problem with the following parameters:

c_j = cost associated with locating a facility at location j.
$a_{ij} = 1$ if the center at j covers vertex i otherwise 0.
$x_j = 1$ if a facility is placed at center j otherwise 0.

The LP formulation is

$$\text{Minimize } \sum_{j=1}^{n} c_j x_j$$

$$\text{subject to } \sum_{j=1}^{n} a_{ij} x_j >= 1 \qquad i = 1, 2, \ldots, m \qquad x_j = 0 \text{ or } 1$$

Applying the model to our data we have

$$\text{Min } Z = 80x1 + 70x2 + 110x3 + 60x4 + 120x5 + 110 \\ x6 + 90x7 + 70x8 + 105x9 + 115x10 + 65x11$$

Subject to

$$\begin{aligned} x1 + x2 + x5 &>= 1 \\ x1 + x2 + x3 + x4 + x7 &>= 1 \\ x2 + x3 &>= 1 \\ x2 + x4 + x7 + x11 &>= 1 \\ x1 + x5 + x6 + x8 + x9 &>= 1 \\ x5 + x6 + x7 + x8 + x9 &>= 1 \\ x2 + x4 + x6 + x7 + x11 &>= 1 \\ x5 + x6 + x8 + x9 + x10 &>= 1 \\ x8 + x9 + x10 &>= 1 \\ x4 + x7 + x11 &>= 1 \end{aligned}$$

$x_j >= 0 \qquad j = 1, 2, \ldots, 11$

The solution gives:

$$Z = 200 \text{ with, } x1 = 0, x2 = 1, x3 = 0, x4 = 1, x5 = 0, \\ x6 = 0, x7 = 0, x8 = 1, x9 = 0, x10 = 0, x11 = 0$$

indicating the centers should be at nodes 2, 4, and 8.

7.2.2 An Alternative Procedure for Fixed Cost Problem

A slight modification to the no-cost procedure is all that is needed to solve the problems of the fixed cost. Follow the steps in a manner similar to the no-cost procedure, except with one modification. Calculate the cost per vertex for each center as the fixed cost divided by the number of vertices it can reach. For example, center at 1 can reach vertices 1, 2, and 5 and costs $80, therefore, its cost per vertex is 80/3. This data is included as the last column in Table 7.8.

Again, select the least total among the columns (i.e. 2 in column 3). Select the least cost per vertex associated among the active cells 2 and 3. In this case it is 70/5 associated with row 2; hence, place a center at 2. Proceeding further, reduce the matrix and recalculate the cost per vertex based on the reduce matrix, as shown in Table 7.9.

TABLE 7.8 Cost per Vertex Calculations

Center	Vertex										
	1	2	3	4	5	6	7	8	9	10	11	Cost/vertex
1	1	1	0	0	1	0	0	0	0	0	0	**80/3**
2	1	1	1	1	0	0	1	0	0	0	0	**70/5**
3	0	1	1	0	0	0	0	0	0	0	0	**110/2**
4	0	1	0	1	0	0	1	0	0	0	1	**60/4**
5	1	0	0	0	1	1	0	1	1	0	0	**120/5**
6	0	0	0	0	1	1	1	1	1	0	0	**110/5**
7	0	1	0	1	0	1	1	0	0	0	1	**90/4**
8	0	0	0	0	1	1	0	1	1	1	0	**70/5**
9	0	0	0	0	1	1	0	1	1	1	0	**105/5**
10	0	0	0	0	0	0	0	1	1	1	0	**115/3**
11	0	0	0	1	0	0	1	0	0	0	1	**65/3**
Total	**3**	**5**	**2**	**4**	**4**	**5**	**5**	**5**	**5**	**3**	**3**	

TABLE 7.9 Vertex Cost Calculations After First Selection

Center	Vertex						
	5	6	8	9	10	11	Cost/vertex
1	1	0	0	0	0	0	**80/1**
3	0	0	0	0	0	0	**110/0**
4	0	0	0	0	0	1	**60/1**
5	1	1	1	1	0	0	**120/4**
6	1	1	1	1	0	0	**110/4**
7	0	1	0	0	0	1	**90/2**
8	1	1	1	1	1	0	**70/5**
9	1	1	1	1	1	0	**105/5**
10	0	0	1	1	1	0	**115/3**
11	0	0	0	0	0	1	**65/1**
	5	**5**	**5**	**5**	**3**	**3**	

TABLE 7.10 Vertex Cost Calculations After Second Selection

Center	Vertex	
	11	Cost/vertex
1	0	**80/0**
3	0	**110/0**
4	1	**60/1**
5	0	**120/0**
6	0	**110/0**
7	1	**90/1**
9	0	**105/0**
10	0	**115/0**
11	1	**65/1**
	3	

Among the least total we have a tie for columns 10 and 11. Here, the procedure is modified, and rather than choosing any column at random, we chose the column that has the least-cost center. The least cost for column 10 is center 8, with the cost of 70/5 and for column 11 is center 7, with the cost of 90/2. We select the location with least cost (i.e., location 8) as the new center and the new table is generated as Table 7.10.

Among the centers that cover vertex 11, the center at 4 has the lowest cost; therefore, it is selected next.

The entire network is covered with centers at 2, 4, and 8, and the minimum cost of 70,000 + 60,000 + 70,000 = \$200,000. The solution is the same as that we had obtained by the linear programming.

7.3 ABSOLUTE p-CENTER PROBLEM

The next set of problems we consider are called p-center problems. The objective is to locate facilities to serve demand points so that the *maximum* value of the criteria measured, such as the distance or the time, is as small as possible or is within a limiting value. This is also referred to as minimax criteria. Locating emergency service facilities to respond to any urgent call within specified time limit, is one such application. Possible sites for locating the new facilities may have been preselected based on the availability of the sites, and the analysis may involve selecting p-specific locations from these, or alternatively, we may be free to choose any site.

7.3.1 Facilities Located on Preselected Sites

If the sites have been preselected, then a slight modification to the procedure in Section 7.2 is all that is needed. A search is performed on the limiting value. Select a limiting value of the measuring entity and develop a 0–1 matrix, similar to Table 7.1, indicating the customers who can be served from each possible site. Determine the number of centers required by applying the procedure in 7.2. If this value is more than the number of centers allowed, then the limiting value of the measuring criteria has to be increased, and the search is performed again. If it is less, then the limiting value could be decreased, and the procedure is repeated. If the centers required are equal to the specified number, continue the search by decreasing the limiting value, until the facilities required exceed what is permitted. The limiting value before the final change is the minimax solution.

7.3.2 Facilities Located on Networks

The procedure for a p-center problem when a facility can be located anywhere on the network may be a little challenging and is explained next by means of an example.

Consider a tree network, shown in Figure 7.2, with 44 nodes and associated branches connecting these nodes. Note, because this is a tree network, there are no cycles. Suppose we wish to place a single facility that will respond to the demand from a node as quickly as possible, which in this case, is directly proportional to the distance between the new facility and existing demand point. The procedure to select one absolute center (location for one new facility) is as follows:

1. Select any point on the network and find a node that is farthest from that point. Call it node A.
2. Determine another node that is farthest away from node A. Call it node B.
3. The center point of a link connecting node A and B is the absolute one-center.

In our example, suppose we select a point, node 13. The node farthest away from this node is node 1, 19 units away. From node 1 the farthest node is node 18, which is 37 units away. Therefore, the absolute one-center is on the link joining node 1 and node 18 at $37/2 = 18.5$ units away from both nodes. This point is shown as ABS1 in Figure 7.3.

To locate 2 absolute centers, start with the one-center solution. Delete the link containing ABS1. This divides the network into two disconnected tree structures. Apply the one-center solution procedure to each of the resulting tree structures.

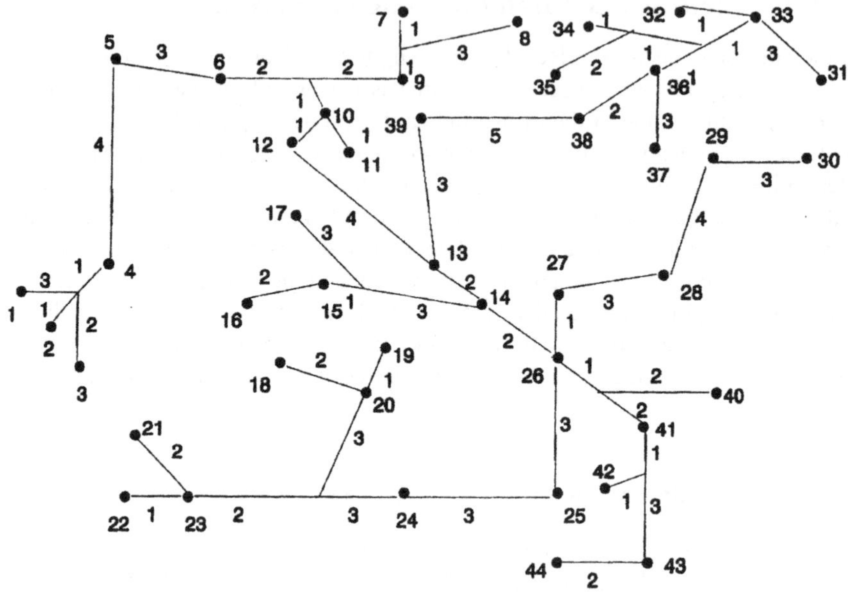

FIGURE 7.2 Tree network.

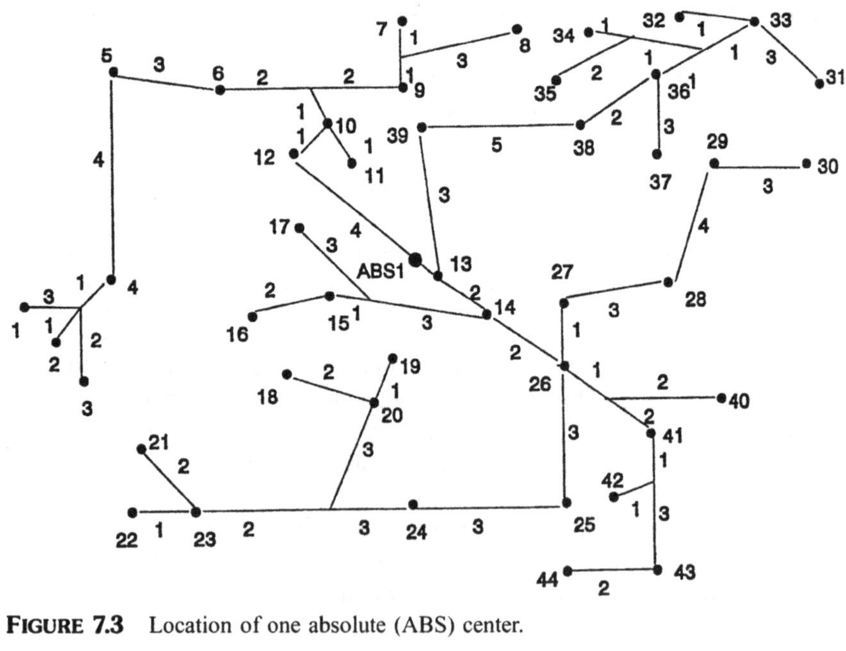

FIGURE 7.3 Location of one absolute (ABS) center.

In our example, these centers are shown as ABS2 and ABS3 in Figure 7.4. The minimax distance in ABS2 network is 9.5 whereas it is 17 for ABS3 network, resulting in overall minimax distance of 17, the larger of two numbers.

Note that in each case, the new facility is located interior in the network. In locating more than 2 facilities we take the advantage of this property. Prefix the maximum permissable value for minimax criteria. Start from an outermost node and proceed interior (therefore, the procedure is called backward procedure), leading toward, and one on it, continuing with, the longest path in the network. Determine the minimum distance that will cover all nodes examined so far, as we go backward. At every branch junction (a point from which more than one branches have stemmed out), apply the backward procedure to the other branch(s) and determine the minimax distance for that branch up to the junction point (if a center is not required earlier). The maximum value among the branches merging from the junction is the distance value that will cover all nodes in all branches emerging from that junction. Continue the procedure. Once that distance is equal to the value set, place a tentative ABS location at that point. Determine all other nodes that this location may serve, making necessary adjustments in ABS location such that the maximum possible nodes are reached. Detach the associated network in an independent tree and delete any empty branches (branch without a node at the end). Continue the procedure on the remaining

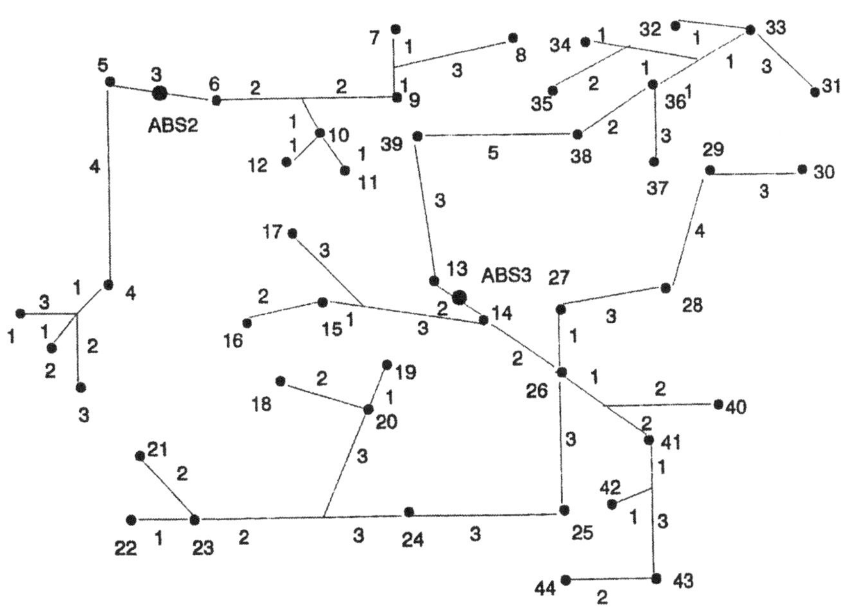

FIGURE 7.4 Location of two absolute centers.

network. When all the nodes are served, the associated number of facilities is the minimum number of ABS required for the present minimax value.

In our example, suppose we fix the minimax value to 13 and wish to determine the least number of centers required to cover all nodes. We can start on an outermost node. In our network the outermost nodes are 44, 21, 30, 31, and 1. These nodes are selected because they are the farthest end nodes in the network. Let us start with a node among these, say node 44. Traveling backward from node 44 we see 2 units will cover both nodes 43 and 44. Going farther back on the main stem (longest path in the network as was determined in one-center analysis), we encounter a branch leading to node 42. So perform the backward analysis starting at node 42. At the junction of branches coming from node 43 and 42, the distance on the backward path of each branch is 5 and 1, respectively. The maximum is 5. This means that if a facility is placed at that junction, a minimax distance of 5 units will cover all the nodes that are examined so far; namely, 44, 43, and 42. Continuing, at the junction of branch from node 41 and 40 the minimax distance is $\max(8, 2) = 8$. At node 26 we have three branches merging; therefore, it is necessary to calculate minimax distance for each branch. For the branch with 30 as the end node, this distance is 11. For the branch with node 44, this distance is 9. For the node 21 branch, again applying the backward procedure, the distance is 14. This distance exceeds the set limit of 13. Therefore, the first ABS center must be at a point A, which measure exactly 13 units in that

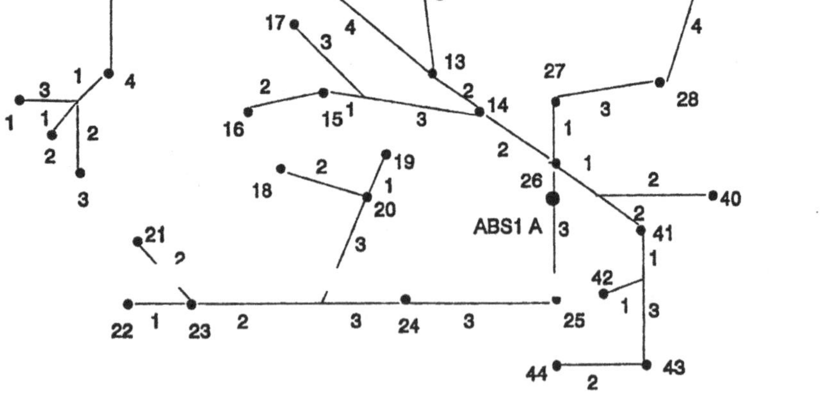

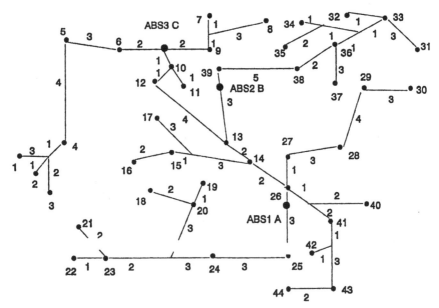

FIGURE 7.5 Location of the three absolute centers.

branch. The other two branches can be attached to this center because the minimax distances would be $9 + 1 = 10$ and $11 + 1 = 12$, respectively.

Check what other nodes can be served from this center. Proceeding on the main stem, node 14 is 3 units away. Minimax distance from branch merging into 14 from nodes 16 and 17 is 7, which is then $7 + 3 = 10$ units from A and thus can be covered from A. Proceeding to node 13, it is 2 units away and is a junction node. So the branches merging into 13 must be checked.

By following the backward procedure from node 31, we determine an ABS2 must be set at location B to serve all nodes in that branch. Similarly a center ABS3 must be established at junction C to serve all outer nodes leading to node 1. Nodes 10, 11, 12, and 13 may be served by multiple ABS nodes. If we follow the policy of assigning nodes to the closest ABS center, nodes 10, 11, and 12, should be served from ABS3, and 13 from ABS2. All the locations are shown in Figure 7.5.

By aplying the procedure for different limiting values of minimax criteria we can also determine the minimum minimax value for a predetermined number of centers.

7.4 ABSOLUTE 1-CENTER ON A WEIGHTED TREE

Now suppose each location has different weight, indicating one customer is more important than another. In establishing a fire station, for example, a nursing home, in which most of the adult population is in the age group of 70, would require a greater weight than a normal household in which most people can escape from fire more easily. Another viewpoint may be a customer who contributes more to the profit or has larger demand or even has more political or social clout may have higher weight than other customers. We wish to place a new facility such as to minimize the maximum demand-weighted distance between the new facility and the existing locations. We assume, as before, that the existing locations are on the nodes and their connecting paths form a tree network.

Suppose the network is as shown in Figure 7.6 and the weights of the nodes are given in Table 7.11.

Before we form details of the procedure, let us observe some properties that are important in its development.

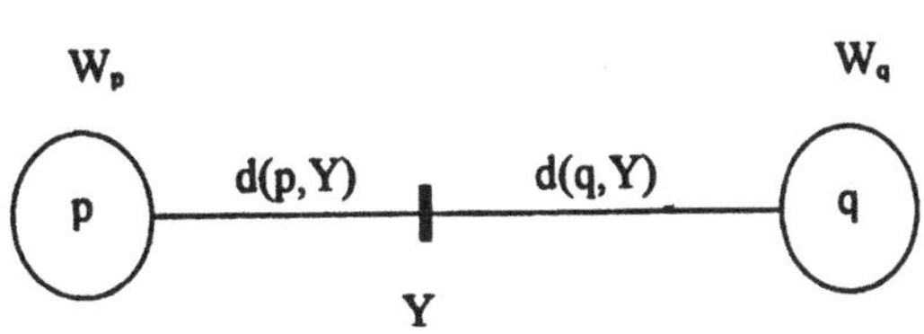

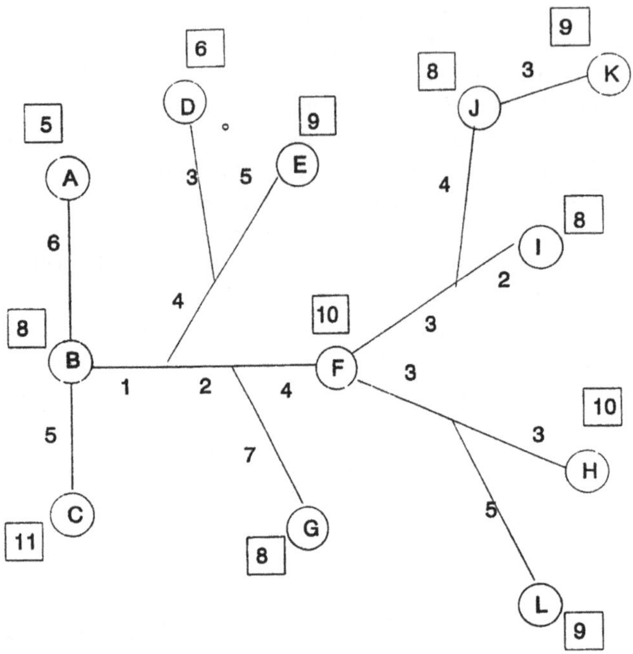

FIGURE 7.6 Absolute one center on a weighted tree.

TABLE 7.11 Weights Associated with Nodes

Nodes	A	B	C	D	E	F	G	H	I	J	K	L
Weights	5	8	11	6	9	10	8	10	8	8	9	9

Let us start by analyzing two nodes connected by a path. Suppose the new facility is to be located between nodes p and q. Let the location of new facility be Y. If the weights for the nodes p and q are W_p and W_q, and the distance between each location to the new facility Y, which is noted as $dt(i,Y)$, then the optimum location Y should satisfy the condition,

$$Y \text{ location} = W_p d(p, Y) = W_q d(q, Y) \tag{1}$$

We also know that

$$d(p, Y) + d(Y, 1) = d(p, q) \tag{2}$$

therefore,

$$W_p[d(p,q) - d(Y,q)] = W_q d(q,Y)$$
$$W_p d(p,q) - W_p d(Y,q) = W_q d(q,Y)$$

And since $d(Y,q) = d(q,Y)$, we obtain

$$\frac{W_p d(p,q)}{(W_p + W_q)} = d(Y,q)$$

Substituting back in condition (1), we note that the optimum location Y has the weight of $W_p W_q d(p,q)/(W_p + W_q)$.

We denote this quantity as WD (weighted distance) then for nodes p and q,

$$WD = W_p W_q d(p,q)/(W_p + W_q)$$

This suggests an approach to determine the optimum location in a network. Select all two-node combinations and calculate the optimum weighted location for each pair. Because each value is associated with the optimum location of the facility if it is to serve only those two nodes, and we are choosing the minimum possible value that covers all the pairs, the best possible location for the entire network is between the pair of nodes that has the largest weighted value. This location serves all nodes with the weighted value that is less than the maximum WD calculated.

A simplified approach is as follows: Starting with any node, say node i, calculate WD to each of the other node. Select the node with maximum value of WD say node j. Next calculate the WD from node j to all other nodes. If now the maximum WD value is associated with node i, then we have determined an unique pair i and j where the WD is maximum in the network. If such is not so, continue the iteration with the new node with the present maximum value of WD. Stop the procedure when we obtain a pair of nodes in which we will go back and forth if we continue the procedure.

In our network let us start arbitrarily at node A. Calculate the value of WD for node from node A. The results are shown in Table 7.12. The maximum is associated with node K which was calculated as follows:

$$W_a W_k d(A,K)/(W_a + W_k) = (5)(8)(23)/(5+9) = 73.92$$

Now beginning with node K determine WD for all other nodes. The results are shown in Table 7.13.

TABLE 7.12 Weighted Distance (WD) from A

From–To	A	B	C	D	E	F	G	H	I	J	K	L
A	—	18.46	37.81	38.18	51.42	46.66	49.23	63.33	67.69	61.58	73.92	67.5

TABLE 7.13 Weighted Distances (*WD*) from A and K

From–To	A	B	C	D	E	F	G	H	I	J	K	L
A	—	18.46	37.81	38.18	51.42	46.66	49.23	63.33	67.69	61.58	73.92	67.5
B											72.0	
C											108.9	
D											82.8	
E											112.5	
F											47.36	
G											97.41	
H											80.52	
I											38.11	
J											12.70	
K											—	
L											81.0	

The node having the largest value of *WD* is node E (with value of 112.5). Because it is not the initial node, node A, start with node E pairing all other nodes and calculate associated values of *WD*s: we obtain Table 7.14.

Here, the largest value of *WD* (i.e., 112.5) is associated with node K, the previous starting node. Because we have reached the cycling stage, we also have the final solution. The new facility lies between the nodes E and K.

The new facility will be located at as distance of $(W_e)d(K, E)/(W_e + W_k)$ from K or $= (9)(25)/(9 + 9) = 12.5$ units away from node K on the path K–E.

TABLE 7.14 Weighted distances from A, K, and E

From–To	A	B	C	D	E	F	G	H	I	J	K	L
A	—	18.46	37.81	38.18	51.42	46.66	49.23	63.33	67.69	61.58	73.92	67.5
B											72.0	
C											108.9	
D											82.8	
E	51.42	42.35	74.25	28.5	—	71.05	76.23	99.47	84.70	93.17	112.5	103.5
F											47.36	
G											97.41	
H											80.52	
I											38.11	
J											12.70	
K											—	
L											81.0	

7.5 ABSOLUTE p-CENTER ON A WEIGHTED TREE

Now suppose we want to place multiple facilities on the tree in Figure 7.6. From the previous section we know that with one facility the minimum value of the weighted distance that will cover all nodes is 112.5. If we are allowed to place two facilities, this value would be smaller than 112.5. However, we do not know what the actual value is until we divide the network into two parts, one served by one facility and other by the second facility. The challenge is to determine how the network should be divided. To this end the following procedure is suggested.

Start with a predefined maximum value of the variable-weighted distance (VWD) that is equal to the maximum weight times the distance of the nodes examined. Initially set the value to be less than maximum WD found earlier, which is less than 112.5, say 70. Start with any end node, let us choose at random node K. Proceed backward from node K. At node J, calculating the weighted value as equal to the weight times the distance we obtain VWD $= 3 \times 9 = 27$. This value is less than 70 so continue in branch J–F. At any point X units from J where $X < 4$, VWD $= \max(9(3+X), 8X) = 9(3+X)$. At $X = 4$ the VWD is 63 for that branch. But another branch I is merging at that point; hence, we must calculate VWD from node I up to the merge point and choose the maximum value as VWD. In this case it is max $(63, 16) = 63$. Because VWD is less than predefined value of 70, continue in the branch toward F. By this time we should note that VWD is the product of weight of only one node, and its distance to the point of interest (it may not be the same node over the entire length of the path). Proceeding backward it is easy to identify which node contributes to VWD, in our case it has been node K.

Determine the location of the first facility at which VWD $= 70$. This point is Y units away from K where Y is $9Y = 70$. Thus the point is 4.77 from J on branch J–F. Determine the nodes that are within 70 units of weighted distance away from this point. Only node F is possible. Cut the network, which can be served by the facility just placed, excluding F because F is still a part of a network that is presently not being served (nodes H and L as well as nodes A, B, C, D, E, and G). Start with another nonserved end node, say L and we see that we need a facility to serve nodes H and L. The remaining nodes are not served; hence, two facilities are not sufficient for maximum VWD $= 70$.

By searching on VWD, when the maximum value of VWD is set to 82, the network can be broken into two distinct networks consisting of K, J, I, F, H, and L and A, B, C, D, E, and G. The *WD* for the first network is 81 between nodes L and K and for the second network 76.23 between G and E. Thus, for two facilities the maximum *WD* is 81. Figure 7.7 shows the facilities placement.

If we have three facilities to place, the problem becomes interesting. With VWD $= 80$, three networks can be formed consisting of A, B, C, D, E, G, and F, K, J, I, and F, and F, H, and L. The problem is to decide which facility will serve

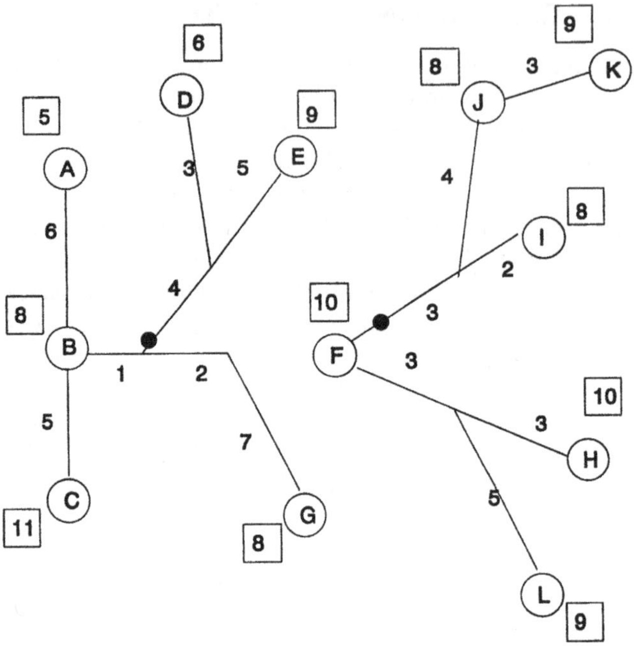

FIGURE 7.7 Absolute two centers on a weighted tree.

F. With F in first group *WD* = 76.23 and without it it is the same. In the third group we observe the same results with *WD* = 37.89. However, in the second group with F, *WD* is 47.36 and without F it is 38.11. Because joining F in either first or third group does not change the *WD*, it should be joined in one of those two groups. The maximum *WD* for three facilities problem is 76.23 between nodes E and G. The results are shown in Figure 7.8. Note that if the final *WD* is greater than the original feasible VWD, then an alternative node allocation to divide the network can be found that will reduce the value of *WD*.

7.6 COMPETITIVE FACILITY PLACEMENT IN SELECT LOCATIONS

Consider a network for which there are already few existing facilities, and a customer (represented by node) receives service from a facility that is closest to it. As a newcomer in the market providing the same services, we wish to establish a similar facility and attract a maximum number of customers to our new facility. Customers will switch to the new facility only if it is closer to them than the

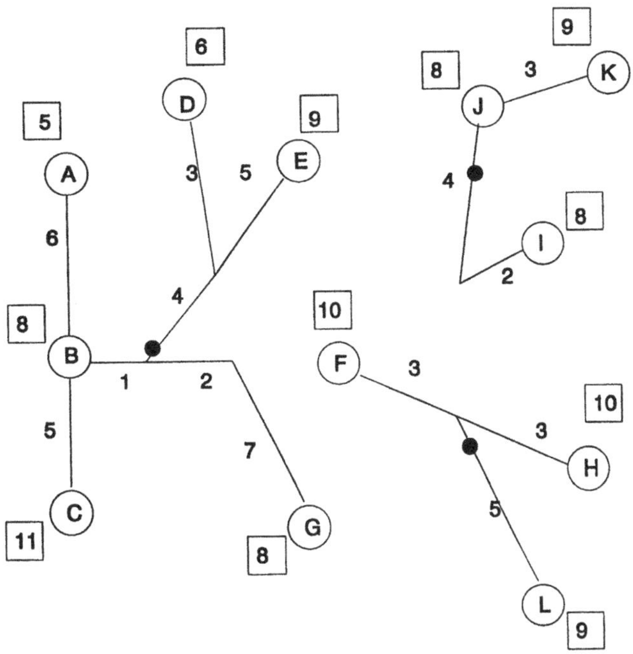

FIGURE 7.8 Absolute three centers on a weighted tree.

existing service center. The problem is to determine the location for the new facility to successfully compete in this environment.

For instance, in Figure 7.9, ten customers are currently being served by a facility located at location O. Furthermore, there is a short list of four possible locations at which the new facility may be located. These locations are 1, 2, 3, and 4, and we wish to select one of these locations as a place for a competing facility that will attract the maximum number of customers. Each customer is equally important and, thus, has equal weight.

The procedure for selection of a competitive site is as follows:

1. A customer travels on a path that has the shortest distance to the existing facility to receive the service. So, as the first step, calculate from existing facility O, the shortest distance to each node. The results for our example are shown in Figure 7.10; they are also displayed in Table 7.15.
2. For a customer to switch to a new facility the distance traveled must be reduced. From each possible location determine the shortest distance to

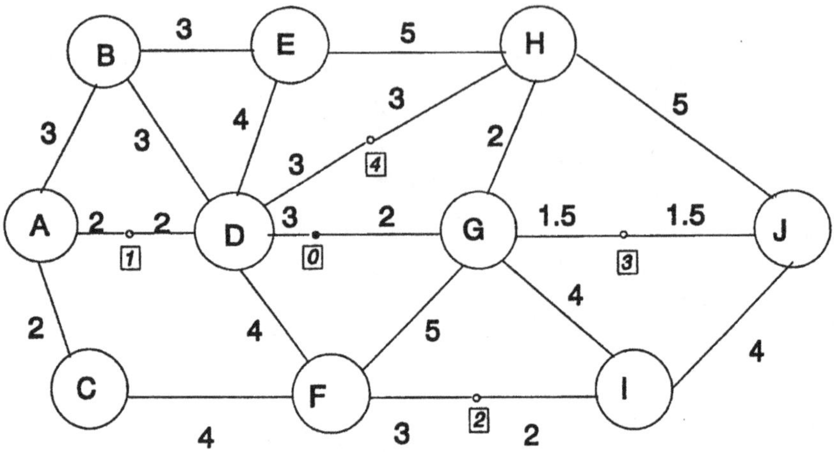

FIGURE 7.9 Ten customers being presently served from 0.

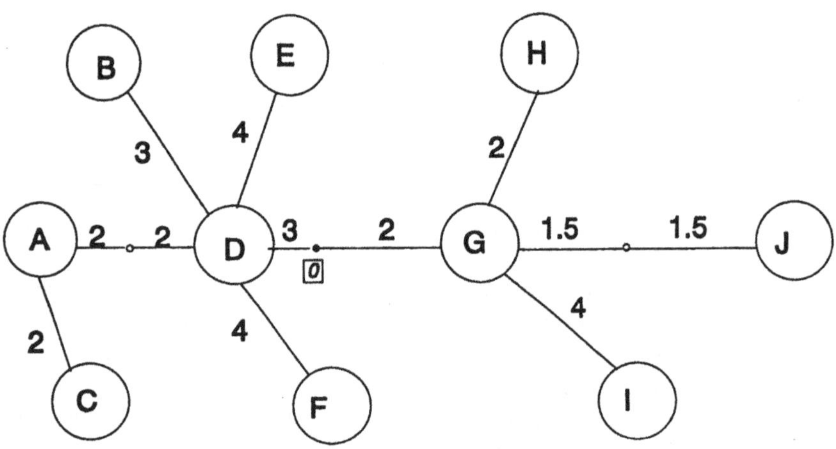

FIGURE 7.10 Shortest distance to a demand node from facility 0.

TABLE 7.15 Shortest Distance from 0

From–To	A	B	C	D	E	F	G	H	I	J
0	7	6	9	3	7	7	2	4	6	5

each demand node in the network. The results are shown in Figures 7.11 through 7.14, and the distances are also tabulated in the Tables 7.16 though 7.19.

All of the tables are compiled together to display the distances for each demand node to the existing and possible new locations. Table 7.20 displays the results.

Now we can convert the distance matrix into the 0–1 format.If a customer has less distance to travel to a new possible facility location than it needs to travel

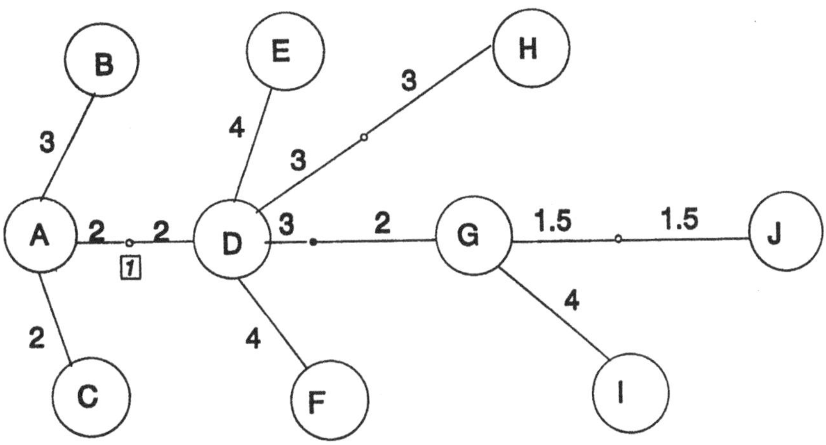

FIGURE 7.11 Shortest distance to a demand node from possible facility 1.

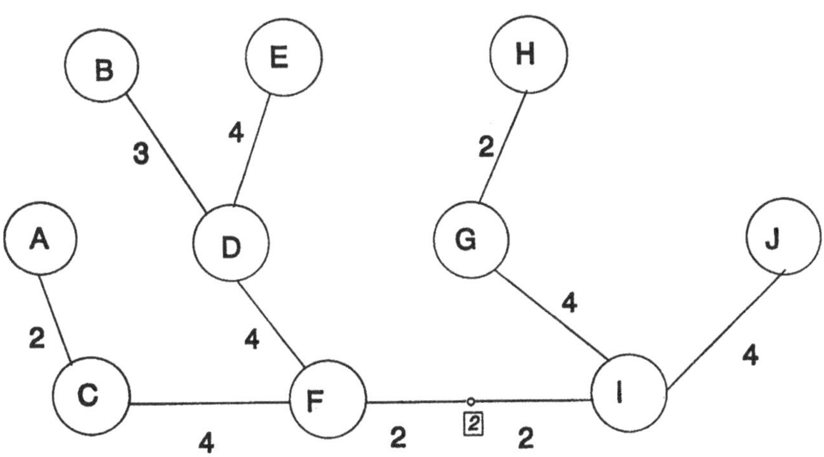

FIGURE 7.12 Shortest distance to a demand node from facility 2.

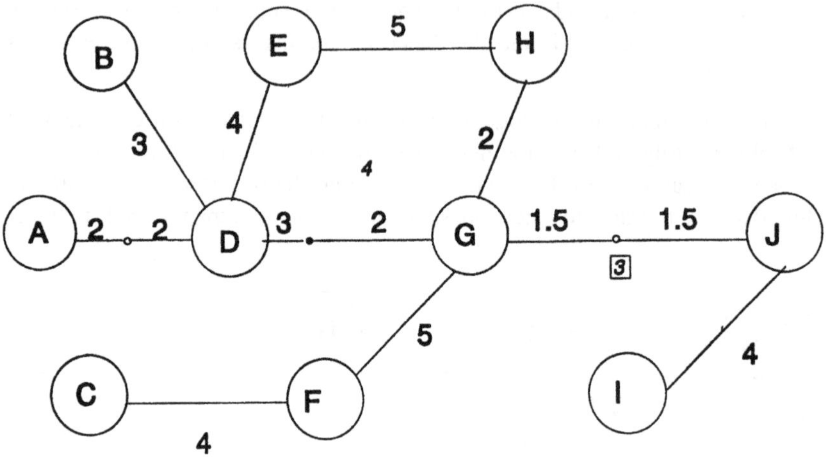

FIGURE 7.13 Shortest distance to a demand node from facility 3.

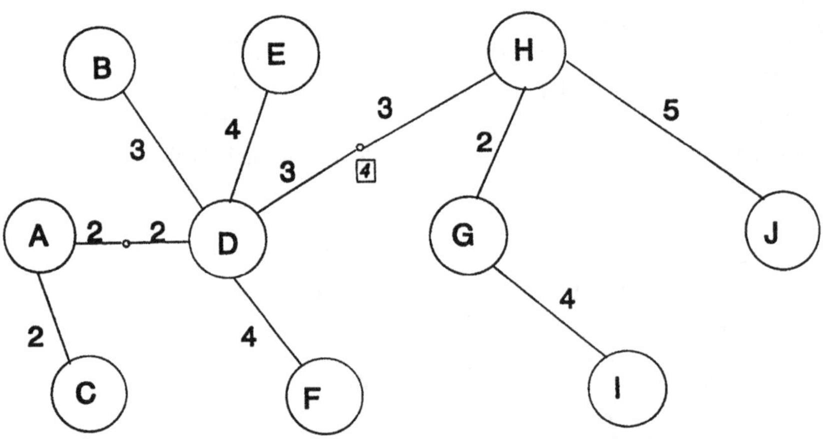

FIGURE 7.14 Shortest distance to a demand node from facility 4.

TABLE 7.16 Shortest Distance from Location 1

From–To	A	B	C	D	E	F	G	H	I	J
1	2	5	4	2	6	6	7	8	11	10

TABLE 7.17 Shortest Distance from Location 2

From–To	A	B	C	D	E	F	G	H	I	J
2	9	10	7	7	11	3	6	8	2	6

TABLE 7.18 Shortest Distance from Location 3

From–To	A	B	C	D	E	F	G	H	I	J
3	10.5	9.5	10.5	6.5	8.5	6.5	1.5	3.5	5.5	1.5

TABLE 7.19 Shortest Distance from Location 4

From–To	A	B	C	D	E	F	G	H	I	J
4	7	6	9	3	7	7	5	3	9	8

TABLE 7.20 Composite Shortest Distance Table

From–To		A	B	C	D	E	F	G	H	I	J
Existing	0	7	6	9	3	7	7	2	4	6	5
	1	2	5	4	2	6	6	7	8	11	10
Possible locations	2	9	10	7	7	11	3	6	8	2	6
	3	10.5	9.5	10.5	6.5	8.5	6.5	1.5	3.5	5.5	1.5
	4	7	6	9	3	7	7	5	3	9	8

to the existing (O) service center, then the associated distance is converted to 1, otherwise it is converted to 0. If the distance is the same we assume that the customer will go either to the new facility or the old existing facility with equal likelihood; therefore, the associated distance is converted to 0.5, to reflect $\frac{1}{2}$ and $\frac{1}{2}$ distributions of the trips. In our illustration the result is Table 7.21.

The total associated with each location is obtained by adding the row values. This total indicates the number of customers who will switch to the associated location if a new facility is placed there. The location with the highest total value is selected for the placement of the new facility. In our example, location 1 on the direct link joining nodes A and D is selected to place a competitive facility.

TABLE 7.21 0–1 Distance Conversion

From–To		A	B	C	D	E	F	G	H	I	J	Total
Existing	0	—	—	—	—	—	—	—	—	—	—	—
Possible	1	1	1	1	1	1	1	0	0	0	0	6
locations	2	0	0	1	0	0	1	0	0	1	0	3
	3	0	0	0	0	0	1	1	1	1	1	5
	4	0.5	0.5	0.5	0.5	0.5	0.5	0	1	0	0	4

7.6.1 Location of a Competitive Facility with One or More Existing Facilities

Suppose that, in the previous network, facilities exists at locations 0 and 1 and we want to place a new competitive facility, how will the procedure change? not by very much. The logic remains the same, except we must determine the present distribution of the customers to the existing facilities and observe which customer may change to the competitive location if a facility is installed there.

From the results of the previous example, we know the nodes or customers covered by the existing facilities 0 and 1. They are

Facility 0	G, H, I, J
Facility 1	A, B, C, D, E, F

By following the same procedure as used for locating one competitive facility, develop a shortest-distance table from the existing and the possible facilities' locations to each customer. Table 7.22 displays the results.

TABLE 7.22 Shortest Distance from Each Location to Each Node

From–To		A	B	C	D	E	F	G	H	I	J
Existing	0	7	6	9	3	7	7	2	4	6	5
	1	2	5	4	2	6	6	7	8	11	10
Possible	2	9	10	7	7	11	3	6	8	2	6
locations	3	10.5	9.5	10.5	6.5	8.5	6.5	1.5	3.5	5.5	1.5
	4	7	6	9	3	7	7	5	3	9	8

TABLE 7.23 0–1 Distance Conversion

From–To		A	B	C	D	E	F	G	H	I	J	total
Existing	0	—	—	—	—	—	—	—	—	—	—	—
	1	—	—	—	—	—	—	—	—	—	—	—
Possible locations	2	0	0	0	0	0	1	0	0	1	0	2
	3	0	0	0	0	0	0	1	1	1	1	4
	4	0	0	0	0	0	0	0	1	0	0	1

Currently, a customer travels to the closest existing facility and will switch only if the new location is closer than any of the existing facility. Convert the distance to 1 if the new location distance is less than the shortest distance to any of the present facility. This is displayed in the zero–one Table 7.23.

The new competitive facility should be located at location 3, the location with the largest total, and will cover nodes G, H, I, and J.

7.7 LOCATION OF COMPETITIVE FACILITY ANYWHERE ON THE NETWORK

Now suppose the competitive facility can be placed in any location, with the objective of attracting as many customers to it as possible. Remember, customers will switch from the existing service facility only if the new facility is closer to them. So in essence we are trying to find a position that is closer to as many customers as possible.

If we observe the present assignments in a network, in Figure 7.13 all customers travel to an existing facility through the end nodes of the branch on which the facility is located. For example, when we have two facilities in locations 0 and 1, the way customers are served is as follows and as shown in Figure 7.15:

Facility at location	Customers served by the facility
0	G, H, I, J
1	A, B, C, D, E, F

The customers coming to 0 come through either node D or G, and customers coming to node 1 come through node A or D. We can develop a list, as displayed in the following, showing the number of customers coming through each of these nodes.

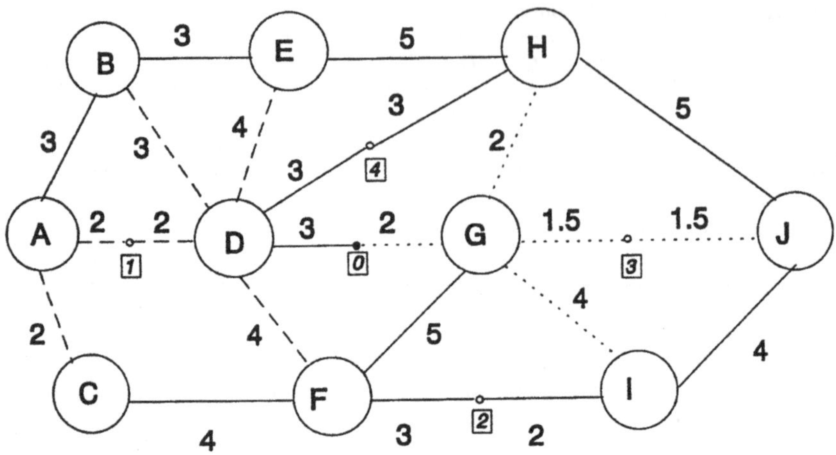

FIGURE 7.15 Customers served by locations 0 and 1 (···· by location 0 and – – – by location 1).

End node	Customers passing through end node	total number of customers passing through the node
G	G, H, I, J	4
D	—	0
A	A, B, C	3
D	D, E, F	3

Choose an end node. Determine the distance between the existing facility that the end node serves and the end node. These distances are as follows:

G 2 units from 0
D 3 units from 0
D 2 units from 1
A 2 units from 1

Now, if the demand coming through any end node is to be attracted to a new site, this site should be on a branch emerging from that end node and should be placed at a distance that is smaller than the distance to the existing facility, as just listed. For example, a competing facility can be placed on any one of the branches GH, GI, GF, and GJ that can attract all the demand passing though G. This facility should be no more than 2 units away from G. After some analysis, we see that if we place the facility on branch GF, a fraction less than two units away from G, it will attract not only nodes G, H, I, J, that are now assigned to 0, but also F which is currently assigned to 1. Similar analysis with other end nodes leads us to conclude that the new location X, on GF is the most attractive alternative as shown in Figure 7.16.

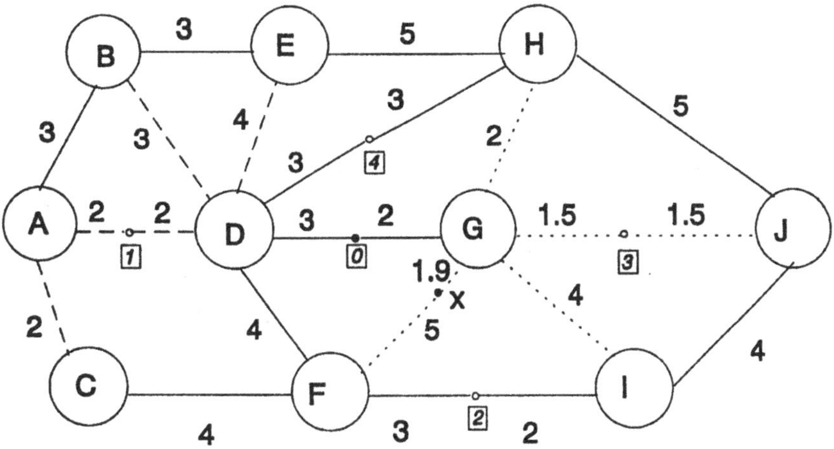

FIGURE 7.16 Customers served by locations x and 1 (···· by location x and – – – by location 1).

7.8 SERVICE AND BACKUP FACILITIES LOCATIONS IN A NETWORK

A simple modification of the backward prorogation procedure results in a solution to another type of problem. Here each demand point requires a facility within a specific distance, and each facility may even need a backup facility within certain proximity. We need to find the minimum number of facilities needed to satisfy such requirements.

Mathematically the problem can be stated as follows:

Let $x = \{xj\}$ denote a set of unknown points for the location of new facilities. Then the objective is to minimize $|x|$.
Subject to

$$d|xi - vj| <= bj \text{ for some } i \text{ and } j$$

$$d|xj - xj| = cj \text{ for some } i \text{ and } i$$

where bj is the maximum distance a facility i can be from demand node vj, and cj are the maximum distance between two facilities, one serving as a backup for the other.

The procedure consists of two steps. Initially, starting with tip (end) nodes and proceeding backward, place a facility in the best location possible. Determine what other nodes this facility can also serve. Once a node is attached to a facility, temporarily detach or ignore the unnecessary branches. In the second phase check the backup facility requirements by adjusting the locations established in the first phase or adding a backup facility wherever it is needed. The procedure is best explained by means of an example.

Location of sewage pumping stations to provide an uninterrupted service within a region is one such example. Suppose the pipeline network is represented by Figure 7.17 in which the nodes represent the population centers. The C_i values indicate the maximum radius on the network from a node center where a facility, in this case a pump station, must be located so that the facility can serve that node. The data are given in Table 7.24. It is assumed that pumps have a large enough capacity that the direction of pumping (or the total amount of flow) is not a factor.

Start with end or node 1. A facility must be placed (see Figure 7.18) at a distance of 1 unit on the branch 1–2, at $x1$, to cover the node 1 (note $C1 = 1$). This facility can also serve node 2, for $C2 = 4$, whereas distance from $x1$ to node 2 is 3. The branch 2–3 is ignored. On branch 3–4, a facility is needed at a distance 3 units from 3. This facility is noted as $x2$, which can also cover node 4, because $c4 = 5$. Branch 6–4 (passing through node 5) joins node 4 to a tip node 6. Thus, node 4 is a branching node. We continue our analysis on branch 6–4. A facility is needed at a distance of 1 unit from node 6 on branch 6–5. Note this as facility $x3$.

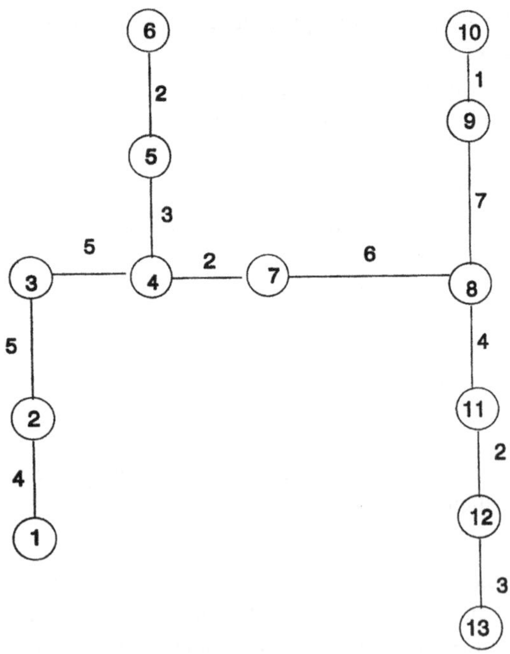

FIGURE 7.17 Tree network.

TABLE 7.24 Population Centers and Associated Maximum Allowable Facility Distances

Node	1	2	3	4	5	6	7	8	9	10	11	12	13
Ci	1	4	3	5	2	1	7	1	4	6	3	3	1

This facility can also cover node 5; although node 4 is already connected, we could also attach node 4 to *x*3. Disconnect the 4–7 link. Start with node 7 toward node 8, because node 7 has now become a pseudo-tip node. C_7 is 7 and distance between 7–8 is 6. So the new facility must be either on branch 8–11 or 8–9. To resolve this, begin the backward analysis from tip nodes 13 or 10. We arbitrarily select node 13. A facility *x*4 is placed 1 unit away from 13; *x*5 also covers node 12. Delete the branch 11–12. Start backward from 11, and the new facility is placed at a distance of 3. The facility is noted as *x*5, which also covers nodes 8 and 7. Node 9 cannot be attached to *x*5; therefore, begin the analysis from the tip node 10. The new facility can be located at a distance of 6 units from node 10. Proceeding to the next node we need a facility no more than 4 units from node 9. If we place *x*6 at 4 units from 9 on branch 9–8 it will serve both nodes 9 and 10 and will also be close

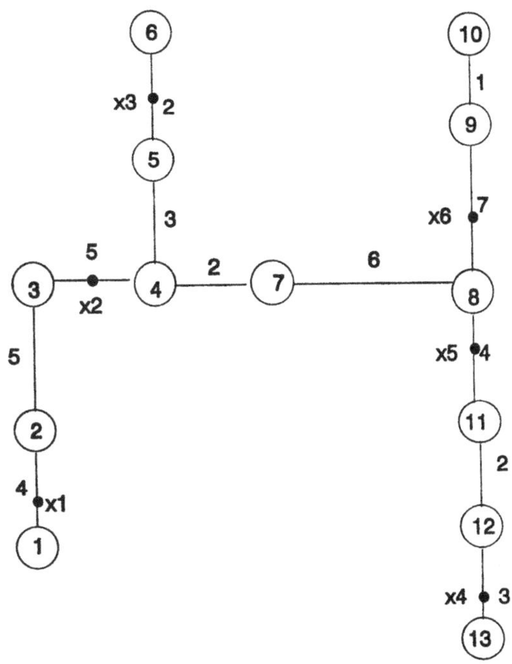

FIGURE 7.18 Initial assignments.

TABLE 7.25 Temporary Assignments for the Required Facilities

Facility x	1	2	3	4	5	6
Location and distance	1 unit from node 1 on 1–2	3 units from node 3 on 3–4	1 unit from node 6 on 6–5	1 unit from node 13 on 13–11	3 units from node 11 on 11–8	4 units from node 9 on 9–8
Nodes assigned	1, 2	3, 4	6, 5 Possibly 4 and 7	13, 12	7, 8, 11	10, 9

to other existing facilities to serve as a backup unit. Thus, location of $x6$ is fixed. Our present temporary assignments are shown in Figure 7.18, with the assignments as shown in Table 7.25.

Now suppose a backup facility is required for each facility within a distance of 7 units. If the existing facilities do not back each other, then any possible modifications to the existing locations and assignments that will make this

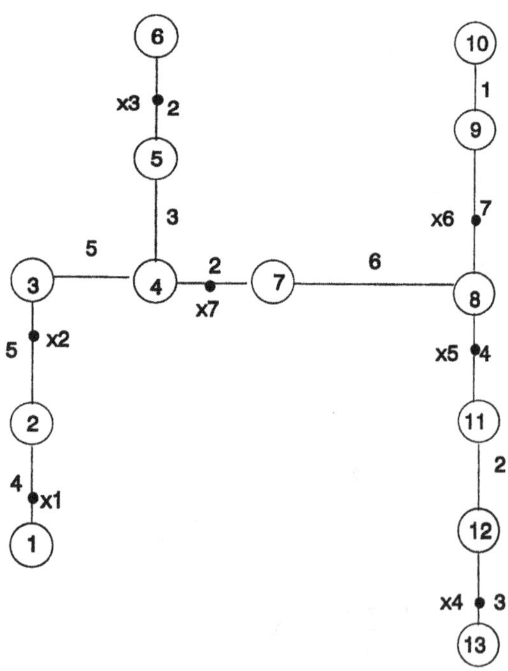

FIGURE 7.19 Final assignments with the backup facility.

possible should be considered. As a last resort, additional facilities may be placed in the network to satisfy the requirement. In our case $x1$ is 11 units away from $x2$. However it is possible to attach node 4 to $x3$ and place $x2$ 3 units away on branch 3–2. The distance between $x1$ and $x2$ is then only 5 units, and each can be backup for the other. Here $x2$ serves only node 3. Facility x cannot be covered by any other facility, and we need to add one additional facility anywhere within 7 units away from $x3$. Node 7 may be reassigned to this facility. The final locations are displayed in Figure 7.19.

7.9 HUB LOCATION PROBLEM

A transportation network frequently represented by set of nodes indicating population centers and branches representing the paths connecting these nodes. In a generalized network each node may represent a flow origination as well as a destination point. In addition, direct paths may be available connecting each node to all other nodes. The objective is to develop a least-cost transportation plan.

Given that there is a flow from each node to every other node, the least distance required to travel between any two nodes is the direct distance between them. As a first thought then, it may appear to be a good solution to transport directly between two nodes. But the cost associated with this type of network may be very high because of two factors: the type and number of vehicles that would be necessary in such scheme and perhaps resulting low utilization of transporters, which is related to the flow between two nodes. For example, if there are N nodes in the network, then there are $N(N-1)$ origin–destination pairs in a fully connected network. For a ten-node problem, we will require $10 \times 9 = 90$ vehicles, or even if we assume that each vehicle can serve say five origin–destination pairs, on average, we still will need $90/5 = 18$ vehicles.

One way to considerably reduce the number of paths is to develop a hub-and-spoke network. In this network, a few nodes are designated as hubs, and other nodes are connected to these hubs, such that a node is connected to only one hub. The flow between the nodes must then go through at least one hub. The economy of scale comes in as all the flow, to and from, can be grouped into one transporter. In addition the flow from one hub to another hub consists of the total for all nodes that are connected to each hub. Thus, the transportation between nodes and hub may take place by a small transporter, whereas between hubs with a large flow, a larger transporter can be used, reducing the cost of transportation per unit. For N nodes there are only $(N-1)$ paths in making such connections.

A typical objective in a hub location problem is to determine the optimum locations for hubs and to determine which nodes should be connected with which hub to minimize the total cost of transportation. There are many examples of hub–spoke arrangements, typically they include airline, train, communication, and computer networks, and trucking and mail delivery systems.

Mathematically, the hub location problem can be formulated as a linear-programming problem, but the number of variables in the problem increase very rapidly, in the order of N^4; therefore, heuristic solutions often provide attractive alternatives.

The following is one such approach to hub location problem with a known covering distance, that is, the maximum radius within which a node must fall such that it can be attached to a hub. This radius is generally determined by the type of vehicle used in transportation. For example, in developing a hub network for an airline, the airline may serve the node and hubs by a small 20-seater plane with a maximum of 40 units, between the hubs, transportation may be with a larger plane with larger passenger and flying time capacities.

Procedure

The following are the steps of the procedure:

1. Given the flow from each node to every other node, formulate the from–to table.
2. We want to select a node as a hub node so that it has the maximum flow in and out of it. By doing so we will minimize the transport flow in the spokes. To attend this objective, add the flow in each column and determine the largest total flow value. This value gives the node with the largest in-and-out flow. Make the selected node as a hub node.
3. From the hub, determine the nodes that are within covering distance and, therefore, can be reached from the hub. Remove the hub and nodes covered by it from the from–to chart. Combine with the remaining data and repeat step 2.

   Repeat the foregoing steps until all the nodes are covered.
4. List the nodes, which are connected to a particular hub, but could also be connected to or reached by another hub (within a given covering distance from both hubs). We will call these dual nodes.
5. From among the dual nodes, connect one to another hub and calculate the total cost. If the cost has increased, then the previous solution is still the best and repeat the step with next dual node. If the cost has decreased, then make this as the solution and reformulate the list and perform step 5 with the new list. When no changes are possible, the associated cost is the optimum cost.

Example

Consider a network of eight nodes shown in Figure 7.20. The flow and distances between nodes, in terms of 100 miles with an appropriate transport mode, are shown in Tables 7.26 and 7.27, respectively. The nodes represent the cities on a

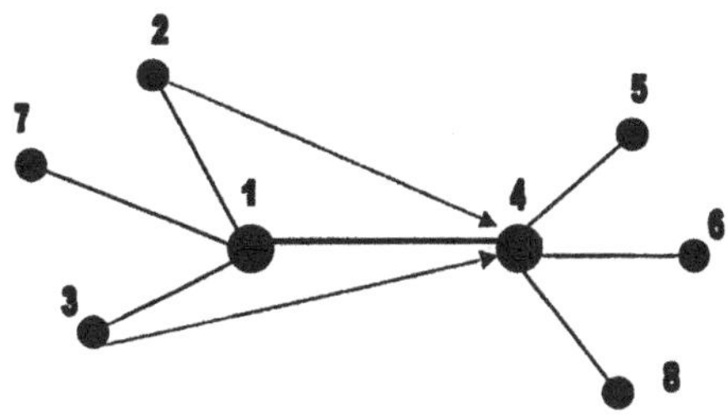

FIGURE 7.20 Initial network.

TABLE 7.26 Flow Between Nodes

		To							
		1	2	3	4	5	6	7	8
From	1	—	25	40	60	38	52	45	50
	2	25	—	10	32	25	15	22	39
	3	40	10	—	58	22	18	54	17
	4	60	32	58	—	41	25	39	43
	5	38	25	22	41	—	33	26	29
	6	52	15	18	25	33	—	21	38
	7	45	22	54	39	26	21	—	31
	8	50	39	17	43	29	38	31	—
	Total	310	168	219	298	214	202	238	247

TABLE 7.27 Distance Between Nodes (in 100)

		To							
		1	2	3	4	5	6	7	8
From	1	—	39	16	50	49	45	25	46
	2	39	—	40	10	31	39	42	48
	3	16	40	—	40	21	46	48	50
	4	50	10	40	—	38	21	49	39
	5	49	31	21	38	—	36	39	41
	6	45	39	46	21	36	—	31	27
	7	25	42	48	49	39	31	—	50
	8	46	48	50	39	41	27	50	—

network for air transport, and the flow between the nodes is the daily passenger flow between those two nodes. The cost of operating a large plane and a small plane over several miles is displayed in Figure 7.21. After a break-even point at 40, the relative cost of the two types remain fairly constant. The cost of operating a small airplane is $0.80 per passenger per unit, whereas a large plane operation cost $0.30 per passenger per unit. Therefore, we will consider the covering distance of a small plane to be 40 units and use the associated operating cost for each type of plane.

It is clear from the total-flow column of the flowchart that the maximum flow is associated with node 1; therefore, node 1 is selected as the first hub location. From hub 1, node 1 in our example, nodes 2, 3, and 7 are within covering distance of 40 units; therefore, they are attached to hub 1.

The next step is to remove the hub and the nodes covered by it from Table 7.26. The resulting modified flowchart is shown in Table 7.28. To determine the possibility of a second hub apply the procedure on the reduced matrix.

The second hub location, from Table 7.28, is node 4, because it has the largest flow of 298. From hub 2 the nodes covered are 5, 6, and 8. As all the nodes are covered, the hub locations are node 1 and node 4.

We need to further evaluate the present hub–spoke network for minimum cost. From the distance table, we see that hub 4 can cover nodes 2 and 3, which are presently linked to hub 1. Hence, the network should be further evaluated for costs with nodes 2 or 3, or both, reassigned to hub 4. Two cases are considered:

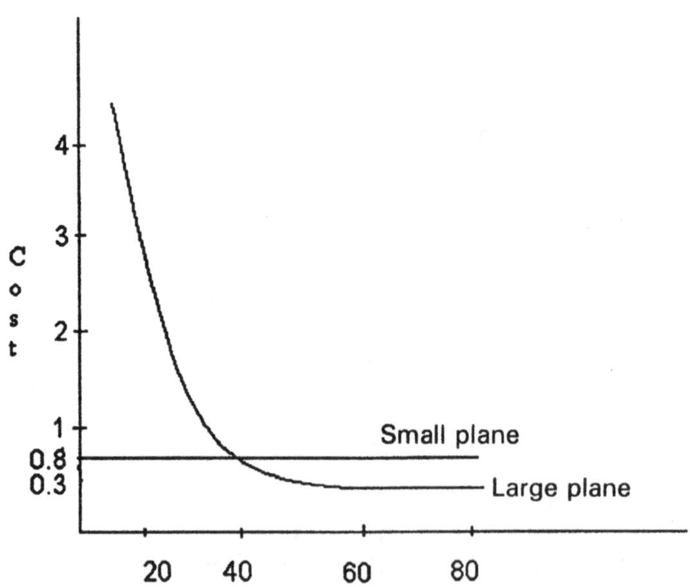

FIGURE 7.21 Cost comparison: large and small planes.

TABLE 7.28 Modified Flowchart

		To			
		4	5	6	8
From	1	60	38	52	50
	2	32	25	15	39
	3	58	22	18	17
	4	—	41	25	43
	5	41	—	33	29
	6	25	33	—	38
	7	39	26	21	31
	8	43	29	38	—
	Total	298	214	202	247

Case 1: Connect node 3 to hub 1, in which case we have two subcases:
Case 1a. Node 2 is connected to hub 1 (Figure 7.22).
The cost associated with the foregoing assignments is

cost between 2 and 1 + cost between 2 and 3 + cost between 2 and 7
+ cost between 2 and 4 + cost between 2 and 5 + cost between 2 and 6
+ cost between 2 and 8 + {cost between 3 and 1 + cost between 3 and 4
+ cost between 3 and 5 + cost between 3 and 6 + cost between 3 and 7
+ cost between 3 and 8}

$$= (25*39*0.8) + (10*39*0.8 + 10*16*0.8) + (22*39*0.8 + 22*25*0.8) + (32*0.8*39 + 32*50*0.3) + (25*0.8*39 + 25*50*0.3 + 25*38*0.8) + (15*0.8*39 + 15*50*0.3 + 15*21*0.8) + (39*0.8*39 + 39*0.3*50 + 39*39*0.8) + \{(40*0.8*16) + (58*0.8*16 + 58*0.3*50) + (22*0.8*16 + 22*0.3*50 + 22*0.8*38) + (18*0.8*16 + 18*0.3*50 + 18*0.8*21) + (54*0.8*16 + 54*0.8*25) + (17*0.8*16 + 17*0.3*50 + 17*0.8*39)\}$$
$$= 16{,}685.2.$$

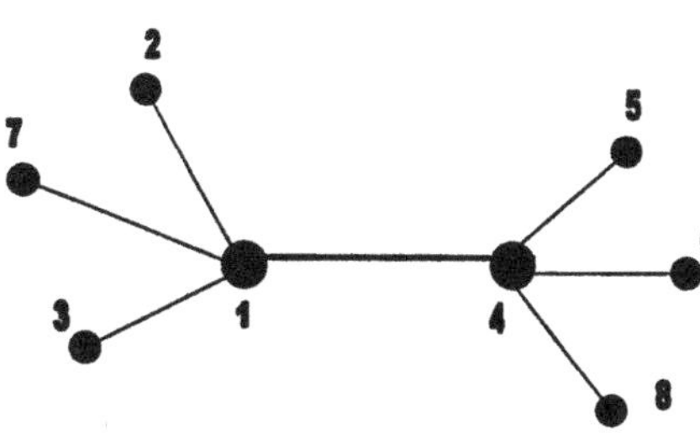

FIGURE 7.22 Case 1a: node 2 and node 3 connected to hub 1.

Similarly, for other cases results are as follows:

Case 1b. Node 2 connected to hub 4 (Figure 7.23):

cost between 2 and 1 + cost between 2 and 3 + cost between 2 and 7 + cost between 2 and 4 + cost between 2 and 5 + cost between 2 and 6 + cost between 2 and 8 + {cost between 3 and 1 + cost between 3 and 4 + cost between 3 and 5 + cost between 3 and 6 + cost between 3 and 7 + cost between 3 and 8}

Cost associated $= 11,977.6$

Case 2. Connect node 3 to hub 4 with subcases as follows:

Case 2a. Node 2 connected to hub 1 (Figure 7.24)
Cost associated = 20,597

Case 2b. Node 2 connected to hub 4 (Figure 7.25)
Cost associated = 15,637.4

The minimum cost is associated with the case when node 3 is connected to hub 1 and node 2 is connected to hub 4. Thus, we connect node 2 to hub 4 and

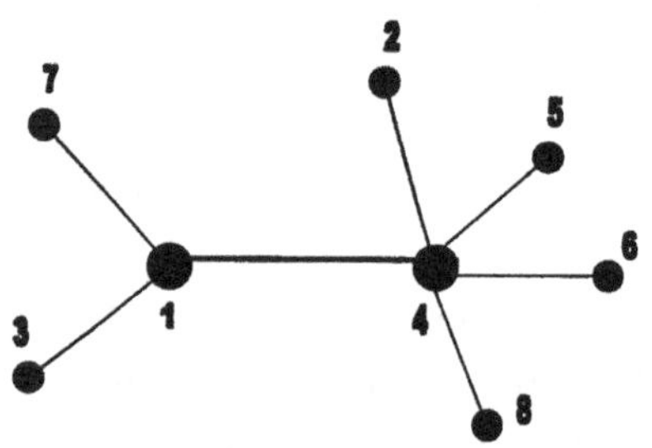

FIGURE 7.23 Case 1b: node 3 connected to hub 1 and node 2 connected to hub 4.

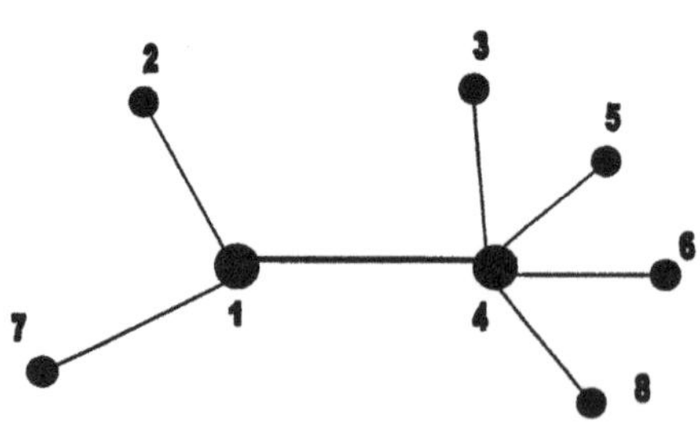

FIGURE 7.24 Case 2a: node 2 connected to hub 1 and node 3 connected to hub 4.

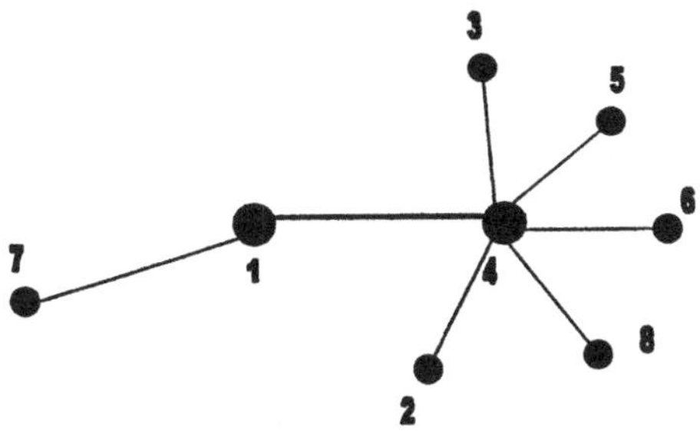

FIGURE 7.25 Case 2b: both nodes 2 and 3 are connected to hub 4.

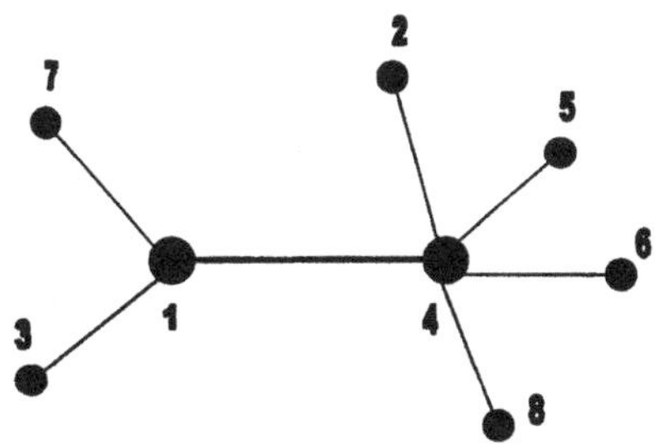

FIGURE 7.26 Final network.

node 3 to hub 1 giving us minimum cost. The optimum network is shown in Figure 7.26.

Exercises

7.1 It has been observed that fish move in groups. Rudolph Fishing Co. is a leader in research on fish in the sea. During its research it found eight probable areas in which these groups may be present. They can throw nets that will cover several different areas at one time. For example, if they throw the net in area 1 it will cover spot 1, 3, and 6 ($S1 = \{1, 3, 6\}$). Determine the minimum number of nets they should throw to cover all areas.

$S1 = \{1, 3, 6\}$
$S2 = \{1, 5, 6, 8\}$
$S3 = \{2, 5\}$
$S4 = \{3, 6, 7\}$
$S5 = \{4, 6, 7, 8, 9\}$
$S6 = \{3, 4, 6, 8\}$
$S7 = \{1, 3, 5, 7, 8\}$
$S8 = \{2, 5, 6, 8\}$

7.2 If traveling to an area and throwing a net has a cost associated with it and that is displayed in the following table, determine the optimum number of nets required to cover all areas.

Area	Fixed cost
1	90
2	100
3	80
4	50
5	70
6	110
7	125
8	130

7.3 A blower is to be located in a network of ducts, as shown in the following figure. Where should it be located so that it will serve all the locations within minimum capacity. The capacity is directly proportional to the maximum distance that the blower is required to serve an opening.

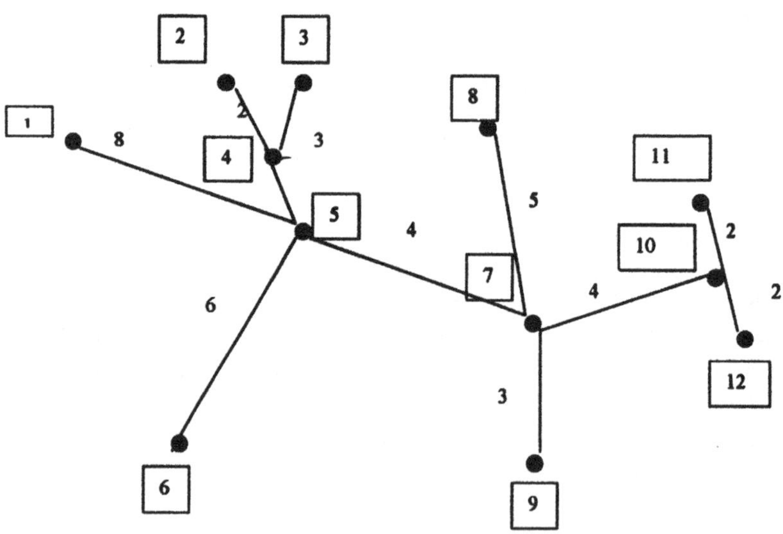

7.4 If instead of one blower, two blowers of smaller capacity are to be located, where will you locate them so that minimum-capacity blowers can be purchased?

7.5 A company wants to improve on its trading business and has selected 10 cities where its wants to extend its market. To do so it needs to locate distributors who will serve these cities. The distributor will serve the cities within 60 miles from its location. Determine the minimum number of distributors required to serve all the ten cities with minimum cost. The network along with the distances are shown in the following figure.

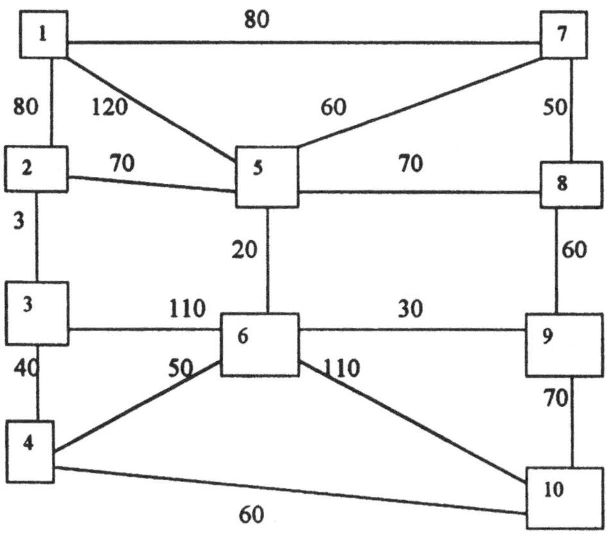

7.6 The ultrasonic sound machine used in breaking kidney stones is expensive and is used by ten large and small hospitals in the region. It is carried to the hospital when needed. The company wants to locate a base for its machine to minimize the maximum distance it has to cover to transport the machine to the hospital. The use frequency for the machine is different for each hospital and is as shown in the following table. The distances between the hospitals are shown in the tree diagram at top of the next page. Determine the location of the base on the tree to satisfy the objective.

Hospital	A	B	C	D	E	F	G	H	I	J
Frequency	6	10	7	9	5	8	6	8	5	10

7.7 A major airline wants to locate a hub terminal for six cities. The passenger flow between the cities and the distances are known. The maximum distance that can be covered by a small plane is 60. The operating cost of the small plane is \$0.75 per passenger per unit and that of the large plane is \$0.25 per

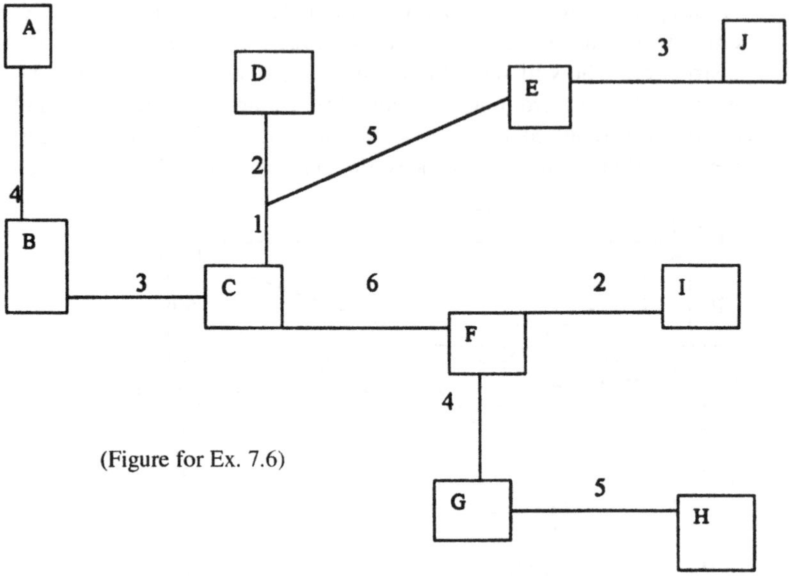

(Figure for Ex. 7.6)

passenger per unit. Determine the location of the hub in such a way that the operating cost is minimum.

Flowchart

	1	2	3	4	5	6
1	—	75	45	65	71	38
2	75	—	52	34	49	64
3	45	52	—	80	46	58
4	65	34	80	—	27	42
5	71	49	46	27	—	42
6	38	64	58	42	42	—

Distance Chart

	1	2	3	4	5	6
1	—	75	64	50	46	38
2	75	—	58	32	38	65
3	64	58	—	56	64	75
4	50	32	56	—	44	66
5	46	38	64	44	—	55
6	38	65	75	66	55	—

7.8 Super-A Store, Inc, an owner of a chain of stores, is contemplating the placement of few stores at different locations around a city. It has divided the market into eight different regions. Sets (S1–S8) show the regions that will be served if the store is placed in a particular region.

$S1 = \{1, 4, 7, 8\}$
$S2 = \{1, 3, 5\}$
$S3 = \{2, 4\}$
$S4 = \{3, 4, 8\}$
$S5 = \{4, 5, 6, 8\}$
$S6 = \{3, 5, 7, 9\}$
$S7 = \{1, 2, 3, 4\}$
$S8 = \{2, 3, 7, 9\}$

Find how many stores it would be optimum to place to minimize the cost and serve customers of all the regions.

7.9 Select the locations for the foregoing set of data considering the fixed cost shown in the following table.

Region	Fixed cost
1	85
2	76
3	80
4	30
5	40
6	90
7	131
8	146

7.10 A restaurant is to be located to serve different customers in a locality. Assume all customers carry a weight of 1. Determine the location of the restaurant to minimize the maximum-weighted distance between the restaurant and the customers. (For Exercises 7.10, 7.11, and 7.12, see diagram at top of page 222.)

7.11 Locate two absolute centers in the attached network. What is the maximum value?

7.12 For the network in 7.11, assume that there exists a service facility at node 14. Where would you place a competing facility if it must be at least 4 units away from the existing facility?

7.13 For the network attached, each demand point requires a facility within a specific distance, as shown in the following table. Also, each facility

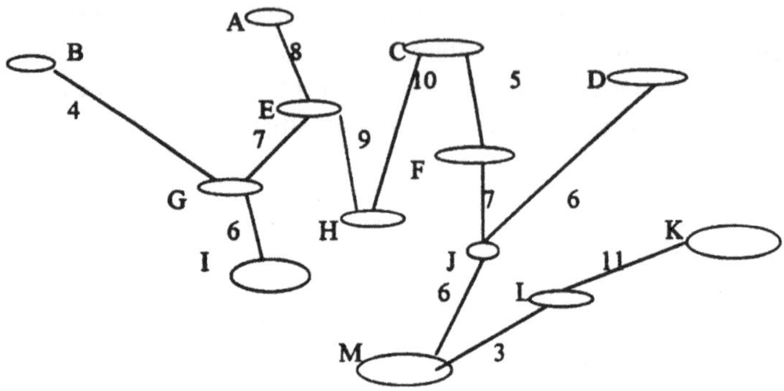

requires a backup facility within a distance of 5 units. Determine the minimum number of facilities required to satisfy the requirements.

Node	1	2	3	4	5	6	7	8	9	10	11	12	13	14
Dist	3	1	2	4	5	7	5	3	6	1	4	2	5	6

7.14 Aesthetics, cleanliness, customer service, and such are some of the factors that attract customers to a shop. Depending on this, an attractiveness factor is associated with each facility for each customer. The attractiveness factor is relative to other existing facilities. A flower shop is to be located in an area where there is an existing shop. But this shop is more attractive to some customers. The attractiveness factor for customers is shown in the following attractiveness table. The distance table shows the distances from existing and possible locations to the customers. Find the optimal location for the new facility. (Hint: divide the distance by the attractiveness factor.)

Attractiveness Table

Customers	Attractiveness factor
A	2
B	0.9
C	3
D	1

Distance Table

	A	B	C	D
Existing facility	10	8	9	5
Location 1	20	6	12	20
Location 2	28	10	33	10
Location 3	12	18	9	14

7.15 A grocery shop is to be located in a particular locality. There is already an existing shop. The new grocery shop will be better, as it will keep fresher vegetables and fruits. Customers like vegetables and fruits to be as fresh as possible (the time they buy should be close to the time of plucking). Each store has a maximum desired number of days (as lead time) of keeping fruits on the stand. The following table shows the data relevant to three customers.

Attractiveness Table

Location	Attractiveness factor
1	4
2	3
Existing	5

The distance to be traveled by each customer to the existing and to the new facility are given in the next table. Two locations are being contemplated for the new grocery shop. Select the optimum location to place the new shop based on these data.

Distance Table

Distance	A	B	C
Existing grocery shop	12	16	8
Location 1	13	17	21
Location 2	22	27	31

PART III

Tour Development Models

8

Logistics in Tour Development

So far we have seen some basic models for locating facilities and assigning customers to these locations. In addition, we will further study a number of new models in subsequent chapters. All these models, however, have a major common thread, which is that all customers are directly served from the facility to which they are assigned.

There are several logistic problems, however, with somewhat modified objectives. For example, we may want to develop a path or a route that connects all customers so that a delivery truck can deliver and collect loads from different customers. The objective here is to minimize the total distance traveled for the tour. Such problems are called tour development problems. Another classic example is the traveling salesman problem. Here, a salesperson must visit many customers, each in different cities. He or she starts from his or her home city and visits each city to which he or she is assigned once and only once, before returning back home. The capacity of the vehicle in which he or she is traveling or what goods are carried, is not a limiting factor; therefore, all customers can be visited in one continuous tour, starting and terminating at home. The distance between all pairs of cities is known, and the problem is to develop the shortest route for the salesperson to complete his or her tour. This problem has proved to be NP-hard, meaning, for large number of customers, the time required to solve the problem increases almost exponentially and, after about 20 customers, there are no procedures that can determine the optimum solution in a reasonable time.

If the distance (cost) between every pair of cities is independent of the direction of travel, the distance matrix is called symmetrical. If, on the other hand, the distance for even one pair varies with the direction of travel, the matrix is called asymmetrical. An asymmetrical matrix may arise because two cities may be at different evaluations and going up is more difficult than coming down, or because of wind direction, if the travel is by plane or by sail. Traveling between two cities may take different times and costs.

Some of the typical problems for which tour development or traveling salesman algorithms are used are the following:

1. School bus routing
2. Postal deliveries
3. Garbage collection within a city
4. Assembly or inspection of printed circuit boards

We illustrate two solution procedures for solving a traveling salesman problem. The first is a simple application of an assignment algorithm, but it may require further observations and deductions. The second, which is developed by Little, uses the branch and bound method. The procedures are illustrated in an example problem.

8.1 ASSIGNMENT PROCEDURE FOR TRAVELING SALESMAN PROBLEM

For a Thanksgiving weekend Mr. Nice, who lives in city A, wants to visit his four friends and relatives in cities B, C, D, and E. Table 8.1 displays the distance (in hundred miles) between the cities where his friends reside. He wants to visit the four locations only once and return home with minimum distance traveled. Thus, we want to find the minimum cost route for Mr. Nice.

8.1.1 Solution Procedure

In developing a tour, we cannot visit any city twice except for the home base. This implies that we cannot go from a city back to the same city again. Therefore,

TABLE 8.1 Distance (Cost) Matrix

	A	B	C	D	E
A	0	1	7	4	3
B	2	0	1	5	4
C	5	2	0	6	1
D	3	1	4	0	2
E	4	3	1	5	0

TABLE 8.2 Resultant of Row Operation

	A	B	C	D	E
A	∞	0	6	3	2
B	1	∞	0	4	3
C	4	1	∞	5	0
D	2	0	3	∞	1
E	3	2	0	4	∞

as the first step, convert the costs of A to A, B to B, C to C, D to D, and E to E to infinity to prevent such returning back.

Apply assignment algorithm, to the resulting cost matrix. To begin, obtain a zero element in each row. Search for the minimum number in each row, and subtract that number from each element of that row. The result is Table 8.2. The cost of reduction so far is the sum of the associated cost elements used in this reduction, that is $1 + 1 + 1 + 1 + 1 = 5$.

Now, to obtain at least one zero element in each column, search for the minimum number in each column and subtract that from the elements of that column. The result is displayed in Table 8.3. The total cost reduction now is $5 + 1 + 0 + 0 + 3 + 0 = 9$.

There is at least one zero in each row and each column; therefore, a path with zero cost may be possible from the resulting table.

Because A is the home base, let us start with A and try to develop a zero-cost tour. From A, the zero-cost cities we can go to are B and D. Suppose we choose D. From D we can go to B with zero cost; from B we can go to A or C. Because all cities are not in the route yet, we do not want to return to A; hence, the entry B–A is marked by ∞ and we proceed to select C as the next city in the tour. From C the only place to which we can go to with zero cost is E, and from E we can go back to C. Thus, our present route is A-D-B-C-E-C. This is not a permissable route because we are visiting C twice and have not yet returned back to A.

If from A we had selected going to B first, our zero-cost route would have been A-B-C-E-C, again an incomplete and nonpermissable route.

TABLE 8.3 Resultant of Column Operation

	A	B	C	D	E
A	∞	0	6	0	2
B	0	∞	0	1	3
C	3	1	∞	2	0
D	1	0	3	∞	1
E	2	2	0	1	∞

So, the present zero elements do not provide a feasible route. Next minimum (nonzero) element in the matrix is 1. Therefore, we want to see if increasing the cost by 1, that is, by incorporating one of the "1"-cost elements with all other zero-cost elements, a feasible route can be developed. But which elements to include? There is no easy answer, and we may have to try a few different routes.

For example, we may develop the following incomplete routes if we include cost 1.

A–B–D–B
A–D–E–C–E
A–B–C–E–D–B

Having found no feasible route so far, we next must increase the permissible cost to the next higher-cost element: in our cost matrix it is 2.

In developing the routes we might choose one 2-cost element and all other zero-cost elements or two 1-cost elements and remaining zero-cost elements. A few of the resulting routes follow. We might notice that some are feasible, whereas others are not.

A–B–C–E–D nonfeasible
A–D–B–C–E–A feasible
A–E–C–E nonfeasible
A–C–E–D–B–A feasible

The cost of each feasible route is 11, obtained by adding the associated cost of each link or adding 2 to the reduction cost for Table 8.3, which was 9.

The method does require a few try outs and for a small dimension distance matrix this is not a major problem, but for a large problem the procedure may become quite time-consuming because of the number of alternatives that we may have to investigate. As an alternative we have Little's branch-and-bound method. This method, described in the next section, can be computerized and, therefore, is used many times as a preferred alternative.

8.2 LITTLE'S METHOD FOR THE TRAVELING SALESMAN PROBLEM

As before, we work with the cost table and sequentially develop the path. The procedure selects one connection at a time in building the route. The steps of the procedure are explained by means of application to the previous example.

1. Reduce Table 8.1 to Table 8.3 by performing row and column operations. The reduction cost in our case has been 9, thus far.
2. Find the penalty of not using each zero-cost cell. If we do not use a zero-cost cell in row i and column j to form a link in a path, then we must use another element in row i and column j to develop the route.

TABLE 8.4 Penalty Evaluation: cycle 1

		A		B		C		D		E
A		∞	0	0		6	1	0		2
B	1	0		∞	0	0		1		3
C		3		1		∞		2	2	0
D		1	1	0		3		∞		1
E		2		2	1	0		1		∞

The best we can do is to use the next minimum cost of cells from row *i* and column *j* for this purpose. Thus, the minimum penalty for not using a zero cell, is the sum of the smallest elements in row *i* and column *j*.

Calculate the penalty for every zero cell and record it in the left-hand corner of the associated cell. In our example, the results are displayed in Table 8.4. For instance, for A–B cell, the penalty is $0 + 0$ (because the next minimum cost in row A is 0 and column B is also 0), for A–D cell the penalty is $0 + 1 = 1$ and so on.

3. Select the zero cell with the largest penalty, let us say it is in row *h* and column *k*; therefore, it is denoted as cell (h, k). We now divide the problem into two subsets, θ(h, k) that contains the link (h, k) and $\theta(\underline{h,\ k})$ that does not contain link (h, k). The cost of not using the link (h, k) is at least equal to the penalty associated in cell (h, k). So the total cost of going in that branch so far is cost plus the penalty in cell (h, k).

   In our example the maximum penalty is associated with cell (C, E). Thus, the cost associated with choosing (C, E), as the link is 9, whereas not choosing is $9 + 2 = 11$. This is displayed in the tree with associated branches.

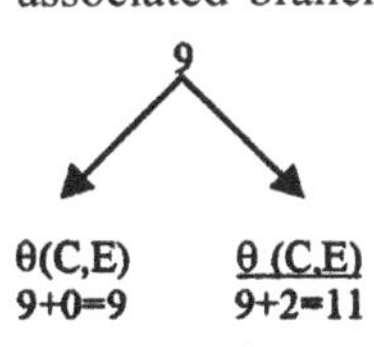

5. Continue with the minimum cost branch so far. Because (C, E) is selected as a link, we cannot have another link that starts at C or ends at E, as it will develop a loop (visiting one of the cities twice without visiting all others at least once). This is achieved by scratching row C and column E, thereby eliminating them from further consideration.

		A		B		C		D		E
A		∞	0	0		6	1	0		2
B	1	0		∞	0	0		1		3
C		~~3~~		~~1~~		~~∞~~		~~2~~	~~2~~	~~0~~
D		1	1	0		3		∞		1
E		2		2	1	0		1		∞

Also change the cost from E to C to ∞, to prevent that link from forming in future route. The resulting table is as shown in Table 8.5.

6. Repeat steps 3 and 4 until a solution is obtained.

Continuing with the procedure, because we do not have at least one zero in each row and column, we must create them. For the rows and columns without a zero element in them, subtract the minimum element associated with each such row and column, from all the elements in that row and column. In our problem, 1 is taken off from row E, increasing the solution value by 1.

In the resulting table, calculate the penalty value for each zero cell entry. The results are displayed in Table 8.6.

The maximum penalty is associated with cell (B, C). The corresponding tree follows.

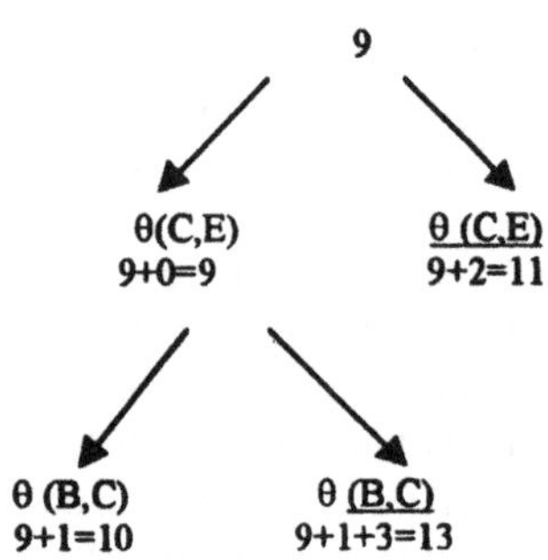

TABLE 8.5 Remaining Cells After Selecting Cell (C, E)

	A	B	C	D
A	∞	0	6	0
B	0	∞	0	1
D	1	0	3	∞
E	2	2	∞	1

TABLE 8.6 Penalty Table: Cycle 2

	A		B		C		D	
A		∞	0	0		6	0	0
B	1	0		∞	3	0		1
D		1	1	0		3		∞
E		1		1		∞	1	0

Continuing with the tree, between all open branches, θ(B, C) has the minimum cost. So, choose θ(B, C) to proceed. Eliminate row B and column C from further considerations. The result is as now shown.

	A		B		C		D	
A			0	0		6	0	0
B	1	0		∞	3	0		1
D		1	1	0		3		∞
E		1		1		∞	1	0

Again to prevent looping, cost of E to B is changed to ∞ in Table 8.7. Subtracting 1 from row E, to get a zero cell and calculating the penalties results in Table 8.7. The total cost of the tour has increased by 1.

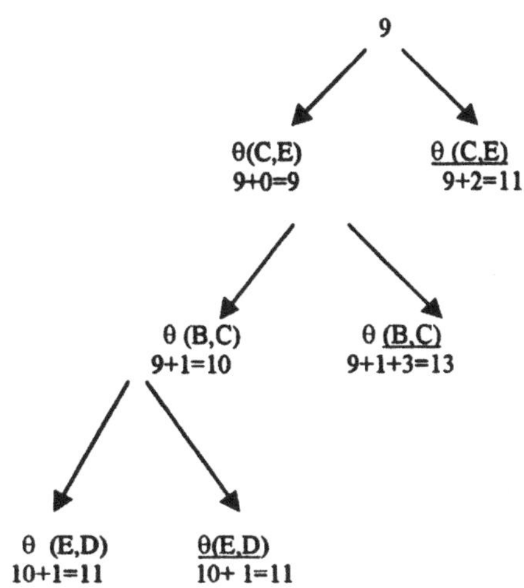

It does not matter whether we select cell (E, D) or not, both branches have penalty of 11.

Let us continue our solution procedure by not selecting cell (E, D). This is achieved by placing the penalty of ∞ in cell (E, D). The resultant table is Table 8.8.

TABLE 8.7 Penalty Table: Cycle 3

	A		B		C	
A		∞	0	0	0	0
D	0	0	0	0		∞
E	0	0		∞	0	0

TABLE 8.8 Penalty Table After not Selecting Cell (E, D)

	A		B		D	
A		∞	0	0	∞	0
D	0	0	0	0		∞
E	0	0		∞		∞

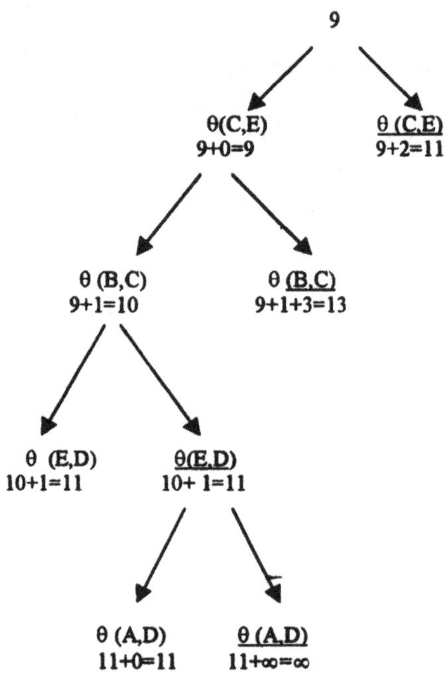

A, D is selected next, leading to following table.

	A		B		D	
A		∞	0	0	∞	0
D	0	0	0	0		∞
E	0	0		∞		∞

The next penalty table is displayed as Table 8.9.

From Table 8.9, either paths, D–B and next E–A may be selected. The associated tree diagram is as shown.

All nodes are assigned with the optimum path being A–D–B–C–E–A. (Trace each branch taken and then make connections. The corresponding distance traveled is 11. This is the same value as we calculated for the cost using the original cost Table 8.1.

TABLE 8.9 Penalty Table After Selecting Cells (A, D)

		A		B
D		∞	0	0
E	0	0		∞

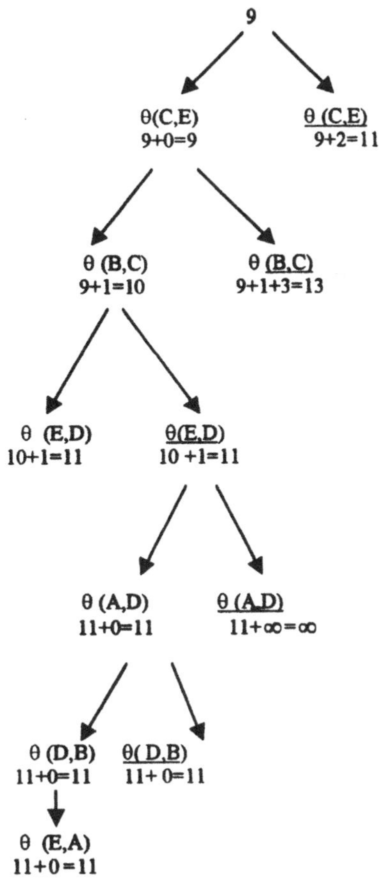

8.3 AN INTEGER-PROGRAMMING FORMULATION FOR TRUCK ROUTING PROBLEM

Next we develop the problem of truck routing. The problem is similar to the traveling salesman problem except it has some additional considerations affecting the way routes are formed. Some of these practical limitations follow:

1. Territory over which the truck is to be operated.
2. The load type and the equipment required for such loading or unloading operation as part of the truck. For example, a heavy load may need a truck fitted with a lift.
3. Truck capacity in weight and volume.
4. Maximum number of stops desired on a route.
5. Maximum travel time or mileage that should be allowed.
6. Any federal, state, or local legal requirements.

The problem of truck routing can be formulated as an integer-programming problem. It requires identifying the possible routes a truck can take and then optimally assigning the trucks to some of these routes to minimize an objective function, such as transportation cost.

Example

Consider a company with two trucks, one with the capacity of 12 tons and another with the capacity of 10 tons. There are six cities A, B, C, D, E, and F from which a load is to be picked up and brought back to the depot at O. The distances between cities and the loads to be picked up from each city are displayed in Table 8.10. We wish to determine truck routes to minimize the total distance traveled by the two trucks.

The first step is to develop different feasible routes and associated mileage for each route. For example, with truck type 1, a single trip can be made to pick up the load from A, as noted by route 1 (R1) in Table 8.11. The associated cost is the distance from O–A, plus A–O which gives $3 + 2 = 5$. Here, the capacity of truck and maximum distance a truck may travel from the constraints. Trip 2 (R2) involves going to O–A, A–B, and B–O with the total load of $6 + 5 = 11$, less than the capacity of truck 1. The distance traveled is $3 + 3 + 3 = 9$. Similar calculations results in 22 different routes from truck 1 and 8 different routes for truck 2. These trips are displayed in Table 8.11. Note that 1s in the first and second row, respectively, which represent the use of truck 1 or 2 in that particular trip.

Now suppose we want to make no more than three trips with truck 1 and four with truck 2. The formulation is as follows:

Let x_i be the possible routes $i = (1, 2, 3, \ldots, n)$

The objective is to minimize $= \{5*x_1 + 9*x_2 + 9*x_3 + 11*x_4 + 7*x_5$

$$+ 8*x_6 + 13*x_7 + 17*x_8 + 15*x_9 + 6*x_{10}$$
$$+ 8*x_{11} + 13*x_{12} + 13*x_{13} + 14*x_{14}$$
$$+ 14*x_{15} + 6*x_{16} + 7*x_{17} + 9*x_{18} + 10*x_{19}$$
$$+ 10*x_{20} + 10*x_{21} + 10*x_{22} + 5*x_{23}$$
$$+ 7*x_{24} + 8*x_{25} + 6*x_{26} + 8*x_{27}$$
$$+ 14*x_{28} + 6*x_{29} + 10*x_{30}\}$$

TABLE 8.10 Distances Between Nodes, Demands from Nodes, and Truck Capacities:

	O	A	B	C	D	E	F
O	—	3	4	2	6	3	5
A	2	—	3	1	4	5	3
B	3	9	—	5	2	6	8
C	4	3	4	—	6	3	4
D	2	6	5	3	—	4	2
E	3	5	3	4	2	—	4
F	5	7	2	6	3	1	—

Demand
A = 6
B = 5
C = 8
D = 5
E = 6
F = 7

Capacity

Truck 1	Truck 2
12 tons	10 tons

Subject to

$$\{x_1 + x_2 + x_3 + x_4 + x_5 + x_6 + x_7 + x_8 + x_9 + x_{10} + x_{11} + x_{12} + x_{13} + x_{14} + x_{15} + x_{16} + x_{17} + x_{18} + x_{19} + x_{20} + x_{21} + x_{22} <= 3\} \quad \text{truck 1 (cap = 12)}$$

$$x_{23} + x_{24} + x_{25} + x_{26} + x_{27} + x_{28} + x_{29} + x_{30} <= 4 \quad \text{truck 2 (cap = 10)}$$

$$x_1 + x_2 + x_3 + x_4 + x_9 + x_{15} + x_{19} + x_{23} = 1 \quad \text{For city A}$$

$$x_2 + x_5 + x_6 + x_7 + x_8 + x_9 + x_{14} + x_{18} + x_{22} + x_{24} + x_{25} + x_{28} = 1 \quad \text{For city B}$$

$$x_{10} + x_{26} = 1 \quad \text{For city C}$$

$$x_3 + x_6 + x_{11} + x_{12} + x_{13} + x_{14} + x_{15} + x_{17} + x_{21} + x_{25} + x_{27} + x_{28} = 1 \quad \text{For city D}$$

$$x_4 + x_7 + x_{12} + x_{16} + x_{17} + x_{18} + x_{19} + x_{29} = 1 \quad \text{For city E}$$

$$x_8 + x_{13} + x_{20} + x_{21} + x_{22} + x_{30} = 1 \quad \text{For city F}$$

TABLE 8.11 Tabular Formulation of the Integer-Programming Model

Truck 1																							Truck 2							
R	1	2	3	4	5	6	7	8	9	10	11	12	13	14	15	16	17	18	19	20	21	22	23	24	25	26	27	28	29	30
	1	1	1	1	1	1	1	1	1	1	1	1	1	1	1	1	1	1	1	1	1	1	1	1	1	1	1	1	1	1
A	1	1	1	1					1						1				1				1							
B		1			1	1	1	1	1					1				1		1		1		1	1			1		
C										1																	1			
D			1			1					1	1	1	1	1		1				1				1		1	1		
E				1			1					1				1	1	1	1										1	
F								1					1							1	1	1								1
Cost	5	9	9	11	7	8	13	17	15	6	8	13	13	14	14	6	7	9	10	10	10	10	5	7	8	6	8	14	6	10

In addition, all variables must take either a 0 or 1 value, 0 if the variable is not a part of the solution, and 1 if it is.

Solving the problem with an integer-programming algorithm results in the following solution.

Objective value: 28.00000

Variable	Value	Reduced cost
X1	0.000000	5.00000
X2	0.000000	9.00000
X3	0.000000	9.00000
X4	0.000000	11.00000
X5	0.000000	7.00000
X6	0.000000	8.00000
X7	0.000000	13.00000
X8	0.000000	17.00000
X9	0.000000	15.00000
X10	**1.000000**	**6.00000**
X11	0.000000	8.00000
X12	0.000000	13.00000
X13	0.000000	13.00000
X14	0.000000	14.00000
X15	0.000000	14.00000
X16	0.000000	6.00000
X17	**1.000000**	**7.00000**
X18	0.000000	9.00000
X19	0.000000	10.00000
X20	0.000000	10.00000
X21	0.000000	10.00000
X22	**1.000000**	**10.00000**
X23	**1.000000**	**5.00000**
X24	0.000000	7.00000
X25	0.000000	8.00000
X26	0.000000	6.00000
X27	0.000000	8.00000
X28	0.000000	14.00000
X29	0.000000	6.00000
X30	0.000000	10.00000

The routes are

Truck 1	**Truck 2**
O–C–O = 6	O–A–O = 5
O–E–D–O = 7	
O–F–A–O = 10	

The method is simple, except it requires formulation of all possible routes beforehand. This could be a difficult task if the number of cities is fairly large. Plus, if there is variation in node demands from time to time, formulation has to be reevaluated each time.

8.4 CLARKE AND WRIGHT PROCEDURE

A very efficient heuristic procedure to develop vehicle routing is presented by Clarke and Wright (CW), which we will study next, with slight modification.

Let truck i, with the capacity of C_i, pick up a load from nodes P_i $i = 1, 2, \ldots n$, and bring it back to the depot, denoted P_0. Let the distances between two such points, y and z be denoted as $d_{y,z}$. Any load greater than one full truck load can be considered as being divided into the number of full loads that are equal to an integer portion of the division of the load and the truck capacity, and a partial load. There would be direct trips from the depot to the point to pick up all full loads and only the partial load will be joined with other points in development of a tour.

Note that when the capacity of a single truck is greater than the total demands from all points, the problem becomes that of traveling salesman problem.

The CW tour-developing procedure is best explained by following an example. Consider a problem where there is one depot and eight demand points. The distance between any pair of points is given in Table 8.12. These distances are displayed in the right hand corner of each cell. Because the matrix is symmetrical, only the lower half is presented. The points are numbered so that they are in the ascending order of their distances from the depot. **Q** is a vector of demand from each point.

Calculate the savings $(d_{o,y} + d_{o,z} - d_{y,z})$. This savings is associated with connecting any two points y and z in a tour, as opposed to making independent

TABLE 8.12 Distance Matrix

	P_0	P_1	P_2	P_3	P_4	P_5	P_6	P_7	P_8	Q
P_1	4	—	—	—	—	—	—	—	—	4
P_2	7	8	—	—	—	—	—	—	—	5
P_3	12	8	15	—	—	—	—	—	—	3
P_4	17	6	16	8	—	—	—	—	—	6
P_5	26	10	20	9	13	—	—	—	—	7
P_6	32	15	21	12	12	6	—	—	—	2
P_7	48	22	18	7	20	10	20	—	—	3
P_8	56	30	35	10	19	17	15	8	—	4

TABLE 8.13 Distance–Savings Matrix 1

	P_0	P_1	P_2	P_3	P_4	P_5	P_6	P_7	P_8	Q
P_1	4	—	—	—	—	—	—	—	—	4
P_2	7	3 8	—	—	—	—	—	—	—	5
P_3	12	8 8	4 15	—	—	—	—	—	—	3
P_4	17	15 6	8 16	21 8	—	—	—	—	—	6
P_5	26	20 10	13 20	29 9	30 13	—	—	—	—	7
P_6	32	21 15	18 21	32 12	37 12	52 6	—	—	—	2
P_7	48	30 22	37 18	53 7	45 20	64 10	60 20	—	—	3
P_8	56	30 30	28 35	58 10	54 19	65 17	73 15	96 8	—	4

trips to y and z from depot o. The savings is noted in the left hand of every cell. For example, the saving associated with joining points 3 and 4 in a tour is, $12 + 17 - 8 = 21$, as noted in the left-hand corner of the (P_4, P_3) cell in Table 8.13. No savings are associated with P_0, the point of origin. Therefore, no other recording except for the distance from P_0 to every other point, appears in the column associated with P_0. In this problem the capacity of the vehicle is 15 tons.

Start with the maximum savings and connect those two points if capacity permits. In our case the maximum saving is 96, associated with P_8 and P_7. The capacity required is $4 + 3 = 7$, which is less than 15. Hence, these two points are joined as P_7–P_8 or P_8–P_7. The order will become clear as we develop the tour. Next maximum savings is 73, between P_8 and P_6. We can join P_6 in the path, if by adding this point to the path, the assigned path capacity does not exceed the truck capacity and the point can be connected to P_8. Remember, any point in the tour can have only one point to come from and one point to go to; thus, only two connecting points. In this example, the capacity permits it, and a chain can be formed; therefore, these two points should be connected next. The total load is $4 + 3 + 2 = 9$. The tour is P_7–P_8–P_6.

The next maximum savings of 65 is between P_5 and P_8. But P_8 already has two connections; therefore, it is not an open node. So, ignore this observation. The next savings is 64, between P_5 and P_7. So we will try to connect P_5 from P_7. The load on P_5 is 7 which when added to the earlier load of the truck 9, exceeds the truck capacity of 15. Therefor, P_5 cannot be connected to this tour. Continuing we can connect P_4 to P_7, and no other load can be connected. Thus tour is completed by adding connection to P_0 at the beginning and at the end. Similarly other routes are developed. The results are

Tour 1: P_0–P_4–P_7–P_8–P_6–P_0
Distance traveled $= 17 + 20 + 8 + 15 + 32 = 92$ miles

Develop the next table, Table 8.14, consisting of P_0 and the points that have not yet been connected in a tour; namely, P_1, P_2, P_3, and P_5

TABLE 8.14 Distance–Savings Matrix 2

	P_0	P_1		P_2		P_3		P_5	Q
P_1	4		—		—		—	—	4
P_2	7	8	3		—		—	—	5
P_3	12	8	8	4	15		—	—	3
P_5	26	20	10	13	20	29	9	—	7

By applying the same logic as in tour 1 development we obtain tour 2 as follows:

Tour 2: P_0–P_1–P_5–P_3–P_0
Distance traveled $= 4 + 10 + 9 + 12 = 35$ miles
Total load $= 4 + 7 + 3 = 14$ tons

The reduced matrix now consists of only P_2; therefore, the third tour is

Tour 3: P_0–P_2–P_0
Distance traveled $= 7 + 7 = 14$ miles
Total load $= 5$ tons

Total distance traveled with all tours is $92 + 35 + 14 = 141$ miles.

Incidentally, it is possible that we may develop two or more paths simultaneously as we examine the savings in descending order of magnitude for developing these connections.

8.5 EXTENSION OF TRAVELING SALESMAN (ETS) PROCEDURE

Consider the distance matrix as shown in Table 8.15. Node O is the depot and nodes A, B, C, D, and E are the load points. The load from each point is also known. The loads from each point are: A = 8, B = 4, C = 6, D = 2, and E = 3.

TABLE 8.15 Distance Matrix

	O	A	B	C	D	E
O	∞	2	1	3	2	4
A	2	∞	3	7	4	1
B	1	3	∞	4	5	7
C	3	7	4	∞	2	1
D	2	4	5	2	∞	3
E	4	1	7	1	3	∞

We now advance a procedure that depends on the traveling salesman solution for developing routes. It is divided into two phases. In phase I, a good initial solution is obtained and in phase II, the initial solution is improved by exchanging the nodes.

8.5.1 Phase I

The first step of phase I is to obtain a traveling salesman solution (TS). We have applied the Little method that results in the tree network as outlined in the following diagram.

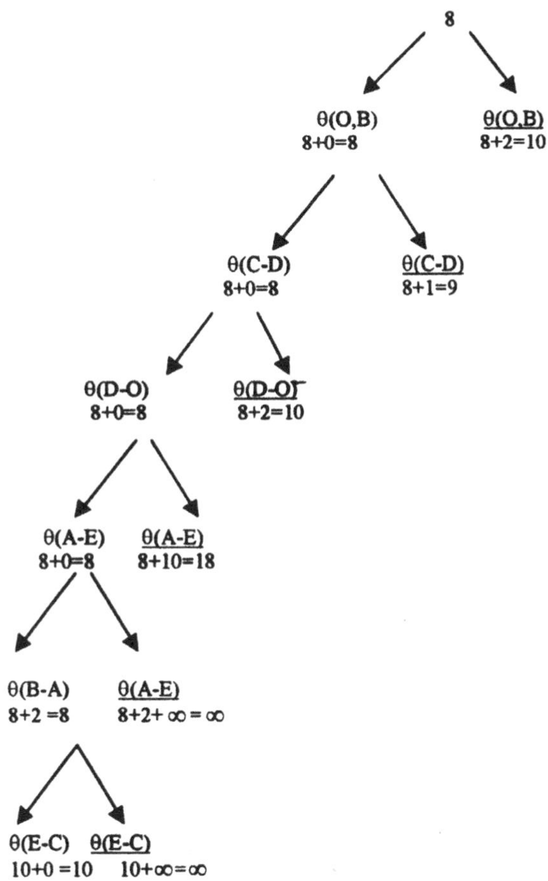

The traveling salesman route is O–B–A–E–C–D–O.

Step 2 of phase I is to write the load associated with each node on the top of each node for simplifying the next phase.

Load at each node is 4 8 3 6 2
TS sequence is O–B–A–E–C–D–O

Next, begin the truck tour from O, connecting the next sequential node of TS, say node X, to the tour if the capacity permits. If the addition of any node in the sequence exceeds the capacity, ignore that node. However, continue to check TS sequentially for another node, say node Y, that may be joined to the route based on its load. Join it in the route if savings are associated with joining the node. This savings is calculated as OX + OY − XY where OX is the distance from home base O to node X, OY is the distance from node Y to O, and XY is the distance between nodes X and Y. Develop the route until no further nodes can be joined to the route.

The reasoning behind the savings formula is straightforward. If we do not go to Y, it will be necessary to return back to O from X. In addition, under the worst conditions, we may have to make an independent trip to node Y to pick up a load, resulting in the travel of 2 × OY. If we could combine the node Y with the present tour we will save coming back to X and going back to Y. But we will encounter additional cost of going from X to Y. Therefore, the maximum possible savings by joining Y to X will be OX + OY − XY.

In our example, with capacity of truck being 10, we can join O to B. We have filled the truck with 4 units. The next node A has a load of 8, and joining A will exceed the truck capacity. The next sequential node that may be joined in this trip is node E, with the capacity of 3. Calculate the savings of joining E: the saving is

$$OB + OE - BE = 1 + 4 - 7 = -2$$

Because the saving is negative, E should not be joined in this path.

Check node C next.
Saving for O-B-C-O is OB + OC − BC = 1 + 3 − 4 = 0.
So we could join node C in the trip. The capacity filled in the truck is 4 + 6 = 10.
Total travel distance from the tip is 1 + 4 + 3 = 8.
The second trip starts at O and goes to A. We may be able to join the next unconnected node in the TS path node D to this path if there are savings.
Savings for joining D is OA + OD − AD = 2 + 2 − 4 = 0.
Capacity filled is 8 + 2 = 10.
Total cost is 2 + 4 + 2 = 8.
The last path is O–E–O.
Capacity filled = 3.
Total cost = 4 + 4 = 8.
Total cost of the routes developed is 8 + 8 + 8 = 24.

8.5.2 Phase II

This is an improvement routine and implemented if lower cost alternative routes can be formed. Check the points in the present tour and see if any of them can be exchanged to an adjacent tour. If it is possible, does it reduce the cost? If the cost is reduced, make the changes. Continue doing so until no changes are profitable.

Consider our example. Our TS path along with the requirement of each node (number at the top of the node) is

  4 8 3 6 2
O–B–A–E–C–D–O, and our touring paths are

  4 6   8 2   3
O–B–C–O, O–A–D–O, and O–E–O.

Let us begin with a trip that has the most slack, O–E–O. The adjacent nodes to E are A and C. A requires 8 units of capacity; therefore, it cannot be conveyed to O–E–O path. C requires 6 units of capacity and may be transferred to the path O–E–O. The associated savings would be that we do not go from B to C, or from C to O, in the first sequence and from E to O in the third sequence. The additional cost is that we do go from B to O in the first sequence and from E to C and C to O in the third sequence. Thus we have the net savings equal to

$$BC + CO + EO - (BO + EC + CO) = (4 + 3 + 4) - (1 + 1 + 3) = 6$$

Because the savings figure is positive, make the change. Our new routes are:

  4   8 2   3 6
O–B–O, O–A–D–O, O–E–C–O

Now let us look at O–B–O with only 4 units of used capacity. Can we join any other node to this route? Going back to the TS sequence, we see that the next node A cannot be joined because of the capacity and next node E was examined earlier and was not profitable to join with B. There is no need to go any farther in the TS sequence because we have obtained a stoppage in search owing to profitability of a next consecutive node.

No other feasible exchange is possible and we stop. Our routes are

O–B–O	Cost = 1 + 1 = 2
O–A–D–O	Cost = 2 + 4 + 2 = 8
O–E–C–O	Cost = 4 + 1 + 3 = 8

For the total cost of 18.

The Clark and Wright method leads to the same results as shown in the following:

Application of CW Method (savings are shown on the right side)

Q	O									
8	2	A	Sav							
4	1	3	0	B	Sav					
6	3	7	−2	4	0	C	Sav			
2	2	4	0	5	−2	2	3	D	Sav	
3	4	1	5	7	−2	1	6	3	3	E

Q = demand; Sav = Savings.

The routes are

O–C–E–O Cost $= 3 + 1 + 4 = 8$
O–A–D–O Cost $= 2 + 4 + 2 = 8$
O–B–O Cost $= 1 + 1 = 2$

The ETS method works well for small examples. Because the distances between demand points do not change, the TS route can be defined only once. If there is a variation in the quantity to be picked up (or delivered) from day to day, alternative routes can be easily formed based on the data for that day, because the TS sequence does not change with demand. This is perhaps a little easier (almost by inspection in some cases) than the CW method, for which the entire procedure needs to be reapplied. For a large set of demand points, however, both methods may require considerable effort. For such a condition, a simpler method called the *Sweep Method*, is presented next.

8.6 SWEEP METHOD

Again, similar to ETS, the method may have two stages. In the first stage, approximate routes are determined and, in the second stage, adjustments are made to improve the routes. In many instances, efficient tours can be developed just by inspection.

Determine the minimum number of trips required by dividing the total demand from different customers by the capacity of the truck. Because an integer number of trips are to be made, there may be some available slack. It can be calculated as: truck capacity × the number of trips required—the total load.

Draw the customers (nodes) and depot locations on an x–y coordinate chart. It is convenient to make coordinates of depot (0,0) and draw cones, with the tip of the cone being at the depot, to include the customers in a tour path. This can be done by starting with a ray, joining some convenient location (say, one closer to the depot) with a depot, and then rotating it clockwise to include the customers in the cone thus created. The cone is expanded, by rotating the ray farther, to include additional customers in the cone, as the capacity of the truck may allow. If the addition of the next customer will exceed the truck capacity, go to the next point to see if it can be joined in this route. If the path becomes too long and unattractive (one can judge this by inspection) then stop this route. Determine the slack in this route, which is the capacity of the truck minus the total load on the planned route. This amount is also the slack consumed by the present route from the total available slack. If the slack consumed is more than what is available, then the trip cannot be terminated with the last added customer. Join the next feasible customer and again evaluate the slack allocation. Continue the process until the route is defined.

If the slack consumed by the route is less than that available, then terminate the route. Recalculate the now available slack in the system and begin with the next route from the last customer in the path of the clockwise-rotating ray that is yet to be joined in a route and repeat the procedure to develop the next tour. Continue the application of this process until all customers are connected.

To improve, or if at some point the additional paths cannot be developed because the slack has been exhausted, then consider the customers who are closest to the existing paths with slacks, and determine which customers can be joined into one of the existing slack paths with the least cost (distance). Add the associated customer to such path, and reevaluate the remaining paths by using the remaining customers. Select the least-cost alternative from among the various alternatives thus created. Repeat the procedure by starting the initial ray with a different customer. Some patterns may repeat themselves, even though the initial ray may have started with different customer, as we will see in the example problem. Examine the total distance traveled in each pattern and select the optimum.

In some examples, the minimum number of trips may not be sufficient; for example, when there are three customers, each with a load of 8, and with a truck capacity of 15. Although, theoretically two trips should be sufficient, practically no two customers can be combined and we will need three trips. In such cases, the minimum trip number needs to be increased and the procedure repeated.

Example

Consider a 12-location problem with the origin being the depot and other locations with the coordinates as given in the following table. The demand from each point is also given in the table.

No.	1	2	3	4	5	6	7	8	9	10	11	12
Location	(−0.5, 0.3)	(0.5, 1.5)	(2.2, −0.7)	(3.5, 0.4)	(−3.3, 3.0)	(0.0, −5.8)	(0.0, 6.5)	(6.4, 4.4)	(−4.5, −8.1)	(6.8, −7.6)	(7.0, 8.7)	(5.0, −11.3)
Demand	3	2	4	5	6	1	3	3	6	2	4	4

The corresponding distance matrix, calculated as a direct distance between two points, is as follows:

P_0												
0.58	P_1											
1.58	1.56	P_2										
2.3	2.87	2.78	P_3									
3.52	4.0	3.19	1.7	P_4								
4.45	3.88	4.08	6.62	7.28	P_5							
5.8	6.12	7.31	5.55	7.11	9.39	P_6						
6.5	6.22	5.02	6.2	7.03	4.81	12.3	P_7					
7.76	8.02	6.57	6.6	4.94	9.8	12.04	6.73	P_8				
9.26	9.3	10.82	9.98	11.67	11.16	5.05	15.27	16.58	P_9			
10.19	10.75	11.06	8.29	8.65	14.64	7.03	15.65	12.0	11.31	P_{10}		
11.16	11.26	9.7	10.55	9.0	11.77	16.1	7.33	4.34	20.35	16.3	P_{11}	
12.35	12.83	13.56	10.96	11.79	16.53	7.43	18.48	15.76	10.02	4.11	20.09	P_{12}

The plot of the customers locations and associated demands, included in ( ) are shown in Figure 8.1.

Assume that the truck has a capacity for 11 units.
The total load = 43.
Hence the minimum number of trips required = 43/11 = 3.90.
Round up the number of trips to the next possible integer, which is 4.
So, the total capacity becomes $11 \times 4 = 44$, leaving behind the slack of 1.

By applying the procedure, the first route is developed by starting with location 1. Next, rotating the ray clockwise, we can join customer 5 to make the total load of $3 + 6 = 9$ units. Location 7 with the load of 3, cannot be added because such addition to this route exceeds the truck capacity of 11. The slack in this route is $11 - 9 = 2$. Maximum available slack is 1. Therefore we must continue with the route. The next node in clockwise rotation is 2 with load of 2. It can be combined with this route. The truck is full, and we go to the second route.

The route starts at location 7 with load of 3. Combine node 11 with load 4 and next node 8 with load 3 to make the total $3 + 4 + 3 = 10$. Next, node 4 with the capacity of 5 cannot be joined. We could stop here, because the slack used in this route is $11 - 10 = 1$, the available slack. But suppose we continue, and join next feasible node to this route, node 6 with the load of 1. The remaining two routes are similarly developed to include nodes 4, 3, and 10 in one route, and 12 and 9 in another route. The details are shown in Figure 8.2 and case 1 table.

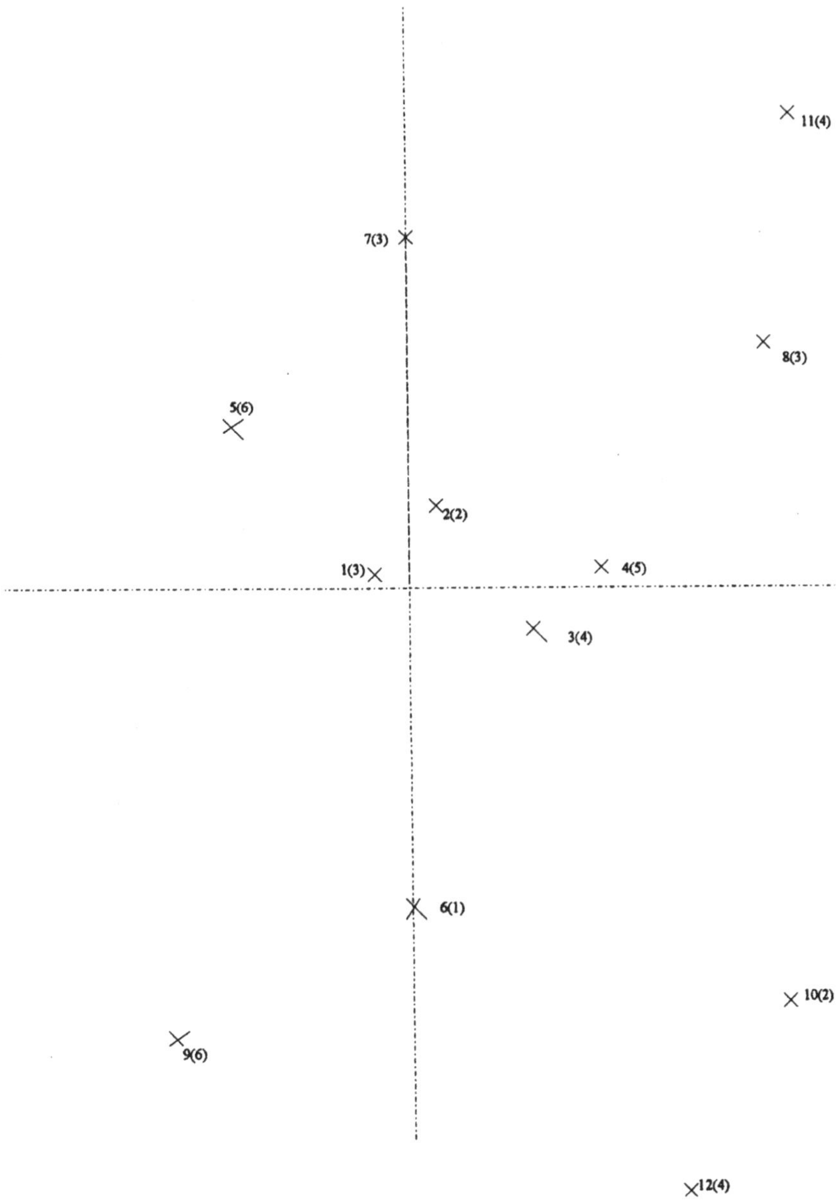

FIGURE 8.1 Customer's locations and associated demands.

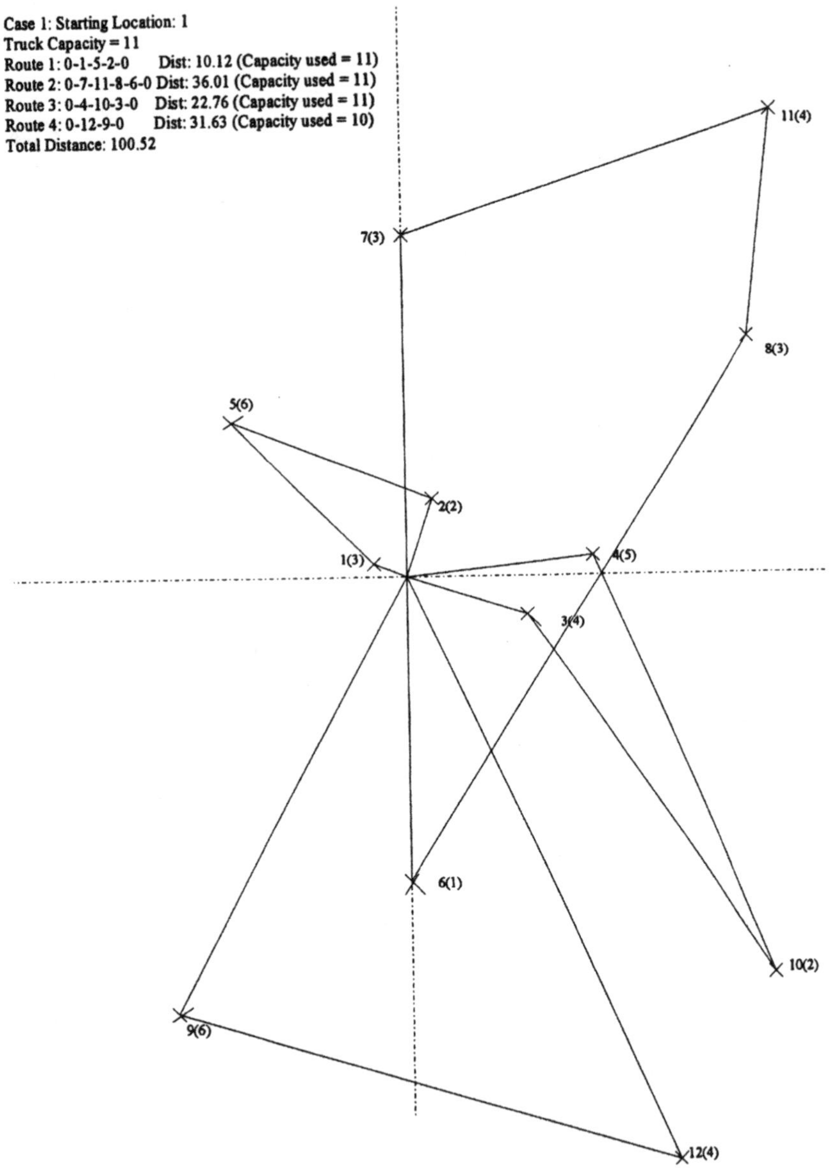

FIGURE 8.2 Route starting with location 1.

Case 1. Starting location 1. Figure 8.3 and case 1, show improvement calculations. Route 0–12–9–0 has 1 unit of slack. We can include point 6 with load of 1 in this route and reevaluate other routes. This results in route 2 using 1 unit of slack, and terminating with node 8. The total distance of all routes has been reduced from 100.52 to 91.52.

Route no.	Route	Slack used	Slack remaining	Distance	Total distance
1	0–1–5–2–0	0	1	10.12	
2	0–7–11–8–6–0	0	1	36.01	100.52
3	0–4–10–3–0	0	1	23.70	
4	0–12–9–0	1	0	31.63	

Case 1. Improvement, starting location 1. Other routes are similarly developed, starting at different points and are shown in Figures 8.4 through 8.6 and associated calculations are shown in cases 2 through 4. We might observe, as in Figure 8.4, that some starting locations give the same routes, reducing the actual calculations. This is true in most examples.

Route no.	Route	Slack used	Slack remaining	Distance	Total distance
1	0–1–5–2–0	0	1	10.12	
2	0–7–11–8–0	1	0	25.93	91.32
3	0–4–10–3–0	0	0	23.70	
4	0–6–12–9–0	0	0	34.09	

Case 2. Starting location 5.

Route no.	Route	Slack used	Slack remaining	Distance	Total distance
1	0–5–7–2–0	0	1	15.86	
2	0–11–8–3–0	0	1	24.40	89.62
3	0–4–10–12–0	0	1	28.63	
4	0–6–9–1–0	1	0	20.73	

Case 3. Starting location 7.

Route no.	Route	Slack used	Slack remaining	Distance	Total distance
1	0–2–7–11–10–0	0	1	40.42	
2	0–1–8–4–0	0	1	17.06	112.97
3	0–3–6–9–0	0	1	22.16	
4	0–12–5–0	1	0	33.33	

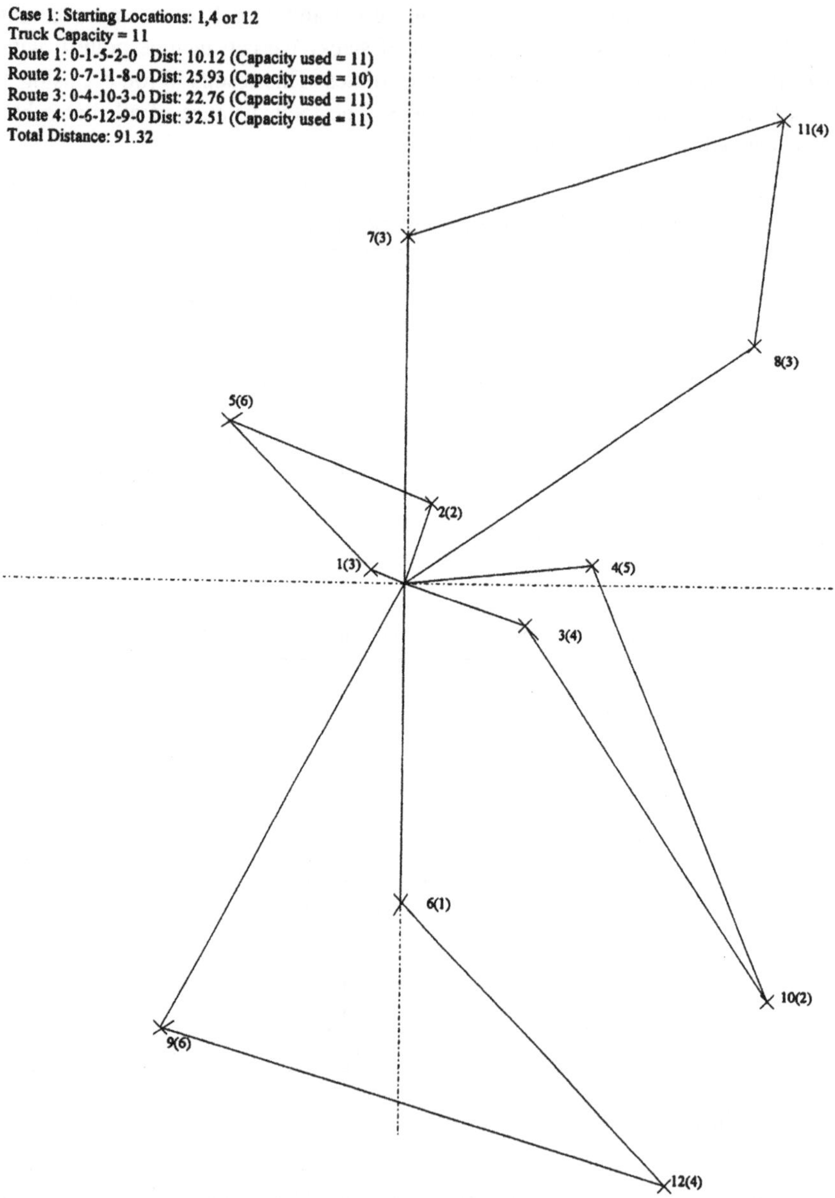

FIGURE 8.3 Routes starting with locations 1, 4, or 12.

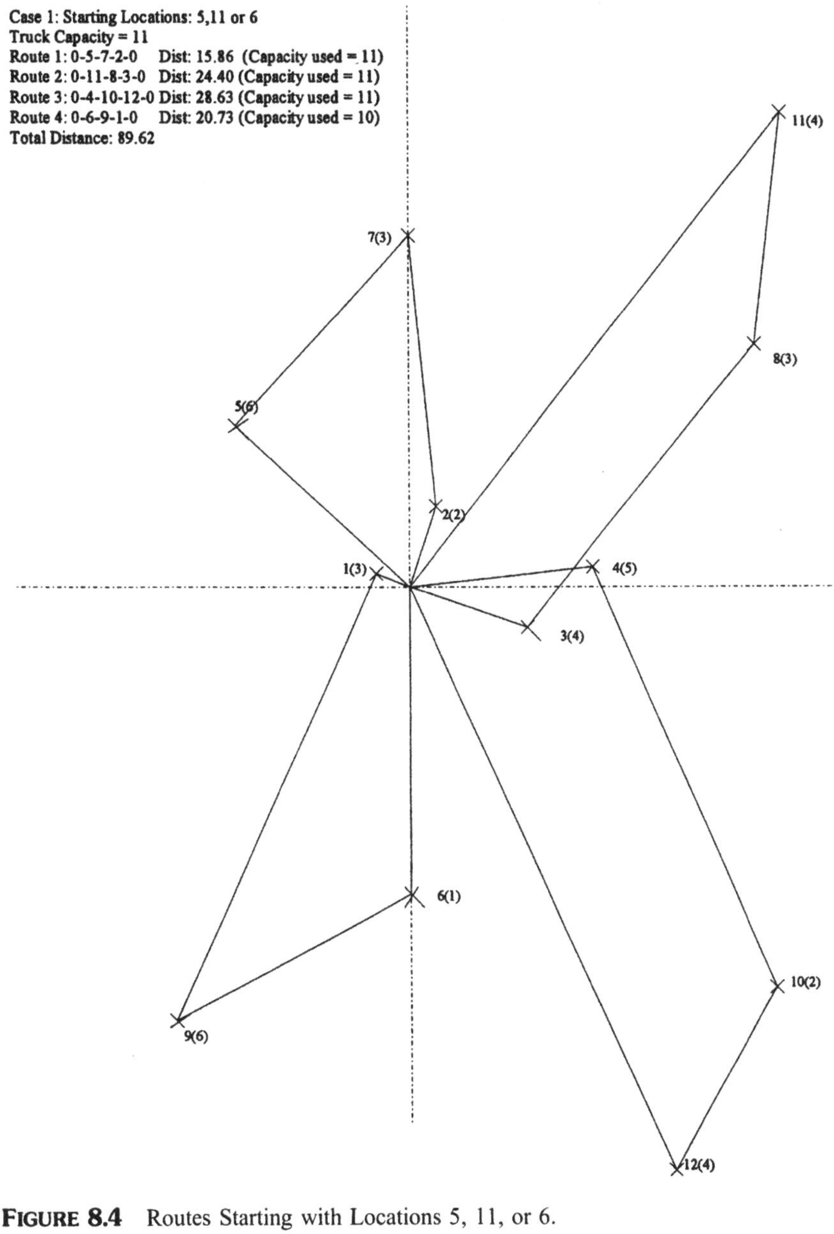

FIGURE 8.4 Routes Starting with Locations 5, 11, or 6.

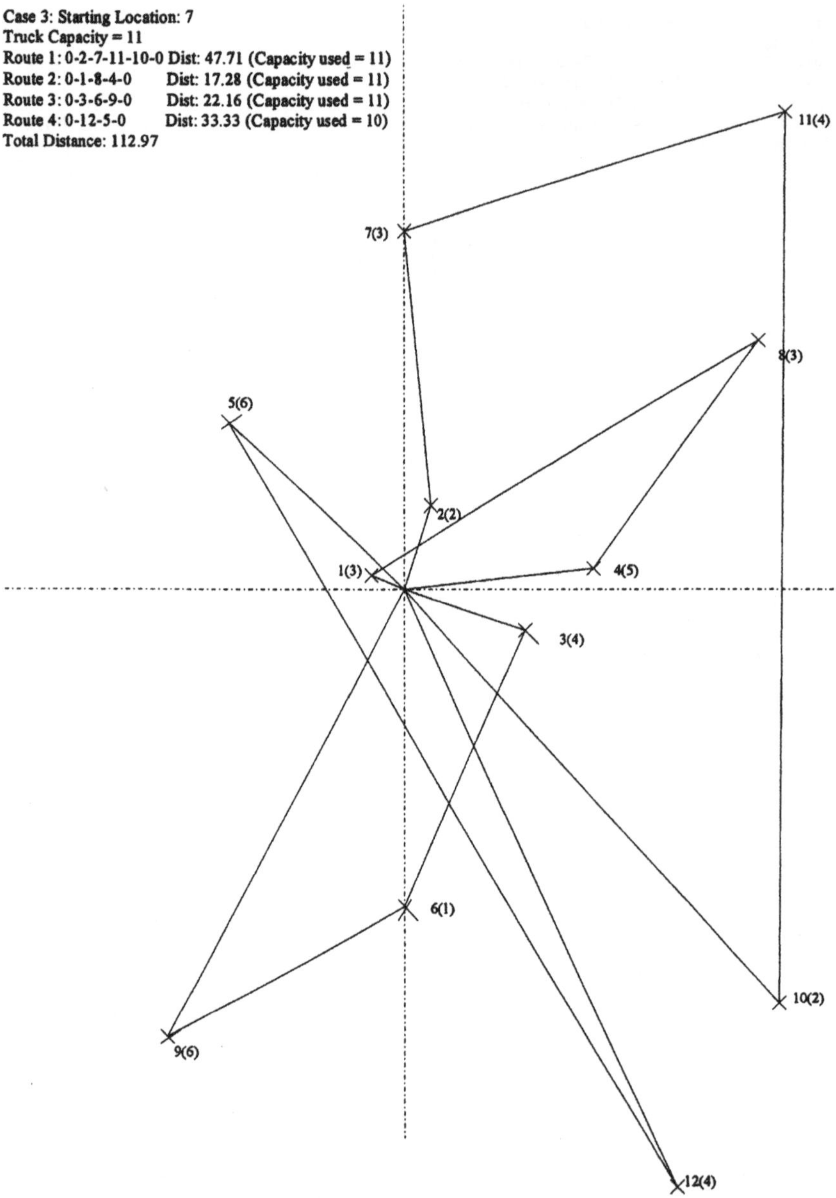

FIGURE 8.5 Routes starting with location 7.

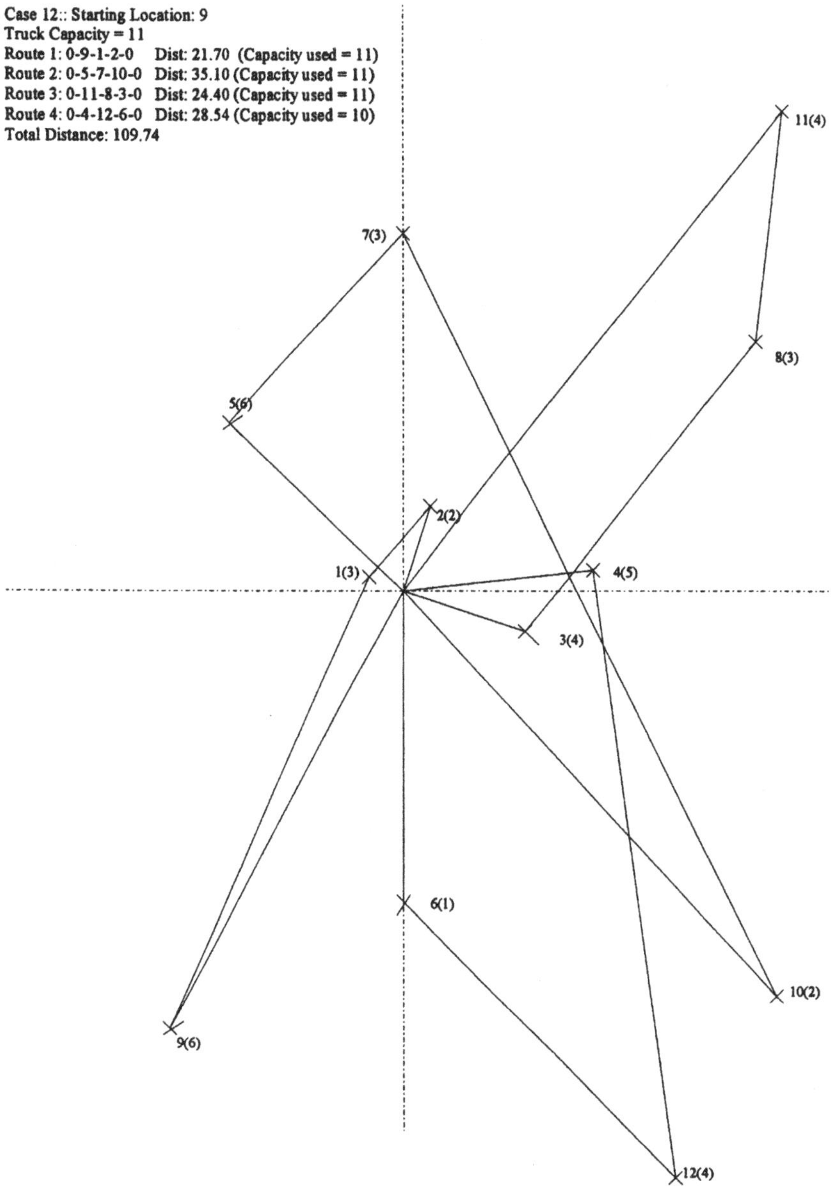

FIGURE 8.6 Routes starting with location 9.

Case 4. Starting location 2.

Route no.	Route	Slack used	Slack remaining	Distance	Total distance
1	0–2–11–4–0	0	1		
2	0–8–3–10–6–0	1	0		

Further assignments are not possible because by selecting any remaining locations, we exceed the truck capacity. And we cannot fulfil the capacity constraint of the truck.

Case 5. Starting location 11: The routes formed are similar to those of case 2.

Case 6. Starting location 8.

Route no.	Route	Slack used	Slack remaining	Distance	Total distance
1	0–4–8–10–6–0	0	1	33.29	
2	0–3–12–1–0	0	1	26.67	119.36
3	0–9–7–2–0	0	1	31.13	
4	0–5–11–0	1	0	27.38	

Case 7. Starting location 4: The routes formed are similar to those of case 1.

Case 8. Starting location 3.

Route no.	Route	Slack used	Slack remaining	Distance	Total distance
1	0–3–10–12–6–0	0	1	27.93	
2	0–9–1–2–0	0	1	21.70	98.70
3	0–5–11–0	1	0	27.38	
4	0–7–8–4–0	0	0	21.69	

Case 9. Starting location 10

Route no.	Route	Slack used	Slack remaining	Distance	Total distance
1	0–10–12–6–1–0	1	0	28.43	
2	0–9–7–2–0	0	0	31.13	99.21
3	0–5–4–0	0	0	15.25	
4	0–11–8–3–0	0	0	24.40	

Case 10. Starting location 12: The routes formed are similar to those of case 1.

Case 11. Starting location 6: The routes formed are similar to those of case 2.

Case 12. Starting location 9.

Route no.	Route	Slack used	Slack remaining	Distance	Total distance
1	0–9–1–2–0	0	1	21.70	
2	0–5–7–10–0	0	1	35.10	109.74
3	0–11–8–3–0	0	1	24.40	
4	0–4–12–6–0	1	0	28.54	

The optimum solution is determined by comparing the total distances for each case and selecting the minimum among them. Here, the optimum routes are observed in case 2, with starting at location 5. The same result can be obtained if we start the route from locations 11 or 6, as seen in cases 5 and 11.

8.7 BACKTRACKING

When we developed the tours in the previous analysis we had assumed that the tours are made either to deliver or to pick up loads from the customers. But when customers require both pickup and deliveries, there are two strategies that may be followed. If unloading or loading can be done at the same time, then routes can be developed that allow us to do that. But unloading and loading may not be possible at the same time because of conditions such as the physical constraint for which we do not want a pickup load in front of delivery load in the truck, or we deliver a load in the morning and pick up the load in the evening, giving the customers an opportunity to work and develop new products during day. To achieve unloading and loading operations backtracking provides an option. Here, we distribute the loads first and, then, drive back on the same route and collect the loads. Note that in this case, we do not go from the last destination in the route to home, but travel back from the last node to the first node in the route, before coming home.

If A and B are two customers and we collect the loads independently, then the total distance traveled would be 2OA + 2OB. If we combine A and B in a backtracking tour then the total distance traveled would be OA + AB + BA + AO. By assuming a symmetrical distance matrix, the savings by combining A and B in a single tour is 2 (OB–AB). Because 2 is constant multiplier, for simplicity, we can consider savings as simply OB–AB. Using these savings we can apply the CW method to develop backtracking routes.

For example, consider the same distance matrix as in Table 8.12, reproduced here as part of Table 8.16. Also shown in the table is the demand and supply from each customer. Demand is what we need to deliver first, and supply is the quantity that we need to pick up on the backtracking trip. The savings by joining locations i and j is now calculated as the distance from (P_0 to P_j) − distance from (P_i to P_j) and is displayed in left-hand corner of Table 8.16.

TABLE 8.16 Distance–Savings Matrix

	P_0	P_1	P_2	P_3	P_4	P_5	P_6	P_7	P_8	Demand	Supply
P_1	4	—	—	—	—	—	—	—	—	3	4
P_2	7	(4) 3	—	—	—	—	—	—	—	2	5
P_3	12	4 8	−3 15	—	—	—	—	—	—	3	1
P_4	17	11 6	1 16	9 8	—	—	—	—	—	4	6
P_5	26	16 10	6 20	15 9	13 13	—	—	—	—	5	4
P_6	32	17 15	11 21	20 12	(20) 12	(26) 6	—	—	—	2	2
P_7	48	26 22	(30) 18	41 7	28 20	38 10	28 20	—	—	3	2
P_8	56	26 30	21 35	(46) 10	37 19	39 17	41 15	(48) 8	—	4	3

An application of CW procedure is illustrated in Table 8.17. The only variation here is that we need to keep track of both demand and supply from each customer. The cumulative total for either should not exceed the truck capacity, in our example 15. The calculations are straightforward and are illustrated in the Table 8.17.

The final routes are

Route 1: 0–5–7–8–3

Distance for route 1: $2(26 + 10 + 8 + 10) = 2(54) = 108$

Route 2: 0–4–6–1–2

Distance for Route 2: $2(17 + 12 + 15 + 8) = 2(52) = 104$

Total distance traveled $= 108 + 104 = 212$

8.8 SIMULTANEOUS UNLOADING AND LOADING OPERATIONS

In this model we study the tour development procedure when unloading and loading is performed at the same time. Thus, with a single stop at the customer, the truck is unloaded first with the demand from the customer, and then loaded with the customer's supply. The truck, therefore, starts at the depot and returns to the depot in each route without going back to the same customer again (no backtracking). The problem is similar to one solved with CW method except for the additional loading and unloading constants. If from any customer, the supply (loading) is greater than demand (unloading), then we must make sure that there is an empty capacity available in the truck to accommodate such overload. If, on the other hand, the customer's demand is more than supply, then we will create a slack in the truck at that point. Our calculations should reflect such variations. The procedure tries to minimize the total distance traveled, at the same time

TABLE 8.17 Backtracking Solution

Savings	Route no.	Between locations	Entering location	From connecting point		Cumulative		Action C, connect NC, cannot connect A, capacity exceeded B, chain cannot be formed	Resulting chain
				Demand	Supply	Demand	Supply		
48	1	7–8	7	3	5	0 + 3 = 3	0 + 5 = 5		
			8	2	3	3 + 2 = 5	5 + 3 = 8	C 7–8	7–8
46	1	8–3	3	3	1	5 + 3 = 8	8 + 1 = 9	C 8–3	7–8–3
41	1	8–6						B	—
41	1	7–3						B	
39	1	8–5						B	
38	1	7–5	5	5	4	8 + 5 = 13	9 + 4 = 13	C 5–7	5–7–8–3
37	1	8–4						B	
30	1	7–2	2	3	2			B	
26	1	6–5	6	3	4			A	
26	1	7–1	1	4	3			A	
20	2	4–6	4	4	6	0 + 4 = 4	0 + 6 = 6		
20	2		6	3	4	4 + 3 = 7	6 + 4 = 10	C 4–6	
17	2	1–6	1	4	3	7 + 4 = 11	10 + 3 = 13	C 6–1	4–6–1
	.	.	.	.	.	.			
	.	.	.	.	.	.			
	.	.	.	.	.	.			
4	1	2–1	2	3	2	11 + 3 = 14	13 + 2 = 15	C 2–1	4–6–1–2

checking the feasibility of the solution. However, there is no guarantee that it will lead to the optimum solution every time.

We again start with the distance Table 8.13 reproduced here as Table 8.18 for convenience, with added columns associated with demand and supply from each customer slightly different from what we used in Table 8.17. This is to illustrate the approaches in detail. Later we will also solve the problem using the same data for demand and supply that we used in Section 8.7.

T is the initial capacity of our truck. We assume that in this example the truck has a capacity of 15; therefore, $T = 15$. The minimum number of trucks required is the maximum (total demand/truck capacity, total supply/truck capacity) rounded upward. In our case it is max $(27/15 \text{ or } 28/15) = 2$ trucks.

The procedure is illustrated in Table 8.19 and is based on the data from Table 8.18. In column 1 (C1) the savings are recorded from Table 8.18. We start with the maximum savings, in our case 96, between nodes 8 and 7. These values are entered in columns 1 and 2, respectively. Column 3 shows the present possible chain. Column 4 shows the next node to join the chain. Associated node's demand and supply are noted in columns 5 and 6. Cumulative demand and supplies for the chain so far, are noted in columns 7 and 8, respectively. The difference, indicated as "Dif," cumulative supply and cumulative demand is noted in column 9. If this difference is positive (i.e., >0), then we calculated a quantity R, which is truck capacity (T) – Dif – cumulative supply. If R is negative the chain is not feasible and is marked by NF. If R is positive or zero, the chain is feasible and we form the route (R#) and connect the node to the chain.

Now, go back to Table 8.18 and scratch the savings element that was just examined. Select the next largest savings element and continue the process. The following rules apply:

1. Begin a new route if none of the pairs of nodes associated with the current savings value have been included in any of the previous routes.

TABLE 8.18 Distance–Savings Matrix 1

	P0	P1	P2	P3	P4	P5	P6	P7	P8	Demand	Supply
P1	4	—								4	3
P2	7	3 8								3	2
P3	12	8 8	4 15							3	1
P4	17	15 6	8 16	21 8						4	6
P5	26	20 10	13 20	29 9	30 13					5	4
P6	32	**21** 15	18 21	37 12	37 12	52 6				3	4
P7	48	30 22	37 18	**53** 7	45 20	64 10	60 20			3	5
P8	56	30 30	28 35	58 10	54 19	**65** 17	73 15	**96** 8		2	3
									TOTAL	27	28

TABLE 8.19 Solution for Simultaneous Unloading and Loading Problem

1 Saving	2 Between nodes	3 Possible chain	4 Added node	5 Demand	6 Supply	7 Cum. demand	8 Cum. supply	9 Dif = C8 – C7, if Dif > 0	10 R = T – Dif – C7	11	12 chain
96	7–8	7–8	7	3	5	3	5	2	15 – 2 – 5 = 8		
			8	2	3	5	8	3	15 – 3 – 8 = 11	R1	7–8
73	8–6	7–8–6	6	3	4	8	12	4	15 – 4 – 12 = –1 NF		7–8
65	8–5	7–8–5	5	5	4	10	12	2	15 – 2 – 12 = 1	R1	7–8–5
64	7–5 NC										
60	7–6	6–7–8–5	6	3	4	3	4				
			7	3	5	6	9				
			8	2	3	8	12				
			5	5	4	13	16E				
58	8–3 NC										
54	8–4 NC										
53	3–7	3–7–8–6	3	3	1	3	1	–2			
			7	3	5	6	6	0			
			8	2	3	8	9	–1			
			5	5	4	14	13	1	15 – 1 – 13 = 1	R1	3–7–8–5
52	6–5 E										
45	7–4 NC										
37	6–4	6–4	6	3	4	3	4	1	15 – 4 – 1 = 10	R2	
			4	4	6	7	10	3	15 – 3 – 10 = 2		6–4
37	7–2 NC										
32	6–3 NC	R1 + R2 E									
30	7–1 NC										
30	8–1 NC										

(continued)

TABLE 8.19 *(continued)*

1 Saving	2 Between nodes	3 Possible chain	4 Added node	5 Demand	6 Supply	7 Cum. demand	8 Cum. supply	9 Dif = C8 – C7, if Dif > 0	10 R = T – Dif – C7	11	12 chain
30	4–5	6–4–5	5	5	4	12	14	2	–1 NF		
29	5–3 NC										
28	8–2 NC										
21	6–1	1–6–4	1	4	3	4	3				
			6	3	4	7	7	0			
			4	4	6	11	13	2	15 – 2 – 13 = 0	R2	1–6–4
21	4–3	R + R2 E									
20	5–1	R + R2 E									
18	6–2 NC										
15	4–1 NC										
13	5–2	3–7–8–5–2	2	3	2	17 E	15				
8	4–2	1–6–4–2	2	3	2	14	15	1	15 – 1 – 15 = –1 NF		
4	3–2	2–3–7–8–5	2	3	2	17 E					
3	1–2	2–1–6–4	2	3	2	3	2				
			1	4	3	7	5				
			6	3	4	10	9				
			4	4	6	14	15	1	15 – 15 – 1 = –1 NF		

NC, no connection possible; E, capacity exceeded; NF, nonfeasible.

2. Ignore the savings value if an associated chain cannot be formed. It is marked as NC (no chain) in the table. For example, for savings of 64 between 7 and 5, we already have a route 7–8–5. Connecting 7–5 is not feasible because both nodes are already in a chain. Similarly, for savings 58, 8–3 connection is not feasible, because 8 has already been connected (7–8–5) on both sides by other nodes.
3. If the next connecting node is at the end of the previously developed route so far, then check for result of addition of a new node by incorporating the data for the present node with the last row that was developed for the route. For example, add node 6 to chain 7–8 to make it 7–8–6. On the other hand, if the entering node is in the front of the chain, then the entire chain needs to be reevaluated (e.g., by adding node 6 to the present chain of 7–8–5 to make chain 6–7–8–5).
4. If the savings between two nodes leads to a pair for which one node is in one route and other in another, but a chain can be formed, then evaluate the effect of combining two routes. Example, for savings 32, between 6 and 3, route R1, with chain of 3–7–8–5, and route R2 with chain of 6–4 could be joined to make 4–6–3–7–8–5. In this case, the combined route is not possible as the total demand of 14 + 7 and supply of 13 + 10, both exceed the capacity of the truck, which is 15.
5. A node cannot be joined to a route if, by doing so, one of the following conditions appear.
   a. R is negative.
   b. The cumulative demand exceeds truck capacity.
   c. The cumulative supply exceeds truck capacity.
6. Continue developing the routes until all savings are examined. If, at this point, we still have some nodes that are not part of any route yet developed, then, an independent route, from origin to the node and back, is needed for each such node.

Route 1: 0–3–7–8–5–0	Distance 12 + 7 + 8 + 17 + 26 = 70
Route 2: 0–1–6–4–0	Distance 4 + 15 + 12 + 17 = 48
Route 3: 0–2–0	Distance 7 + 7 = 14
	Total distance: 132

8.9 MAXIMUM PERMISSIBLE TIME FOR A ROUTE

The routes developed in the previous analysis had only truck capacity as the limiting factor. At times, however, there are some other considerations that also must be taken into account in developing the truck routes. For example, the maximum or minimum distance a truck should travel, the travel time available for

TABLE 8.20 Time Matrix

Demand	Supply	P_0								
4	3	0.3	P_1							
3	2	0.7	0.5	P_2						
3	1	2.0	0.4	0.9	P_3					
4	6	2.2	0.3	1.1	0.2	P_4				
5	4	1.1	0.9	1.8	0.3	1.0	P_5			
3	4	1.3	1.2	2.2	0.8	0.8	0.3	P_6		
3	5	2.8	1.8	1.2	1.0	1.6	1.0	2.2	P_7	
2	3	2.3	2.3	3.1	0.9	2.1	1.3	1.8	0.6	P_8

making a trip, the load distribution among trucks, and other such considerations may form one or more additional constraints on the problem. They can be handled easily by expanding on Table 8.19.

Suppose the time required to travel between two nodes for the distance data in Table 8.18 is displayed in Table 8.20. Note that the times are not directly proportional to the distances, as the driving condition between any two cities may also influence the travel time. Now we wish to develop routes so that the maximum travel time on any route is restricted to 5 units.

The solution is shown in Table 8.21 and the associated time calculations are shown in Table 8.22. At each step, the complete route time must be considered. For example, if nodes 7 and 8 can be joined in a chain, the route would be 0–7–8–0, where 0 is the depot. This time exceeds the limit of 5; therefore, such connection is not feasible. As a node is added the previous route time is modified. This may require just adding and subtracting a few time elements from the previously feasible route, depending on whether the new node is added at the end or at the beginning of the previously feasible chain. For example going from 0–8–5–0 to 0–8–5–7–0 requires subtracting time for 5–0 from time for 0–8–5–0 and adding times for 5–7 and 7–0. The solution is displayed at the bottom of Table 8.21.

8.10 TO MINIMIZE THE NUMBER OF VEHICLES USED

Another objective may be to use the minimum number of vehicles to satisfy demand and supply requirements of all customers without very much concern for the distance traveled. The objective is to develop clusters of customers, such that a vehicle can support the customers, as far as possible, with even loads of total demands and supplies. The steps of a heuristic procedure are as follows:

1. Calculate total demand and supply for all customers. Divide the larger value by the capacity of the vehicle. This gives the minimum number of vehicles that are needed to fulfill the total distribution and collection requirements.

TABLE 8.21 Solution for Maximum Permissible Time of 5 Units

1 Saving	2 Between Nodes	3 Possible chain	4 Added node	5 Demand	6 Supply	7 Cum. demand	8 Cum. supply	9 Dif = C8 – C7, if Dif > 0	10 R = T – Dif – C7	11	12 Chain	13 Time
96	7–8	7–8	7	3	5	3	5	2	8			
			8	2	3	5	8	3	4	R1	7–8	5.7
												*TE
73	8–6	8–6	8	2	3	2	3	1	11			
			6	3	4	5	7	2	6	R1	8–6	5.4
												*TE
65	8–5	8–5	8	2	3	2	3	1	11			
			5	5	4	7	7			R1	8–5	4.7
64	7–5	8–5–7	7	3	5	10	12			R1	8–5–7	7.4
												*TE
60	7–6	7–6	7	3	5	3	5	2	8			
			6	3	4	6	9	3	3	R2	7–6	6.3
												*TE
58	8–3	3–8–5	3	3	1	3	1					
			8	2	3	5	4					
			5	5	4	10	8	2	5	R1	3–8–5	5.3
												*TE
54	8–4	4–8–5	4	4	6	4	6	2	7			
			8	2	3	6	9	3	3			
			5	5	4	11	13	2	0	R1	4–8–5	6.7
												*TE
53	7–3	7–3	7	3	5	3	5	2	8			
			3	3	1	6	6			R2	7–3	5.8
												*TE

(*continued*)

TABLE 8.21 *(continued)*

1 Saving	2 Between Nodes	3 Possible chain	4 Added node	5 Demand	6 Supply	7 Cum. demand	8 Cum. supply	9 Dif = C8 − C7, if Dif > 0	10 R = T − Dif − C7	11	12 Chain	13 Time
52	6–5	8–5–6	6	3	4	10	11	1	3	R1	8–5–6	5.2
												*TE
45	7–4	7–4	7	3	5	3	5	2	8			
			4	4	6	7	11	4	0	R2	7–4	6.6
												*TE
37	6–4	6–4	6	3	4	3	4	1	10			
			4	4	6	7	10	3	2	R2	6–4	4.3
37	7–2	7–2	7	3	5	3	5	2	8			
			2	3	2	6	7	1	7	R3	7–2	4.7
32	6–3	3–6–4	3	3	1	3	1					
			6	3	4	6	5					
			4	4	6	10	11					
30	5–4	R1 + R2										
		E										
30	8–1	1–8–5	1	4	11	10				R1	1–8–5	5
30	7–1	R1 + R3										
		E										
29	5–3	1–8–5–3	3	3	1	14	11			R1	1–8–5–3	6.2
												*TE
28	8–2											
	NC											
21	4–3	6–4–3	3	3	1	10	11			R2	6–4–3	4.3

NC, no connection possible, E, capacity exceeded; NF, nonfeasible; *TE, time exceeded.

TABLE 8.22 Time Calculations for the Routes in Table 8.21

Route	Chain	Calculation	Final time
R1	0–7–8–0	2.8 + 0.6 + 2.3	5.7
R1	0–8–6–0	2.3 + 1.8 + 1.3	5.4
R1	0–8–5–0	2.3 + 1.3 + 1.1	4.7
R1	0–8–5–7–0	(4.7) − 1.1 + 1.0 + 2.8	7.4
R2	0–7–6–0	2.8 + 2.2 + 1.3	6.3
R1	0–3–8–5–0	(4.7) − 2.3 + 2.0 + 0.9	5.3
R1	0–4–8–5–0	(4.7) − 2.3 + 2.2 + 2.1	6.7
R2	0–7–3–0	2.8 + 1.0 + 2.0	5.8
R1	0–8–5–6–0	(4.7) − 1.1 + 0.3 + 1.3	5.2
R2	0–7–4–0	2.8 + 1.6 + 2.2	6.6
R2	0–6–4–0	1.3 + 0.8 + 2.2	4.3
R3	0–7–2–0	2.8 + 1.2 + 0.7	4.7
R1	0–1–8–5–0	(4.7) − 2.3 + 0.3 + 2.3	5.0
R1	0–1–8–5–3–0	(5.0) − 1.1 + 0.3 + 2.0	6.2
R2	0–6–4–3–0	(4.3) − 2.2 + 0.2 + 2	4.3

Solution
Route 1: 0–1–8–5–0 Distance: 4 + 30 + 17 + 26 = 77
Route 2: 0–6–4–3–0 Distance: 32 + 12 + 8 + 12 = 64
Route 3: 0–7–2–0 Distance: 48 + 18 + 7 = 73
Total distance: 77 + 64 + 73 = 214

2. Subtract the cumulative demand from the cumulative supply and divide by the maximum number of vehicles calculated in the earlier step. This gives the ideal "maximum slack" in the cumulative (demand–supply) of each vehicle.
3. Start with the location that is nearest the origin. See what the demand–supply for that point is. The next location to be served by the vehicle, would be the location that is nearest the first point, and has a (demand–supply) such that the cumulative (demand–supply) for the locations is approaching the targeted maximum slack calculated in step 2.
4. Repeat step 3. If a condition is reached so that only one more location can now be added to the route, then the last location is selected from all the remaining locations in such a way that vehicle capacity is used to the maximum possible. Thus, we have developed a cluster of locations to be served by a single vehicle.
5. After a cluster of locations are selected, if the cumulative (demand–supply) for the vehicle is equal to the maximum permissible slack for that vehicle, then the maximum slack for the next vehicle is the same as that set earlier. Otherwise, the difference between the obtained cumulative (demand–supply) difference and the actual permissible maximum slack, is added to the maximum permissible slack for the next vehicle. This becomes its maximum permissible slack.

TABLE 8.23 Distance Matrix

Loc	0	1	2	3	4	5	6	7	8	9	10	D	S	D − S
0	—													
1	33	—										2	1	1
2	36	28	—									3	2	1
3	61	82	76	—								2	6	−4
4	60	88	95	74	—							2	3	−1
5	66	100	91	29	52	—						5	3	2
6	152	178	154	95	165	114	—					1	3	−2
7	141	175	185	91	110	76	110	—				4	1	3
8	115	138	135	76	63	45	132	43	—			3	4	−1
9	97	100	123	140	74	125	235	181	133	—		2	1	1
10	80	97	113	112	40	91	205	144	96	37	—	3	6	−1

$\Sigma D = 27$; $\Sigma S = 30$; Σ; (Total demand) − (Total supply) = −3

6. Proceed in the similar manner for the remaining customers to determine the clusters of the customers to be served by each vehicle. This should give the minimum value for the total vehicles used.

The locations served by the individual vehicles from the groups. The order in which the individual vehicles serve the locations is decided by following the same procedure that was explained in the earlier part of the chapter for minimizing the total distance traveled for a given number of locations.

Example

Consider an example with the data displayed in Table 8.23.

Solution

Vehicle with capacity 15:

$$\Sigma D = 27$$

$$\Sigma S = 30$$

Hence, the minimum number of vehicles that may be required = $30/15 = 2$ to serve all the locations

$$\Sigma D - \Sigma S = 30 - 27 = -3$$

Hence, maximum cumulative *D*–S for each vehicle = $-3/2$,

That is, "maximum slack" is approximately −1 for one vehicle and −2 for the other vehicle. Application of the procedure is straightforward and is shown in Table 8.24.

Groups obtained: Vehicle 1→1, 3, 5, 8, 9
Vehicle 2 →2, 4, 10, 7, 6

TABLE 8.24 Solution to Truck Minimization Problem

Next nearest dist. to previous location	Location	D	S	$(D-S)$	ΣD	ΣS	Cum. $(D-S)$
	0						
33	1	2	1	1	2	1	1
28	2	3	2	1	5	3	2 (NIG)
82	3	2	6	−4	4	7	−3
91	5	5	3	2	9	10	−1
45	8	3	4	−1	12	14	−2
43	7	4	1	3	16 (E)		
63	4	2	3	−1	14	17 (E)	
96	10	3	6	−3	15	20 (E)	
133	9	2	1	1	14	15	−1
	Vehicle 1 goes with 14						
	2nd vehicle can go with a cumulative $(D-S)$ of −2						
	0						
36	2	3	2	1	3	2	1
95	4	2	3	−1	5	5	0
40	10	3	6	−3	8	11	−3
144	7	4	1	3	12	12	0
110	6	1	3	−2	13	15	−2
	Vehicle 2 goes with 13						

NIG→Not included in group as $(\Sigma D - \Sigma S)$ is not in the direction of the maximum permissible slack.
E→Capacity of the vehicle is exceeded; hence, is infeasible.

To obtain the sequence in which the locations will be served by the individual vehicles, Tables 8.25 through 8.28 show the route development for two vehicles. The procedure is the same as was explained earlier and, therefore, only the results are presented.

TABLE 8.25 Saving Matrix for Locations Assigned to Vehicle 1

Location	0	1	3	5	8	9	Demand	Supply	D − S
0	—								
1	33						2	1	1
3	61	82 12	—				2	6	−4
5	66	100 −1	29 98	—			5	3	2
8	115	138 10	76 100	45 136	—		3	4	−1
9	97	100 30	140 18	125 38	96 116	—	2	1	1

TABLE 8.26 Route Development for Vehicle 1

1 Saving	2 Between nodes	3 Possible chain	4 Added node	5 D	6 S	7 Cum. Demand	8 Cum. supply	9 Dif = C8 − C7, if Dif > 0	10 R = T − Dif − C7	11 Route no.	12 Chain
136	8–5	8–5	8	3	4	3	4	1	11		
			5	5	3	8	7	−1		R1	0–8–5–0
116	9–8	9–8–5	9	2	1	2	1	−1			
			8	3	4	5	5	0			
			5	5	3	10	8	−2		R1	0–9–8–5–0
100	8–3 NC										
98	5–3	9–8–5–3	3	2	6	12	14	2	1	R1	0–9–8–5–3–0
38	9–5 NC										
30	9–1	1–9–8–5–3	1	2	1	2	1	−1			
			9	2	1	4	2	−2			
			8	3	4	7	6	−1			
			5	5	3	12	9	−3			
			3	2	6	14	15	1	0	R1	0–1–9–8–5–3–0

TABLE 8.27 Savings Matrix for Locations Assigned to Vehicle 2

Location	0	2		4		10		7		6	D	S	$D-S$
0	—												
2	36	—									3	2	1
4	60	95	1	—							2	3	−1
10	80	113	3	40	100	—					3	6	−3
7	141	185	−8	110	91	144	77	—			4	1	3
6	152	154	34	165	47	205	27	110	183	—	1	3	−2

Vehicle 1

Final route for vehicle 1: 0–1–9–8–5–3–0
Total distance traveled: $33 + 100 + 96 + 45 + 29 + 61 = 364$

Vehicle 2

Final route for vehicle 2: 0–2–6–7–4–10–0
Total distance traveled: $36 + 154 + 110 + 110 + 40 + 80 = 530$
Total distance traveled by both vehicles: $364 + 530 = 894$

Both methods illustrated in Sections 8.7 and 8.8 can also be applied in conjunction with the sweep method, with minor modifications. We will not illustrate these here, but leave it to the reader for further analysis.

Also, it is fairly straightforward to develop a truck route if the trucks have different capacities. Because the procedure connects the customers with maximum savings first, it is profitable to fill the capacity of a larger truck first before starting to develop the tour for the next smaller-sized truck. If the trucks are ranked in descending order of their capacities, then we should develop a tour for each truck in the same sequence.

8.11 VARIABILITY IN THE DEMANDS

Oftentimes the demands from the customers are not constant and vary from day to day (period to period). How can we take care of such variability? How will the variability influence the touring paths? If the demands from all customers are known in advance, say in the morning of the service day, then new tours can be formed by using the available new data for the day using the procedures described earlier. However, if the demands are not known until the truck visits the customers, then it becomes a challenging issue.

For example, if there is variability in nodes connected in a tour as shown in Figure 8.7, with the mean demand as noted beside each node and with a truck capacity of 18 units, is it possible to judge the route the truck should take? O–4–3–2–1–O or O–1–2–3–4–O? If the distances are approximately proportional to

TABLE 8.28 Route Development for Vehicle 2

1 Saving	2 Between nodes	3 Possible chain	4 Added node	5 *D*	6 *S*	7 Cum. demand	8 Cum. Supply	9 Dif = C8 − C7, if Dif > 0	10 R = T − Dif − C7	11 Route no.	12 Chain
183	6–7	6–7	6	1	3	1	3	2	12		
			7	4	1	5	4	−1		R1	0–6–7–0
100	10–4	10–4	10	3	6	3	6	3	9		
			4	2	3	5	9	4	6	R2	0–10–4–0
91	7–4	R1 + R2				10	13	3	2	New R1	0–6–7–4–10–0
77	7–10 NC										
47	6–4 NC										
34	6–2	2–6–7–4–10	2	3	2	3	2	−1			
			6	1	3	4	5	1	10		
			7	4	1	8	6	−2			
			4	2	3	10	9	−1			
			10	3	6	13	15	2	0	R1	0–2–6–7–4–10–0

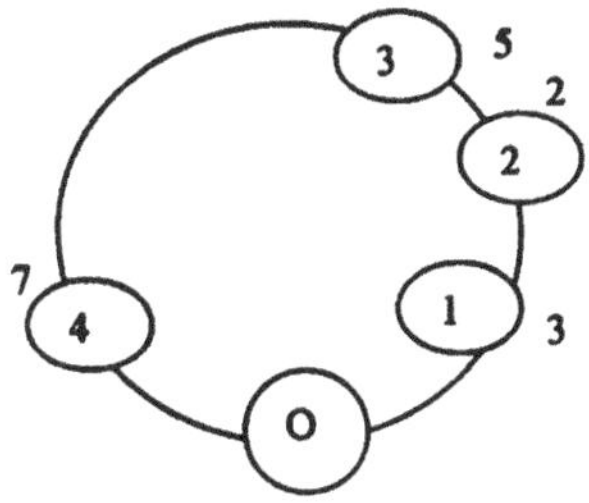

FIGURE 8.7 Travel routes.

the arc shown, then it is obvious that going counter-clockwise from O–1–2–3–4–O is preferable over going clockwise, O–4–3–2–1–O. This is because there is fair degree of certainty, assuming variations are not extraordinary, that one truck can pick up the loads from nodes 1, 2, and 3, and if full or cannot accommodate the normal load from 4, it can come back and we can make a separate trip to node 4 and back. On the other hand, if the route is 4–3–2–1, it is possible that we are at node 3 and find insufficient capacity to continue with the trip. Another truck will have to visit nodes 1 and 2 for comparatively small loads. With such variability in mind, the distance traveled by route O–1–2–3–4 would, in the long run, be smaller than O–4–3–2–1 route.

But such clear and observable path directional identification may not be possible every time. There are statistical alternatives we can follow that may give us sufficient confidence that on any trip most of the demand attached to that trip will be collected.

For example, suppose in Figure 8.7, after an interval of time we observe that the distribution of the demand from each customer follows a normal distribution with following parameters.

Customer	1	2	3	4
Mean	3	2	5	7
Variance	1	0.5	1.5	3.0

Thus the range of each customers demand may be expected to be between mean ±3 standard deviations for the customer.

The total mean demand of the path is the sum of mean demand of each customer which is $3 + 2 + 5 + 7 = 17$ and, similarly, the variance of the path is equal to the sum of the variances of the customers on the path, or in our case, it is $1 + 0.5 + 1.5 + 3 = 6$

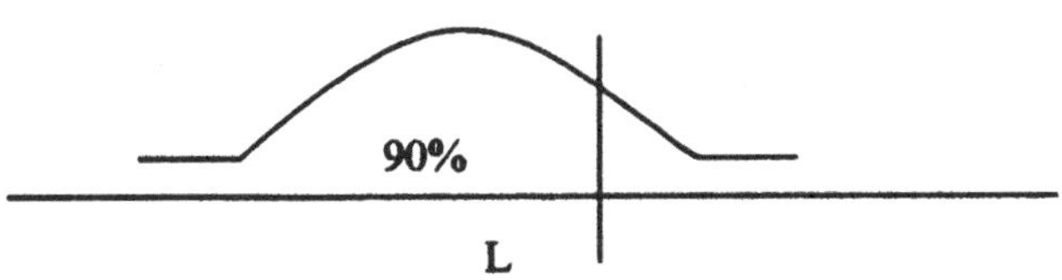

If we want to be 90% certain that a trip will pick up all loads and we are free to choose the capacity of the truck L, then we can apply the property of normal distribution to obtain,

$$Z = \frac{L - 17}{\sqrt{6}} = 1.282 \text{ (from normal table)}$$

and hence $L = 20.14$.

Similar calculations can be used in other problems. For example, given the truck capacity, we can determine the maximum load on the route we should have or the probability that we are able to pick all the loads from a given route.

Exercises

8.1 For a rural delivery, it is required to visit five locations B, C, D, E, and F that a post office located in location A, is serving. A postal van visits all these locations only once and returns back to A. Develop a route for the delivery system with minimum distance traveled. The following table displays the distance in miles between the locations.

	A	B	C	D	E	F
A	—	10	2	3	5	3
B	10	—	3	8	4	6
C	2	3	—	1	4	9
D	3	8	1	—	6	4
E	5	4	4	6	—	2
F	3	6	9	4	2	—

8.2 Consider a company that wants to start a new plant in one of the five available cities B, C, D, E, and F. The company wants to conduct a survey and forms a team for this purpose. The team wants to visit all cities, but only once. The table shows the distances (in hundred miles) between cities. The main plant is in city A. Develop the optimum route.

	A	B	C	D	E	F
A	—	9	12	18	14	18
B	9	—	15	8	16	20
C	12	15	—	8	22	17
D	18	8	8	—	16	15
E	14	16	22	16	—	16
F	18	20	17	15	16	—

8.3 Consider a plant that produces five different flavors of beverages. The flavors are produced and bottled at five different locations and stored at location

A, the warehouse, for further delivery. A vehicle is used to collect the bottles from the locations and bring them back to the warehouse. The required production level of different flavors is known and is given in the following table. The vehicle can carry no more than 1000 bottles at a time. The distances between the locations and the production from each machine are given in the following tables. Determine the optimum trips to make.

Distance Matrix (in 10s of meters)

	A	B	C	D	E	F
A	—	10	6	4	5	4
B	10	—	2	4	6	3
C	6	2	—	3	4	4
D	4	4	3	—	5	4
E	5	6	4	5	—	2
F	4	3	4	4	2	—

Production (in hundreds)

B	C	D	E	F
3	8	4	5	6

8.4 A company promises a free repair for the product in case of any damage within the first 6-month period. Usually there are complaints from five different locations in the city. The number of complaints for the present week from each location is known and is shown in the following table. A technician cannot repair more than ten products in a week. The distances (in miles) between locations are given in the next table.

Find the minimum number of repairmen needed to repair all the products and the routes they should take to minimize the total distance traveled.

	A	B	C	D	E	F
A	—	8	9	4	6	5
B	8	—	3	4	2	3
C	9	3	—	5	6	4
D	4	4	5	—	5	3
E	6	2	6	5	—	4
F	5	3	4	3	4	—

Location	B	C	D	E	F
Number of complaints	6	5	3	4	4

8.5 Consider a school van with a maximum carrying capacity of 15. There are eight stops from which children are to picked up and brought back to the school situated at location O. The distances between the locations are shown in the table. Find out the routes to minimize the total distance traveled.

Q	O									
7	2	A								
5	5	3	B							
4	6	5	4	C						
6	7	7	8	2	D					
7	6	6	5	4	3	E				
6	4	5	7	6	5	3	F			
5	5	8	6	7	5	5	4	G		
4	7	10	9	8	6	3	4	2	H	

8.6 Consider a company, which produces a spare part for a copier. There are eight well-known copier-manufacturing companies that use the part produced by company O. Every month company O supplies the part to these different companies as per their demand. Demands for the next month (in thousands of units) by these companies and the distances (in hundreds of miles) are given in the following table. It is necessary to find the routes to minimize the total distance traveled. The capacity of the truck used for the delivery of parts if 15,000.

D	O									
6	2	A								
4	5	3	B							
3	6	5	4	C						
8	7	7	8	2	D					
5	6	6	5	4	3	E				
9	4	5	7	6	5	3	F			
4	5	8	6	7	5	5	4	G		
6	7	10	9	8	6	3	4	2	H	

8.7 Consider a 12-location problem with the depot at (0,0). The coordinates and the demands of the other locations are given in the following table. The total capacity of the truck is 20 tons. Develop efficient tours, if no backtracking is permitted.

O	A	B	C	D	E	F	G	H	I	J	K	L
(0,0)	(−2, 3)	(−1, 6)	(−1, 8)	(1, 4)	(3, 6)	(5, 1)	(5, −5)	(3, −2)	(−2, −6)	(−6, −7)	(−2, −2)	(−3, −6)
	8	6	12	3	5	9	7	6	4	3	7	9

8.8 A cargo carrier will be used for shipping goods from a factory located at city O to 12 other cities. The cities are located in such a way that they encompass city O where the factory is located. The locations of the cities are given in terms of coordinates as shown, and the demand from each city is given in hundreds of tons. Determine the minimum number of possible routes so that the total distance covered by the carrier is minimum. The maximum capacity of the carrier is 2000 tons.

Cities	O	A	B	C	D	E	F	G	H	I	J	K	L
Location	(0, 0)	(−5, 6)	(−2, 0)	(−3, 6)	(1, 7)	(2, 1)	(6, 6)	(2, −4)	(4, −1)	(−3, −5)	(−3, −1)	(−5, −7)	(−2, −6)
Demand		4	7	10	5	3	6	9	4	8	2	6	7

8.9 A trucking firm, located in O, must deliver and pick up units from cities A, B, . . . , H. The demand d, from each city (units required at the city) and supply, S, from each city are as given in the following table, along with the distances between the cities. The trucking firm has vehicles with capacity of 15.

S	D	O								
7	6	2	A							
5	4	5	3	B						
2	3	6	5	4	C					
4	8	7	7	8	2	D				
8	5	6	6	5	4	3	E			
7	9	4	5	7	6	5	3	F		
5	4	5	8	6	7	5	5	4	G	
7	6	7	10	9	8	6	3	4	2	H

Develop tours if

a. Simultaneous loading and unloading is permitted.
b. If simultaneous loading and unloading is not permitted.

8.10 Solve problem 8.9, if the trucking firm has two 20-unit capacity trucks and remaining 10-unit capacity trucks.

8.11 How will you accommodate an uncertain requirements from each customer, if both demand and supply from each customer have a normal distribution with the same mean as in 8.9, and a variance that is equal to 1.

8.12 For the 4 locations given in the following table, using the distance matrix that corresponds to the cost of going on those routes, find the route with the minimum cost to serve all the locations with no location being visited more that once.

Use the Little's method.

Distance (Cost) Matrix

	A	B	C	D
A	0	5	4	1
B	3	0	8	9
C	1	7	0	2
D	4	5	6	0

8.13 For the solution obtained in the foregoing problem, if the loads on the 4 locations are known, apply the ETS procedure and find the most optimum route. The loads on the different locations are A = 5, B = 3, C = 2, D = 8.

8.14 For the four locations, a store manager has to find the most optimum route to serve all the locations using the two trucks with capacities of 7 and 9 available to him. Formulate the equations and solve them using an integer-programming algorithm with the aim of minimizing the total distance traveled to serve all the locations.

	0	1	2	3	4	Demand	Capacities
0	—						
1	5	—	6	8	2	8	Truck 1–7
2	1	9	—	1	1	7	
3	3	7	2	—	4	13	Truck 2–9
4	8	5	9	4	—	6	

8.15 A major meat-processing unit has to supply meat to eight different retailers. Each of the locations has a different demand. As the meat has to reach the retailers soon, the manufacturer must find the shortest route to serve all the locations with the vehicles available. The vehicles available have a capacity of 8. Find the shortest routes possible for all the locations to be served using the sweep algorithm.

Location	1	2	3	4	5	6	7	8
Demand	2	6	4	3	5	7	1	4
Position	(.0, 3.4)	(8.3, 3.6)	(−1.3, 2.9)	(−5.4, 6.2)	(−6.0, −3.0)	(−5.0, −0.7)	(5.3, −0.9)	(0.7, −4.3)

8.16 For the following ten locations and given supplies and demands for individual locations, using the algorithm for the simultaneous drop and pickup, find the minimum distance required to serve all the locations.

Assume that the vehicle has a capacity of 15.

Loc	0	1	2	3	4	5	6	7	8	9	10	*D*	*S*
0	—											—	—
1	36	—										0	2
2	93	43	—									2	4
3	68	28	44	—								3	5
4	63	83	89	43	—							4	3
5	18	71	118	41	35	—						5	6
6	111	165	200	98	42	77	—					1	4
7	120	142	212	127	69	55	20	—				5	3
8	6	34	104	38	72	25	99	95	—			2	7
9	26	74	165	92	136	59	151	101	37	—		3	7
10	58	53	125	81	151	95	183	153	54	44	—	1	2

8.17 Solve the foregoing problem with the aim to minimizing the number of vehicles. Compare the two solutions on the basis of the final cost, if it is given that each vehicle can travel a maximum of 50 units per day. The cost for renting the vehicle is $100, and the rate per unit of distance traveled is given as $0.60.

PART IV

Additional Quantitative Models

9

Dynamic Facility Locations

The facilities location and allocation procedures developed so far assume that the decisions are based on data presented in a single or a short term period and that this is sufficient. Thus, the assumption is that the data do not change with time or, if they do, the change is not very significant and will not influence the decisions. For most problems this is a reasonable assumption because the objective is to solve the immediate problem, and future data changes are either difficult to predict or are insignificant, or all location data change in about the same proportion relative to time, making relative data fairly stable. On other hand, some changes over a time period could be significant. For instance, a demand from a customer may change as a result of a trend, or the cost of delivery may change as the transportation mode changes (e.g., trucks vs. trains). Even the cost of operation may vary from period to period in each location with changes in economic conditions that affect the regional operating and wage structures. The problem is then to plan a schedule of services optimally for a predetermined time period. Over a long time location decision that is optimum in one period may or may not be optimum in the following period. This is especially true when there are limitations on the number and capacities of facilities that we may have. This type of problem is best handled by dynamic programming.

9.1 DYNAMIC PROGRAMMING

Dynamic programming is a mathematical technique for optimizing a multistage decision process. A problem requiring n decisions simultaneously is broken down into n sequential problems requiring only one decision at a time. It is much easier to work with one variable at a time than to manipulate n variables simultaneously. The basic principles of dynamic programming were illustrated in Section 7.1 and a reader needing a quick overview may refer back to this section.

9.2 TRANSPORTATION ALGORITHM

The initial data analysis may require use of transportation algorithm, and the reader familiar with transportation methods may proceed directly to Section 9.3; however, in this section we briefly review the transportation algorithm.

Transportation problems are well known in operations research and business areas, where they have many uses. The range of applications varies from production and financial planning to personnel staffing and transportation activities. There are numerous solution procedures available that are well documented in many standard text books in quantitative methods and operations research. In this chapter, however, we will illustrate the procedure that we believe is the easiest and quickest to follow. In the following section, we present the mathematical formulation of the transportation problem, and the alternative procedure will be demonstrated next.

9.2.1 Mathematical Formulation

If we define x_{ij} as the quantity shipped from the facility j to the customer i and c_{ij} as the cost of shipping 1 unit between these two and, furthermore, denote a_i as the total demand from the customer i $(i = 1, 2, \ldots, n)$, and b_j as the capacity of warehouse j $(j = 1, 2, \ldots, m)$, then the problem can be formulated as follows:

$$\text{Min} \sum_{j=1}^{b} \sum_{i=1}^{n} c_{ij} x_{ij}$$

subject to

$$\sum_{j=1}^{m} x_{ij} = a_i \qquad \text{(demand requirement of customer } i\text{) for } i = 1, 2, \ldots, n$$

$$\sum_{i=1}^{n} x_{ij} = b_j \qquad \text{(supply restriction of warehouse } j\text{) for } j = 1, 2, \ldots, n$$

The problem can be solved mathematically by application of linear programming; however, an efficient special technique called the transportation technique is illustrated next by means of an example.

Sample Problem

Consider a situation, illustrated in Figure 9.1 in which a service facility with a capacity of 250 units is available at each of three sites and we are faced with the problem of assigning four customers, each with a demand as shown in Table 9.1, to receive service from these facilities. The table also shows the unit cost of transportation between each customer and each of the facilities in the right-hand corner of the corresponding cell. The objective is to assign the customers to the facilities to minimize the total cost of transportation.

Note: A unique feature of the transportation problem is that the total demand from all customers is equal to the total capacity of the system (i.e., the sum of the capacities of all the facilities). What if this were not true? A modification to the problem, that of adding a "dummy" customer or a "dummy" facility would allow us to satisfy this requirement. We will discuss this concept again later in the chapter.

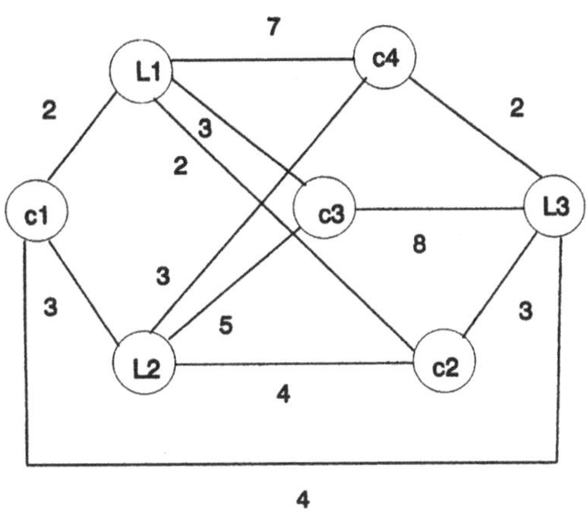

FIGURE 9.1 Network of customers and locations and their connecting paths.

TABLE 9.1 Transportation Cost and Demand

	Location			
Customer	1	2	3	Demand
1	2	3	4	150
2	2	4	3	200
3	3	5	8	100
4	7	3	2	150
Capacity	250	250	250	750

To obtain an optimum solution we apply the transportation algorithm, which is divided into two phases: (1) the initial phase, where a basic feasible solution is found, and (2) the improvement phase, where the final solution is determined.

Initial Phase, Basic Feasible Solution. There are three generally accepted methods for determining the initial solution. They are (1) the Northwest Corner Rule, (2) the Least-Cost Assignment, and (3) Vogel's Approximation. As Vogel's Approximation is quick and usually results in a starting solution which, if not the best, is very close to the optimum solution, we will demonstrate it, rather than either of the other two methods.

For convenience, we will show the number of units transported from a facility to a location by assigning the value to a "cell," designated by the row associated with the customer and the column associated with the facility. For example, if cell 2,3 has 100 units assigned, it means that 100 units are to be delivered to customer 2 from the facility in location 3.

The steps involved in Vogel's Approximation are as follows:

1. Calculate the "difference element" for each available row and column: these elements are the difference between the minimum cost and the next lowest cost for each row and column that may have unfilled demand.
2. Determine the maximum difference element; that is, the difference element with the highest value. If there is a tie, break it arbitrarily. Also note the row or column associated with the maximum difference.
3. Go to the row or column found in step 2 and assign as much demand as possible to the location with the least-cost element. The limitation may occur as a result of either the demand (row total) or the available capacity (column total).
4. Scratch each row (customer) or column (facility) as it is filled (reached its limit), from any further consideration in this phase. If all the customers' demands have been assigned, proceed to step 5; otherwise, return to step 1.
5. This is the initial solution.

Let us apply the Vogel's Approximation to the data in Table 9.1 by employing a series of demand assignment tables. In the first, Table 9.2, no demand has actually been assigned as yet; the table simply presents the initial data and lists the difference elements for each row and each column.

In row 1 of Table 9.2, the minimum cost element is 2, and the next smallest element is 3; the difference is $3 - 2 = 1$, as indicated for that row under "Diff elem." Similarly, for column 1, the two lowest-cost elements and both 2, resulting in a net difference of zero, shown in the "Diff elem" row for column 1.

TABLE 9.2 Demand Assignment 1

	Location					
Customer	1	2	3	Remain demand	Total demand	Diff elem
1	2	3	4	150	150	1
2	2	4	3	200	200	1
3	3	5	8	100	100	2
4	7	3	2	150	150	1
Available capacity	250	250	250			
Total capacity	250	250	250		750	
Diff elem	0	0	1			

The maximum of all the difference is 2, found in row 3. The minimum cost element in this row is 3, and we will assign as much demand as possible to this cell. The table also shows for each customer and each location, the remaining demand and the available capacity, respectively. As a check, the total of all the partial demand assignments plus the demand yet to be assigned must be equal to the total demand from the customer. Similarly, in each column, the capacity used plus that still available must be equal to the total capacity of the particular location. In Table 9.2 we see that the demand of customer 3 that is still to be assigned is 100 units, and the capacity still available at location 1 is 250 units. The lesser of these two is 100 units; therefore, that amount is assigned to cell 3,1. Because this assignment satisfies the requirement of customer 3, he is scratched in this phase. Table 9.3 indicates this by removing all cost element for customer 3 and showing the 100 units in the first column. The table also shows the difference elements for the rows and columns still being considered.

TABLE 9.3 Demand Assignment 2

	Location					
Customer	1	2	3	Remain demand	Total demand	Diff elem
1	2	3	4	150	150	1
2	2	4	3	200	200	1
3	3					
	100			0	100	–
4	7	3	2	300	300	1
Avail cap	150	250	250	650		
Total cap	250	250	250		750	
Diff elem	0	0	1			

The new maximum difference is 1, and it appears in four places. Choose any corresponding row or column; for instance, row 1. The minimum cost element there is 2. Assign as much as possible to the cell; namely, 150 units, satisfying both row 1 and column 1, simultaneously. Table 9.4 updates Table 9.3 in the same manner as the latter modified Table 9.2.

Following the same procedure, the final demand assignment is given in Table 9.5, with the total cost of $2050. This cost is the sum of individual cost obtained by multiplying each cell cost with the corresponding assignment. It should be noted that different entries are possible (giving multiple solutions) depending on how a tie for an entry was broken; however, the total cost of each solution would be $2050.

TABLE 9.4 Demand Assignment 3

	Location					
Customer	1	2	3	Remain demand	Total demand	Diff elem
1	2	–	–			
	150			0	150	–
2	–	4	3	200	200	0
3	3	–	–			
	100			0	100	–
4	–	3	2			
				150	150	1
Avail cap	0	250	250			
Total cap	250	250	250		750	
Diff elem	–	0	1			

TABLE 9.5 Final Demand Assignment

	Location				
Customer	1	2	3	Remain demand	Total demand
1	2	3	4		
	150			0	150
2	–	4	3		
		200		0	200
3	3	–	–		
	100			0	100
4	–	3	2		
		50	250	0	300
Avail cap	0	0	0	0	
Total cap	250	250	250		750

Improvement Phase, Final Solution. The first approach we will follow in seeking improvement is called the "Stepping Stone Method." Improvement can occur only if units can be moved to an empty cell (one without any assignment) that will reduce the overall cost. For such a check, we must find the overall effect of moving a unit into an empty cell. For example, consider the cell in row 2, column 3. Moving 1 unit in cell 2,3 will require removing 1 unit presently assigned to cell 2,2 from that cell to keep the row total the same. But it will mean adding 1 unit to cell 4,2 to keep the column total the same. That move however, will change the row 4 total; therefore, 1 unit must be taken away from column 3 (i.e., cell 4,3). Because 1 unit was originally added in column 3, the net change in the number of units assigned per row and column is zero. Table 9.6 graphically demonstrates the steps from stone to stone (cell to cell) that we just discussed. The +1 or −1 in the lower right-hand corner of a cell indicates that 1 unit will be moved into or out of the cell, respectively.

Now let us determine the effect of such a move. Adding 1 unit in cell 2,3 will increase the cost by $3, whereas removing 1 unit from cell 2,2 will decrease the cost by $4. Similarly, a unit in cell 4,2 will add $3, and taking a unit away from cell 4,3 will decrease the cost by $2. The net effect is $3-4+3-2=\$0$ (i.e., there is no increase or decrease in the cost because of this move). A similar check must be made for each of the unfilled cells to see if any improvement is possible.

This example also illustrates another important point. If we trace the "loop" that has been formed in making the moves, we will see it is a closed loop. The loop consists of only one "empty" cell, one under examination, and all others are the filled cells. The direction of the path is changed only at a filled cell, and then only at a 90-degree angle. In the path, cells are alternatively positive and negative (i.e., a unit is added, and then one is subtracted in alternate cells).

TABLE 9.6 Stepping Stone Loop 1

	Location			
Customer	1	2	3	Demand
1	2 150	3	4	150
2	2	3 200 −1	3 +1	200
3	3 100	5	8	100
4	7	3 50 +1	2 250 −1	300
Capacity	250	250	250	750

If there are savings, which are in terms of per unit moved, then we want to move as many units as possible. Looking throughout the loop, we see that moving 200 units away from cell 2,2 makes the assignment in that cell 2,3, while moving 250 units from cell 4,3 will reduce the assignment there to zero. Other cells in the loop, namely cells 2,3 and 4,2, increase in their assignments as the units are moved in the loop. Our objective is to move as much as possible. If we move 250 units, that will leave −50 units in cell 2,2, but a cell cannot have negative assignment because it would mean transporting from a customer to a facility. That is not possible; hence, the maximum amount that can be moved is 200 units, the smaller of the 250 and 200 units associated with each of the "negative cells" (cells in the loop with "−1").

Can there be multiple paths to the same empty cell? The answer is, no. We must, however, have exactly $m + n - 1$ assignments in proper positions; where m is the number of customers and n is the number of facilities. In our example we should have $4 + 3 - 1 = 6$ assignments. In Table 9.6 we see that we have only five filled cells, or only five assignments. In this case, we will not be able to form a loop for certain cells, for example, cell 2,1. An additional assignment must be created in a "proper cell," the value of which must be zero so that the cost of the solution is not changed. A "proper cell" is any cell that, when filled, will allow a path to form to a cell presently not in any loop (e.g., cell 2,1 or cell 1,2). It is not difficult to recognize proper cells; however, if there is still a doubt, the improvement routine called the "U–V Method" will suggest a way of recognizing proper cells.

The U–V Method leads to the optimum solution by simplifying the selection of the cells that would improve on the initial solution. The improvements are in stages; therefore, this is an interative process.

For each row and for each column, we create a measuring index called the "shadow price." We can denote the shadow price for the ith row by U_i, and the shadow price for the jth column by V_j. The shadow prices are calculated by using the expression $C_{ij} - U_i - V_j = 0$ for each filled cell, where C_{ij} is the unit cost of cell i, j. There are $m + n$ total values of shadow prices; however, only $m + n - 1$ assignments are in our table. Essentially, it means we have $m + n$ variables, but only $m + n - 1$ independent equations; therefore, we can choose a value for one of the variables and calculate all others in terms of that first value. It is most convenient to select a zero value for either one of the U_i or of the V_j. If at any point the next value of U_i or V_j cannot be found, then we do not have $m + n - 1$ assignment, and an additional assignment is created so that the process of calculating U_i and V_j can continue.

In our example, Table 9.7 shows the values of shadow prices starting with $U_i = 0$.

For an empty cell, $C_{ij} - U_i - V_j$ indicates the cost of moving a unit in that cell. To learn whether the present solution is optimal (best), we must check each of the empty cells and see if the condition, $C_{ij} - U_i - V_j > 0$, is satisfied for

TABLE 9.7 U–V Shadow Prices

	Location				
Customer	1	2	3	Demand	U_i
1	2	3	4		
	150			150	0
2	2	4	3		
	0	200		200	0
3	3	5	8		
	100			100	1
4	7	3	2		
		50	250	300	−1
Capacity	250	250	250	750	
V_j	2	4	3		

each. For example, for cell 1,2, $C_{ij} - U_i - V_j = C_{12} - U_1 - V_2 = 3 - 0 - 4 = -1$. That is, if we move one unit to cell 1,2, it will save us \$1.

Let us move as many units as possible in cell 1,3. The results are shown in Table 9.8. Because the assignments are changed, we must recalculate all U_i and V_j values. Table 9.8 shows the result again starting with $U_i = 0$.

All of the empty cells have volume $+C_{ij} - U_i - V_j$ either positive or zero, indicating no further improvement is possible and the assignments in Table 9.8 are also optimum assignments. The corresponding cost is $150 * 3 + 150 * 2 + 50 * 6 + 100 * 3 + 50 * 3 + 250 * 2 = \$1800$.

TABLE 9.8 Final Solution

	Location				
Customer	1	2	3	Demand	U_i
1	2	3	4		
		150		150	0
2	2	4	3		
	150	50		200	1
3	3	5	8		
	100			100	2
4	7	3	2		
		50	250	300	0
Capacity	250	250	250		
V_j	1	3	2		

Now let us take up the question of total demand not being equal to total supply. As mentioned earlier, this requires the introduction of a dummy customer or a dummy supplier. If the total supply is greater than total demand, we obviously have excess capacity, which will not be shipped because it is not needed. But in transportation problems, all units from the warehouses must be transported to the customers. For calculation purposes, it is then necessary to "ship" the excess units at no cost; consequently, a dummy customer is created. The cost of shipping a unit from any warehouse to this customer is zero, and the demand from this customer is equal to the excess capacity (i.e., the total capacity then equals the total demand).

Similarly, if the demand exceeds the supply, a column is created that indicates a dummy variable supply source. The excess demand is satisfied by this supplier, and the cost of transportation is again marked as zero because this addition is only made so that the transportation method can be applied. In reality, this supplier is not a part of the actual solution, because the corresponding units are never shipped.

9.3 MULTIPERIOD FACILITY LOCATION PLANNING

Consider a typical problem in dynamic facility allocation. There are six customers who are to be served from facilities installed in two of the four locations. We are planning for four time periods (years). The data for costs and demands for each period are given in Tables 9.9 through 9.12. A facility can be moved from one location to another at the end of any time period; however, it involves a changeover cost, as shown in Table 9.13.

Each facility has the capacity to serve 700 units of demand. Determine the facility locations and customer assignments in each period to minimize the total cost of operation for 4 years.

TABLE 9.9 Year 1 Data

	Location				
Customer	1	2	3	4	Demand
1	7	6	5	4	70
2	6	7	8	4	80
3	3	2	4	6	100
4	2	3	5	2	50
5	5	7	8	10	150
6	4	7	4	9	200
Fixed cost of operation in the year	400	450	300	200	

TABLE 9.10 Year 2 Data

Customer	Location 1	2	3	4	Demand
1	7	6	5	4	80
2	6	3	8	5	200
3	3	2	4	6	90
4	2	3	5	2	80
5	8	7	8	10	150
6	4	7	4	9	200
Fixed cost of operation in the year	400	450	300	200	

TABLE 9.11 Year 3 Data

Customer	Location 1	2	3	4	Demand
1	8	7	6	5	100
2	7	3	9	5	300
3	4	3	5	7	110
4	3	4	6	3	80
5	6	8	9	11	270
6	2	8	5	10	220
Fixed cost of operation in the year	450	500	350	250	

TABLE 9.12 Year 4 Data

Customer	Location 1	2	3	4	Demand
1	8	7	6	5	120
2	7	5	9	5	350
3	4	3	5	7	130
4	3	4	6	3	90
5	6	8	9	11	310
6	2	8	5	10	200
Fixed cost of operation in the year	450	500	350	250	

TABLE 9.13 Changeover Cost

	Location			
Location	1	2	3	4
1	0	70	50	30
2	70	0	80	9
3	50	80	0	55
4	30	9	55	0

9.3.1 Data Analysis

As the first step, determine for each possible location sets in each time period, the best possible customer assignments. The number of such sets depends on the available locations and the number of facilities needing placement. There are C_n^m distinct combinations that can be formed, where m is the number of locations and n is the number of facilities. In each combination, we must assign customers in a manner that will result in a feasible solution with minimum cost.

In our example, there are four locations in which two facilities are to be placed, resulting in C_2^4 possible combinations: namely, (1,2), (1,3), (1,4), (2,3), (2,4), and (3,4). Select one combination, for instance, combination 1,2 for period 1. This means that we have one facility placed in location 1 and another in location 2. To determine the least cost for this combination we apply the least-cost rule, assigning customers to a location that has the least cost of such assignment. This results in the assignments as shown in Table 9.14.

Customers 2, 4, 5, and 6 will be serviced by location 1, whereas customers 1 and 3 will be serviced by location 2. The total demand assigned to location 1 is $80 + 50 + 150 + 200 = 480$ units, while the demand on location 2 is $70 + 100 = 170$ units. Both assignments are within the capacity of each facility; therefore, it is an acceptable solution. The resulting variable cost is

TABLE 9.14 Least-Cost Rule, Combination 1,2

	Location		
Customer	1	2	Demand
1	7	6	70
2	6	7	80
3	3	2	100
4	2	3	50
5	5	7	150
6	4	7	200
Fixed cost	400	450	

$70(6) + 80(6) + 100(2) + 50(2) + 150(5) + 200(4) = 2750$, to which we must add fixed cost for location 1 and 2; namely 400 and 450, to obtain the value of the solution, which is \$3600.

In some cases the demand assignment to a location may exceed the capacity of the facility. If that is so, the transportation algorithm is applied to find the solution. To illustrate, consider combination 2,3 for period 3. Table 9.15 shows the application of the least-cost rule.

The total demand assigned to location 2 is $300 + 110 + 80 + 270 = 760$ units, which exceeds the 700-unit capacity of the facility. The total demand from all customers in the system is 1080 units; whereas two facilities, each with 700-units capacity, can provide 1400 units. To apply transportation method, the total demand must be equal to the total capacity; therefore, a dummy customer must be created with the demand of $1400 - 1080 = 320$ units. The customer will have no

TABLE 9.15 Least-Cost Rule Application to Locations 2–3, Period 3

	Location		
Customer	2	3	Demand
1	7	6	100
2	3	9	300
3	3	5	110
4	4	6	80
5	8	9	270
6	8	5	220
Fixed cost	500	350	

TABLE 9.16 Transportation Algorithm Assignments and Costs

	Location			
Customer	2	3	Demand	Cost
1		100	100	600
2	300		300	900
3	110		110	330
4	80		80	320
5	210	50	270	2,220
6		220	220	1,100
7 (dummy)		320	320	0
Total	700	700	1,400	5,470

cost for transportation. Continuing with the transportation algorithm, the final solution is given in the Table 9.16, with a cost of \$5470.

The fixed cost of \$850 for the two locations is added to total transportation charge, giving us the total of \$6320. It can be noted that the location 2 has a total assignment of 700 units, and location 3 has an assignment of 380 units, excluding that of the dummy customer. Similar analyses are performed on the other combination for each year. Table 9.17 shows the minimum cost associated with a feasible solution in each period for each combination.

The next step is to develop changeover cost table. This table shows the minimum cost of moving facilities that are in one combination to another combination. For example, the changeover cost for 1–2 involves the following. In combination 1, the facilities are in location 1 and 2; in combination 2, the facilities are in location 1 and 3. Thus, the facility in location 1 is not moved, whereas facility in location 2 is moved to 3, the changeover cost is then $0 + 80 = \$80$.

In some other changes, the changeover cost are not that obvious. For example, going from locations 1–2 to locations 3–4, involves a decision on which facility should be moved from which location to where. For example, the facility in location 1 can go either to location 3 or 4. The decision must be to find which is most economical move.

The answer is obtained by applying to the assignment the algorithm for fixed cost. Complete description of assignment algorithm is provided in Chapter 10. At this point it is sufficient to note that for a 2×2 table we could also obtain the best policy by simply adding the diagonals and selecting the least value.

Changeover costs for our example are shown in Table 9.18.

With the minimum cost to move a facility from location 1 to 3 and another from location 2 to location 4, as shown in Table 9.19, with a total changeover cost of \$59.

Similar analysis for all other changeovers are performed. The resulting minimum changeover costs are shown in Table 9.20.

TABLE 9.17 Minimum Cost for Each Combination

	Periods			
Combination	1	2	3	4
1–2	3,600	4,120	5,180	6,460
1–3	3,480	4,730	6,690	7,650
1–4	3,150	4,350	5,540	6,100
2–3	3,860	4,020	6,320	7,730
2–4	4,000	4,360	7,040	7,840
3–4	3,600	4,340	6,920	7,660

TABLE 9.18 Location Changeover Cost Data

To/from	3	4
1	50	30
2	80	9

TABLE 9.19 Final Changeover Assignment from 1–2 to 3–4

From/to	3	4
1	x	
2		x

TABLE 9.20 Minimum Changeover Costs (Combination)

From/to	1–2	1–3	1–4	2–3	2–4	3–4
1–2	0	80	9	50	30	59
1–3	80	0	55	70	110	30
1–4	9	55	0	59	70	50
2–3	50	70	59	0	55	9
2–4	30	110	70	55	0	80
3–4	59	30	50	9	80	0

9.3.2 Application of Dynamic Programming

Continuing with our primary objectives of determining the locations for facilities and the assignment of customer in each year. We apply dynamic programming and break the problem into four stages, with each year being a separate stage. We will proceed with applying the backward pass, starting with the fourth year.

The cost of operation for different combination in year 4 is given in Table 9.21. Because for each combination there is no other alternative, the listed cost is the minimum cost.

Moving one stage back to year 3, we find the total cost of operation (both years 3 and 4) given in Table 9.22. For example, if we are in combination 1–4 and decide going to combination 1–3 for year 4, the total cost of operation would be the cost of operation in year 3 with combination 1–4, plus the changeover from 1–4 to 1–3 plus cost of operation in year 4 with combination 1–3 (i.e., 5540 +

TABLE 9.21 Year 4 Cost

Combination	1–2	1–3	1–4	2–3	2–4	3–4
Cost	6,460	7,650	6,100	7,730	7,840	7,660

TABLE 9.22 Year 3 Total Cost (Years 3 and 4)

To/from (beginning of year 3)	1–2	1–3	1–4	2–3	2–4	3–4	Min	Dec
1–2	11,640	12,910	11,289	12,960	13,050	12,889	11,289	1–4
1–3	13,230	14,340	12,845	14,490	14,640	14,380	12,845	1–4
1–4	12,009	13,270	11,640	13,329	13,450	13,250	11,640	1–4
2–3	12,830	14,080	12,479	14,050	14,215	13,989	12,479	1–4
2–4	13,530	14,800	13,210	14,825	14,880	14,780	13,210	1–4
3–4	13,439	14,600	13,070	14,659	14,840	14,580	13,070	1–4

$80 + 7650 = 13,270$). For each incoming combination, the minimum cost alternatives is selected from the table and the corresponding combination to go the next stage is shown as the decision variable. For example, for incoming combination 1–2, in year 3, the minimum cost is \$11,289, associated with the decision of going to combination 1–4 in year 4.

The total cost for year 2 is given in the Table 9.23. Again, the cost is the sum of cost and operation in year 2 plus the changeover cost and the cost of operation from there on (years 3 and 4), obtained from Table 9.22. For example, if we have facilities in the locations associated with combination 1–3, the cost for year 2 is \$4730. If we plan to go to combination 1–4 in year 3, the minimum changeover cost is 55. The optimal cost of continuing in year 3 on is \$11,640, for a total cost of $11640 + 4730 + 55 = \$16,425$, an entry shown for cell 2–3 in Table 9.23.

TABLE 9.23 Year 2 Total Cost (Years 2, 3, and 4)

To/from (beginning of year 2)	1–2	1–3	1–4	2–3	2–4	3–4	Min	Dec
1–2	15,409	17,045	15,769	16,649	17,360	17,249	15,409	1–2
1–3	16,099	17,575	16,425	17,279	18,050	17,830	16,099	1–2
1–4	15,648	17,250	15,990	16,888	17,630	17,470	15,648	1–2
2–3	15,359	16,935	15,719	16,499	17,285	17,099	15,359	1–2
2–4	15,708	17,315	16,070	16,894	17,570	17,510	15,708	1–2
3–4	15,688	17,215	16,030	16,878	17,630	17,410	15,688	1–2

TABLE 9.24 Year 1 Total Cost (Years 1, 2, 3, and 4)

To/from (beginning of year 1)	1–2	1–3	1–4	2–3	2–4	3–4	Min	Dec
1–2	19,009	19,779	19,257	19,009	19,338	19,347	19,009	1–2 or 2–3
1–3	18,969	19,579	19,183	18,909	19,298	19,198	18,909	2–3
1–4	18,568	19,304	18,798	18,568	18,928	18,888	18,568	1–2 or 2–3
2–3	19,319	20,024	19,567	19,219	19,623	19,557	19,219	2–3
2–4	19,349	20,209	19,718	19,414	19,708	19,768	19,414	2–3
3–4	19,068	19,729	19,298	19,257	19,388	19,288	19,068	1–2

Applying the same procedure for the first year results in the figures listed in Table 9.24.

The overall minimum cost is 18,568. Tracing backward from Table 9.20 through Table 9.17 results in the following optimum policy.

Year	Combination
1	1–4
2	1–2 or 2–3
3	1–2
4	1–4

The dynamic programming application results in development of optimum strategy. The number of calculations involved are reduced considerably from exhaustive enumeration. In our example, there are six combination each year. Had we chosen to examine each route (a route is a path consisting of one combination from each year), there would have been $6 * 6 * 6 * 6 = 1296$ routes to examine. By applying dynamic programming, we reduced this number to, at most, $6 + 6 + 6 + 6 = 24$ paths. Even this number could be quite large for a problem involving combinations. A procedure is suggested in the next section that will provide a good solution.

9.4 REDUCTION TECHNIQUE

9.4.1 No Changeover

A very quick estimate of one value of the operation cost for all four periods could be obtained by assuming there is no changeover taking place in any period. This may also lead us to the optimum cost when the changeover cost is very high, which prevents or minimizes any changeover. Table 9.17 is reprinted for convenience as Table 9.25 to show the costs for each combination in our problem, these combinations have also been numbered for each of display.

Combination 6 can be seen to have the minimum total cost of $18,614.

9.4.2 With Changeover

The next step of the procedure is to select a certain number of states (combinations) in each period and consider only them in our dynamic-programming application. This will permit changeover from period to period. We must decide which states to select. As a rule, we should rank the combination in each period in the ascending order of their costs and select from these as many as is practical; at the same time we should be careful not to enlarge the states in each year to an unmanageable level.

For instance, if we restrict ourselves to selecting only three combinations in each period, we will have reduced the problem (Table 9.25) to the combinations displayed in Table 9.26 for our considerations.

TABLE 9.25 Minimum Cost for Each Combination in Each Period

		Periods				
Combination	Comb Number	1	2	3	4	Total cost
1–2	1	3,600	4,120	5,180	6,460	19,360
1–3	2	3,480	4,730	6,690	7,650	22,550
1–4	3	3,150	4,350	5,540	6,100	19,140
2–3	4	3,860	4,020	6,320	7,730	21,930
2–4	5	4,000	4,360	7,040	7,840	23,240
3–4	6	3,600	4,340	6,920	7,660	18,614

TABLE 9.26 Three Combinations with Ascending Order of Cost in Each Period

Periods							
1		2		3		4	
Comb	Cost	Comb	Cost	Comb	Cost	Comb	Cost
3	3,150	4	4,020	1	5,180	3	6,100
2	3,480	1	4,120	3	5,540	1	6,460
1	3,600	6	4,340	4	6,320	2	7,650

The dynamic-programming application is shown in Tables 9.27 through 9.30. The calculations are similar to those for the tables developed in the previous analysis; therefore, no explanation of the individual tables is given here. In each table, only the states that are selected previously in Table 9.26 are permitted.

The cost in each table includes the cost of preparation for the period in the state we are presently in plus the cost of changeover into the next time period to the state we could be in, plus the optimum cost of operation from then to the end of planning period.

The minimum overall cost from Table 9.30 is $18,568. Tracing the solution backward from there through Table 9.23 shows the following optimum selection.

TABLE 9.27 Year 4 Cost

Combination	Cost
*3	6,100
1	6,460
2	7,650

TABLE 9.28 Year 3 Cost

To/from	3	1	2	Min	Dec
*1	11,289	11,640	12,910	11,289	3
3	11,640	12,009	13,245	11,640	1
4	12,479	12,830	14,040	12,479	3

TABLE 9.29 Year 2 Cost

To/from	1	3	4	Min	Dec
*4	15,359	15,719	16,499	15,359	1
*1	15,409	15,769	16,649	15,409	1
6	15,688	16,030	16,828	15,688	1

TABLE 9.30 Year 1 Cost

To/from	4	1	6	Min	Dec
*3	18,568	18,568	19,230	18,568	4,1
2	18,909	18,969	19,540	18,909	4
1	19,009	19,009	19,689	19,009	4,1

Year	Combination
1	3
2	4 or 1
3	1
4	3

Incidentally, this solution is the same as we obtained in the previous section.

9.5 OPTIMUM TIME FOR LOCATION CHANGE FOR A FIXED PERIOD, TIME-DEPENDENT PROBLEM

Consider the following problem. We have four locations each with x, y coordinates as shown in Table 9.31. Each location generates a demand for a new facility that is expected to change relative to time. The demand functions are linear and are given by W.

We wish to determine the optimum location for a service facility for an operational time period of t units. Suppose in our problem $t = 10$.

9.5.1 No Change Allowed

If a facility is to be placed in a location permanently to operate over t time units, what should be the location for the facility?

We need to determine the optimum location that minimizes the cost over the planning duration. Start by examining the cost if the facility is placed at the best location at time $t = 0$. This location is determined by applying 50th percentile method (the details are as in Chapter 3). Ranked by ascending x-values, we have the following: For the x value the calculations are Table 9.32.

The 50th percentile is $38/2 = 19$. The associated x-coordinate is 5.

Similarly in ascending order of y-coordinates, we have y value calculation in Table 9.33.

TABLE 9.31 Data for Time-Dependent Demand Problem

Point number	x	y	W_i
1	1	3	$2 + 3t$
2	5	8	$20 - 2t$
3	7	9	$4 + 2t$
4	0	15	$12 - t$

TABLE 9.32 Optimum x Value Determination

x	Point number	W_i	Cumulative W_i
0	4	12	12
1	1	2	14
5	2	20	34
7	3	4	38

TABLE 9.33 Optimum y Value Determination

y	Point number	W_i	Cumulative W_i
3	1	2	2
8	2	20	22
9	3	4	26
15	4	12	38

The 50th percentile y-coordinate value is 8. Therefore, the optimum location for a facility at time $t = 0$ is $x^{(1)}, y^{(1)} = (5, 8)$.

Also determine the optimum location for $t = 10$, using the 50th percentile method again. This time apply it to the data associated with time $t = 10$. If the optimum locations at $t = 0$ and $t = 10$ are the same, then there is no other location that needs examination. This is because the linear changes in data over time assures us that the same location will remain optimum throughout the time period. If this is not so we will have to examine a few alternative locations as explained later.

For $t = 10$ the data are in Table 9.34, and the x-coordinate calculations are shown in Table 9.35. The 50th percentile value is 29, corresponding to an x coordinate of 1.

Similarly, the y coordinate is given by Table 9.36.

We obtain the y coordinate as 3. Hence, the optimum location of the facility at time $t = 10$ is $(x^{(2)}, y^{(2)}) = (1, 3)$.

TABLE 9.34 Weight Values at $t = 10$

Number	x	y	W_i
1	1	3	$2 + 3(10) = 32$
2	5	8	$20 - 2(10) = 0$
3	7	9	$4 + 2(10) = 24$
4	0	15	$12 - 10 = 2$

TABLE 9.35 Optimum x Value Calculations

x	Customer number	W_i	Cumulative W_i
0	4	2	2
1	1	32	34
5	2	0	34
7	3	24	58

TABLE 9.36 Optimum y Value Determination

y	Customer number	W_i	Cumulative W_i
3	1	32	32
8	2	0	32
9	3	24	56
15	4	2	58

The change in optimum position for the location from $t = 0$, (5,8) to $t = 10$, (1,3) signifies that there might be other location besides these two where cost could be minimum. Based on 50th percentile method, the optimum location will change only if the demand changes, so that the 50th percentile is no longer with the old coordinate. Because the demands are changing relative to time, we can determine the time at which the 50th percentile changes significantly to change the optimum location. For example, the x-coordinate goes from 5 to 1. The time at which this shift is possible is when total demand from the points above x coordinate 1 is equal to the total demands from the points consisting of an x-coordinate of 1 and below (50th percentile). To determine the time at which the shift takes place, set the cumulative demands or W_i equations, higher than the 50th percentile point of $x = 1$ equal to the W_i equations after the 50th percentile, and solve for t. Thus, we obtain an equation as follows:

$$20 - 2t + 4 + 2t = 2 + 3t + 12 - t$$

which gives $t = 5$.

Similarly, for y-coordinates, we have demands from points with a y-coordinate of 3 and below that are equated to demands from a y-coordinate of 8 and higher.

$$2 + 3t = 20 - 2t + 4 + 2t + 12 - t \qquad \text{or} \qquad t = 8.5.$$

Thus the points we need to check for costs are the best point at $t = 0$, the best point at $t = 10$, the best point at $t = 5$, and the best point at $t = 8.5$. These points are

Time to change	Optimum point
0	5,8
5	1,8
8.5	1,3
10	1,3 (no change)

Because the facility is installed at time 0 to serve until time 10, the cost calculations will require integration of the cost function relative to the location of the facility, between 0 and 10. The cost for locating at (5,8) is

$$\text{Cost} = \int_{t=0}^{t=10} w_i * (|x^{(1)} - x_i| + |y^{(1)} - y_i|)\mathrm{d}t$$

$i = 1, \ldots, n$ the number of existing facilities.

Substituting the values:

$$\text{Cost} = \int_{t=0}^{t=10} [\{|5-1| + |8-3|\}(2+3t) + \{|5-5| + |8-8|\}(20-2t)$$
$$+ \{|5-7| + |8-9|\}(4+2t) + \{|5-0| + |8-15\}(12-t)]\mathrm{d}t$$
$$= \int_{t=0}^{t=10} [174 + 21t]\mathrm{d}t$$
$$= [174t + 21t^2/2] \int_{t=0}^{t=10}$$
$$= 2790$$

Similarly, cost for the location $(x^{(2)}, y^{(2)}) = (1, 3)$.

$$\text{Cost} = \int_{t=0}^{t=10} w_i * (|x^{(2)} - x_i| + |y^{(2)} - y_i|)$$
$$= 3490$$

and cost for location at $(x^{(3)}, y^{(3)}) = (1, 8)$ is

$$\text{Cost} = \int_{t=0}^{t=10} w_i * (|x^{(3)} - x_i| + |y^{(3)} - y_i|)$$
$$= 2790.$$

Hence, select either (5,8) or (1,8) as the optimum location.

9.6 GRAPHICAL METHOD

Consider another somewhat expanded example with seven customers. Their locations and demand distributions are given in Table 9.37.

Following the procedure from the previous example, we apply the 50th percentile method to develop optimum times at which facility locations should be changed, if the location of the facility can be changed as many times as required. They are displayed in Table 9.38.

Depending on the time, there are five optimum locations for the new facility.

If a facility is to be placed in a single location and no change in the location is permitted, then we can check the cost of placing a facility in each location over the entire time period and select the location that gives the minimum cost to place the facility.

If, however, one or more location changes are allowed, the problem becomes somewhat difficult. There are numerous combinations of locations over different time periods that need examination. The number of combinations

TABLE 9.37 Data for Example 2

Customer number	Location	Demand
1	0,3	$22 - t$
2	3,8	$5 + t$
3	5,9	$8 + t$
4	1,10	$2 + t$
5	2,15	$4 + 2t$
6	6,13	$3 + t$
7	8,12	$10 + 3t$

TABLE 9.38 Optimum Time to Change (for Unlimited Location Changes) and Associate Location

Point	Optimum time to change	Optimum location
1	0	2,9
2	0.5	3,9
3	2.66	3,10
4	5.0	3,12
5	6.0	5,12

increase rapidly as the number of intervals increase. Thus, a simpler method is needed to recognize the optimum locations.

The graphical method is one such way to resolve the problem. It requires the following steps:

1. Calculate the cost at the beginning and end of each optimum time change interval, based on the optimum facility location in that time period.
2. Draw the time–cost diagram. It consists of plotting for each start and end of a time period, associated cost and then connecting these points by a straight-line. The graph thus created shows the cost envelope, from which minimum cost locations can be judged.

In our example, the optimum location over a time period is given in Table 9.39.

The cost values at different breakpoints are as Table 9.40.

The plot of a time–cost diagram is shown in Figure 9.2. The total cost is given by the area under the curve (straight-line in our case) associated with

TABLE 9.39 Time Period and Optimum Location

Time period	Optimum location
0–0.5	2,9
0.5–2.66	3,9
2.66–5.0	3,10
5.0–6.0	3,12
6.0–10.0	5,12

TABLE 9.40 Optimum End Point and Associated Cost

Time	Optimum point	Cost
0.0	2,9	352
0.5	2,9	375
0.5	3,9	375
2.66	3,9	465
2.66	3,10	465
5.0	3,10	550
5.0	3,12	550
6.0	3,12	578
6.0	5,12	578
10.0	5,12	674

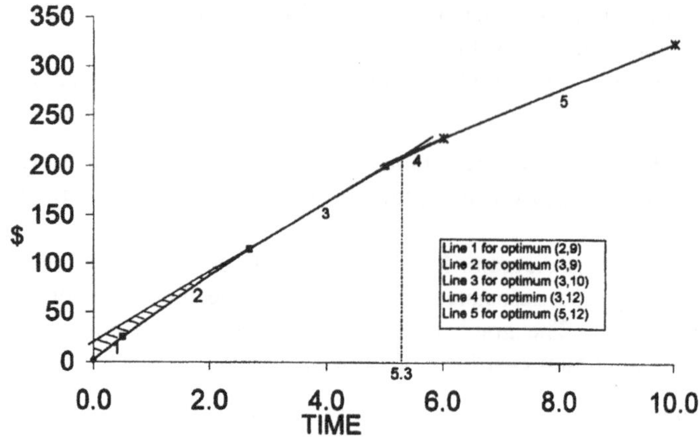

FIGURE 9.2 One change allowed.

placement of a facility. For example, if one change is allowed initial placement of the facility at (3,10) and changing it to (5,12) at time 4.6, leads to minimum cost over 10 units of time. The shaded area indicates the cost above the minimum possible if there is no limitation on the number of allowed changes. In trying to decide the optimum facility locations, given a restriction on the number of changes we try out different straight-line extensions to minimize this area. For example, if two changes are allowed the associated optimum solution is given in Figure 9.3. The facility is placed at location 3,9 first and changed to 3,10 at time 2.66. The second change from 3,10 to 5,12 occurs at time 5.3. Again these points

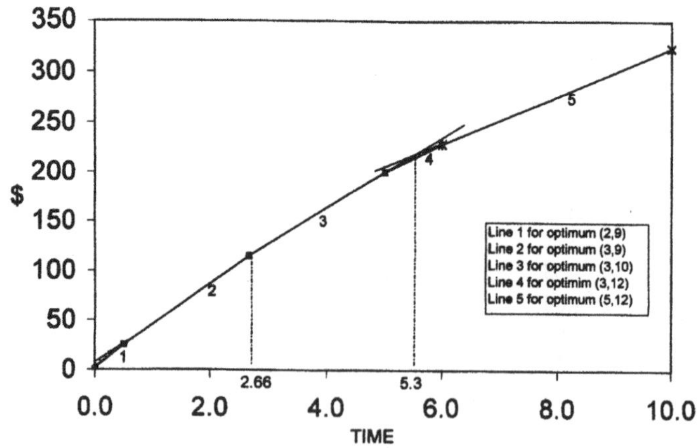

FIGURE 9.3 Two changes allowed.

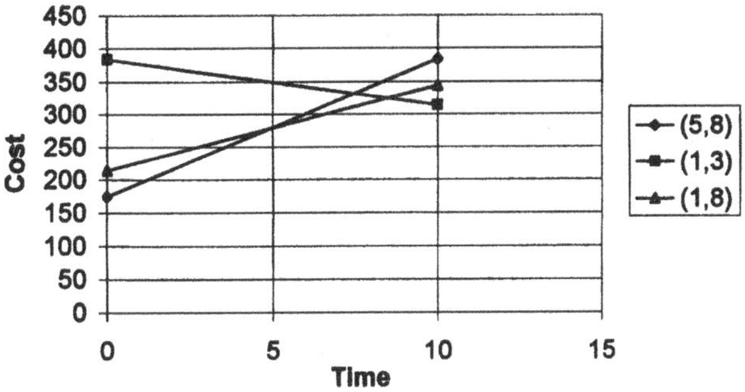

FIGURE 9.4 No change allowed.

are obtained by finding the intersections of cost curve associated with locating the facility in a position as shown in Figure 9.3.

Figure 9.4 shows graphical solution to earlier no-change problem. The area under the cost curve for both 5,8 and 1,8 location is the same, indicating either location can be used as an optimum facility location.

9.7 TABULAR METHOD

We can also represent the information in a tabular form and make the location decisions. For example, for no-change condition the facility must be placed in one of the existing demand points. The total cost associated with each point is as follows:

1. (2,9) Cost $= 62t + 352$
2. (3,9) Cost $= 42t + 354$
3. (3,10) Cost $= 36t + 370$
4. (3,12) Cost $= 28t + 410$
5. (5,12) Cost $= 24t + 434$

As these are linear curves, the cost can be calculated as average cost multiplied by the planning time period. Our planning time horizon is from 0 to 10 and we can calculate the cost at $t = 0$, $t = 10$, and take the average to represent the average cost over the period. Table 9.41 presents the results. Select the location that gives the minimum cost; in this case, point (3,10) or (3,12), with the cost of 5500.

9.7.1 One Change Allowed

We have determined in Table 9.38, the times when the change should be made if unlimited changes are allowed. But when the number of changes are restricted,

TABLE 9.41 No-Change Calculations

Point	End times	Time interval	Cost at end times	Average cost	Total cost
(2,9)	0	$10-0=10$	352	662	6,620
	10		972		
(3,9)	0	$10-0=10$	354	564	5,640
	10		774		
(3,10)	0	$10-0=10$	370	550	5,500
	10		730		
(3,12)	0	$10-0=10$	410	550	5,500
	10		690		
(5,12)	0	$10-0=10$	434	554	5,540
	10		674		

then it is required to find the actual time of change by equating the côst at the point of change. For example, the first possible shift is from (2,9) to (3,9). We have seen from our graphical solution procedure that the optimum time of such change would be when the cost associated with having a facility at (2,9) is equal to cost associated of having it at (3,9). Find t such that

$$\text{Cost}(2,9) = \text{cost}(3,9) \qquad \text{or} \qquad 62t + 352 = 42t + 354$$

$$t = 0.1$$

TABLE 9.42 Optimum Time to Shift from Preceding Location to Succeeding Location in the Pair

Pair	Time
(2,9), (3,9)	0.1
(2,9), (3,10)	0.69
(2,9), (3,12)	1.70
(2,9), (5,12)	2.15
(3,9), (3,10)	2.66
(3,9), (3,12)	4
(3,9), (5,12)	4.44
(3,10), (3,12)	5
(3,10), (5,12)	5.33
(3,12), (5,12)	6

Similarly, we can calculate the optimum shift time for each consecutive pair. The results are shown in Table 9.42. Note the table contains all the combinations at which one change in location can be made.

Table 9.43 is constructed in a manner similar to Table 9.41, except now we must calculate cost at three points, beginning, change point, and end point. Because we do not know the change point we must consider all combinations, the same combinations shown in Table 9.42. The calculations in the table are self-explanatory and are similar to Table 9.41.

TABLE 9.43 Cost Calculations for One Change

Points	Significant time	Time interval t_i	Minimum cost at significant time	Average cost a	Sector cost $a * t_i$	Total cost
(2,9), (3,9)	0	0.1	352	355.1	56.61	5,661
	0.1	9.9	358.2	566.1	5,604.39	
	10		774			
(2,9), (3,10)	0	0.69	352	373.39	257.639	5,538.48
	0.64	9.39	394.78	562.39	5,280.84	
	10		730			
(2,9), (3,12)	0	1.7	352	404.7	697.99	5,449.7
	1.7	8.3	457.4	573.7	4,761.71	
	10		690			
(2,9), (5,12)	0	2.15	352	418.65	900.09	5,450.35
	2.15	7.85	485.3	574.65	4,550.25	
	10		674			
(3,0), (3,10)	0	2.66	354	438	1,165.08	5,553.37
	2.66	7.34	465.72	597.86	4,388.29	
	10		730			
(3,9), (3,12)	0	4	354	438	1,752	5,388
	4	6	522	606	3,636	
	10		690			
(3,9), (5,12)	0	4.44	354	447.24	1,985.74	5,361.99
	4.44	5.56	540.48	607.24	3.376.25	
	10		674			
(3,10), (3,12)	0	5	370	460	2,300	5,400
	5	5	550	620	3,100	
	10		690			
(3,10), (5,12)	0	5.33	370	645.94	2,483.46	5,350.70
	5.33	4.64	561.88	617.94	2,867.24	
	10		674			
(3,12), (5,12)	0	6	410	494	2,964	5,468
	6	4	578	626	2,504	
	10		674			

The minimum cost in this case is \$5350.70. Thus, if we are allowed to change once, our policy should be, start with (3,10) until time 5,33 then change to (5,12).

One Change with Fixed or Permanent Changeover Cost/Savings

If a changeover cost is associated with moving from one location to other, then a slight modification to the previous procedure is needed to develop the optimum solution.

If the changeover cost is a one-time cost, then we just need to add that cost to the appropriate cost in the last column of Table 9.43 and select the minimum. But sometimes changeover cost is in the form of additional operational cost needed to shift personal from one location or other, or is the differential cost of running the facility at the second location compared with the first one. Such cost continues to be present throughout the duration of changeover. This can be both a saving or a cost. Suppose in our example, the permanent changeover costs are as given in Table 9.44.

The only modification we need from Table 9.43 is to add the changeover cost at both ends of the cost at the point of changeover and then determine the average cost for the associated time period. The results are shown in Table 9.45. Other calculations are again similar to Table 9.43.

The minimum cost is at pair (2,9), (3,9); that is, place facility at (2,9) until time 0.1 and then move to (3,9). The minimum cost is cost \$4744.71.

Change at Fixed Time

Sometimes restrictions may apply to a facility, that allows one to change its location only at a certain fixed time. Change after or before that time is not permitted. Consider the same example 2, with data in Table 9.37 with a restriction that change is allowed only at time 3. Which location should be move to obtain minimum cost? This problem can be solved with the same steps as given in the foregoing except one of the significant time in each case is the time when change can be made (i.e., time 3). The calculations are shown in Table 9.46.

TABLE 9.44 Permanent Cost of Changeover

	(2,9)	(3,9)	(3,10)	(3,12)	(5,12)
(2,9)	0	−90	200	150	80
(3,9)	−90	0	−50	250	20
(3,10)	200	−50	0	150	200
(3,12)	150	250	150	0	200
(5,12)	80	20	200	200	0

TABLE 9.45 With Permanent Changeover Charge

Points	Significant time	Time interval t_i	Minimum cost at significant time	Average cost a	Sector cost $a * t_i$	Total cost
(2,9), (3,9)	0	0.1	352	313.2	31.32	
	0.1		358.2			4,744.71
	0.1	9.9	268.2	476.1	4,713.39	
	10		684			
(2,9), (3,10)	0	0.69	352	373.39	257.64	
	0.69		394.78			7.416.48
	0.69	9.39	594.78	762.39	7,158.84	
	10		930			
(2,9), (3,12)	0	1.7	352	404.7	687.99	
	1.7		457.4			6,902.2
	1.7	8.3	607.4	748.7	6,214.21	
	10		890			
(2,9), (5,12)	0	2.15	352	418.65	900.1	
	2.15		485.3			6,078.25
	2.15	7.85	565.3	659.65	5,178.25	
	10		754			
(3,9), (3,10)	0	2.66	354	409.86	1,090.22	
	2.66		465.72			5,111.51
	2.66	7.34	415.72	547.86	4,021.29	
	10		680			
(3,9), (3,12)	0	4	354	438	1,752	
	4		522			6,888
	4	6	772	856	5,136	
	10		940			
(3,9), (5,12)	0	4.44	354	447	1,984.68	
	4.44		540.48			5,472.13
	4.44	5.56	560.48	627.24	3,487.45	
	10		694			
(3,10), (3,12)	0	5	370	460	2,300	
	5		550			6,150
	5	5	700	770	3,850	
	10		840			
(3,10), (5,12)	0	5.33	370	465.94	2,483.46	
	5.33		561.88			6.303.23
	5.33	4.67	761.88	817.94	3,819.77	
	10		874			
(3,12), (5,12)	0	6	410	494	2,964	
	6		578			6,268
	6	4	778	826	3,304	
	10		874			

TABLE 9.46 Fixed Time Change

Points	Significant time	Time interval t_i	Minimum cost at significant time	Average cost a	Sector cost $a * t_i$	Total cost
(2,9), (3,9)	0	3	352	445	1,335	
	3		538			5,724
	3	7	480	627	4,389	
	10		774			
(2,9), (3,10)	0	3	352	445	1,335	
	3		538			5,563
	3	7	478	604	4,228	
	10		730			
(2,9), (3,12)	0	3	352	445	1,335	
	3		538			5,479
	3	7	494	592	4,144	
	10		690			
(2,9), (5,12)	0	3	352	445	1,335	
	3		538			5,465
	3	7	506	590	4,130	
	10		674			
(3,9), (3,10)	0	3	354	446	1,338	
	3		480			5,566
	3	7	478	604	4,228	
	10		730			
(3,9), (3,12)	0	3	354	446	1,338	
	3		480			5,587
	3	7	524	607	4,249	
	10		690			
(3,9), (5,12)	0	3	354	446	1,338	
	3		480			5,468
	3	7	506	590	4,130	
	10		674			
(3,10), (3,12)	0	3	370	424	1,272	
	3		478			5,521
	3	7	524	607	4,249	
	10		690			
(3,10), (5,12)	0	3	370	424	1,272	
	3		478			5,402
	3	7	506	590	4,130	
	10		674			
(3,12), (5,12)	0	3	410	467	1,401	
	3		524			5,531
	3	7	506	590	4,130	
	10		674			

Exercises

9.1 A naval base is supposed to protect five locations, as given in the table. The importance of these places is forecasted to change with time. Thus, the weight given to these stations varies with time. Determine the location of the base such as to reduce fear of attack, if it is possible to change two times, within next 10 years.

No.	x	y	Weight
1	2	4	$5+2t$
2	3	1	$-3t$
3	7	6	$20+4t$
4	15	25	$-15-2t$
5	10	8	$5-10t$

9.2 An electronic company wants to target five areas as new customers and offer a variety of electronic products. It will serve from two of the four locations

Data for First Year

	Location				
Area	1	2	3	4	Demand
1	10	7	8	15	70
2	12	11	10	8	120
3	13	5	15	7	80
4	20	9	11	8	90
5	11	7	17	10	150
Fixed cost of operation	400	300	220	150	

Data for Second Year

	Location				
Area	1	2	3	4	Demand
1	13	8	12	10	85
2	10	16	12	3	32
3	13	8	18	6	80
4	20	10	8	16	100
5	15	10	18	17	175
Fixed cost of operation	450	320	200	250	

Data for Third Year

Area	Location 1	2	3	4	Demand
1	16	8	5	14	80
2	13	12	9	14	90
3	17	6	16	8	60
4	16	8	12	9	120
5	10	10	15	8	160
Fixed cost of operation	300	350	200	250	

Changeover Cost

Location	Location 1	2	3	4
1	0	80	20	100
2	80	0	60	50
3	20	60	0	70
4	100	50	70	0

that the company is planning for 3 years hence. The demand from these areas and fixed cost of operation for the 3 years is represented in the following tables. Whenever there is a change in the location, it involves a changeover cost shown in the table. At the same time the company wishes to minimize the total cost of operation. If so, determine the facility locations and customer assignments in each period.

9.3 A jet pump manufacturing company wants to acquire new customers from four areas it has targeted. The company is proposing to locate two service centers

Data for First Year

Area	Location 1	2	3	4	Demand
1	12	8	8	13	120
2	10	15	11	14	100
3	17	17	23	12	180
4	12	10	13	7	210
Fixed cost of operation	200	300	170	190	

Data for Second Year

	Location				
Area	1	2	3	4	Demand
1	15	12	12	17	90
2	12	19	14	12	61
3	14	12	10	9	99
4	12	12	13	16	130
Fixed cost of operation	120	410	370	290	

Data for Third Year

	Location				
Area	1	2	3	4	Demand
1	12	17	12	15	89
2	13	16	19	23	97
3	12	16	14	13	78
4	11	19	21	28	46
Fixed cost of operation	210	290	300	270	

from the available four locations. It is proposing a plan for the next 3 years. The data for costs and demands for each period are known. The cost of shifting the service centers from one location to another is shown in the fourth table. Determine the location of the service centers and customer assignments to minimize the total cost of operation.

Changeover Cost

	Location			
Location	1	2	3	4
1	0	76	80	97
2		0	43	72
3			0	91
4				0

9.4 Discuss possible situations when demand may change from period to period.

9.5 A farm equipment manufacturer wishes to target two new regions to improve its sales. It will base its operations in one of the two locations. The sales have been forecasted for the next two time periods as shown in the following tables. Also the cost of serving these customers from each location is represented in the tables. If there is a change of location the cost associated with it is shown in the changeover cost table. Determine the location to minimize the total cost of operation.

Time Period 1

	Location		
Region	1	2	Sales forecast
1	3	2	10
2	1	3	15
Fixed cost	100	120	

Time Period 2

	Locations		
Region	1	2	Sales forecast
1	2	2	10
2	3	2	10
Fixed cost	90	110	

Changeover Cost

Locations	1	2
1	0	20
2	17	0

9.6 Software firm wants to expand its business to three growing industrial cities in the country. It wishes to serve clients from these cities from one of the two possible locations. Initially the firm is planning for 3 years. The cost of serving the clients from each of the locations and the expected number of clients they might acquire for each year is shown in the tables. Cost is associated with changing the location as represented in the changeover cost table. Determine the location to minimize the total cost of operation to the software firm.

Year 1

	Location		
Cities	1	2	Number of clients
1	9	12	100
2	10	15	150
3	13	11	125
Fixed cost	50	80	

Year 2

	Location		
Cities	1	2	Number of clients
1	9	10	110
2	10	12	170
3	10	10	125
Fixed cost	60	80	

Year 3

	Location		
Cities	1	2	Number of clients
1	9	10	115
2	9	12	175
3	10	10	130
Fixed cost	60	80	

Changeover Cost

Locations	1	2
1	0	5
2	4	0

9.7 A company is to supply its trucking fleet with foul weather equipment from two of four possible locations. The cost data and demand for the months of January, February, and March are shown in the first three tables. The distribution point can be changed in any month, but with the changeover costs shown in the last table. Determine the best locations.

January

	Location				
Customer	1	2	3	4	Demand
1	5	13	12	9	200
2	7	4	8	13	300
3	9	10	5	4	250
4	11	8	7	6	175
5	3	6	10	11	280
Fixed cost	1,100	1,075	950	1,000	

February

	Location				
Customer	1	2	3	4	Demand
1	5	10	9	9	175
2	7	4	8	11	250
3	9	8	5	4	200
4	7	7	7	6	140
5	3	6	10	10	220
Fixed cost	1,100	1,075	950	1,000	

March

	Location				
Customer	1	2	3	4	Demand
1	5	8	7	9	150
2	7	3	8	10	210
3	9	8	5	4	175
4	7	7	7	6	100
5	3	6	10	10	
Fixed cost	950	975	900	880	

Changeover Cost

	Location			
Location	1	2	3	4
1	0	15	40	65
2	55	0	20	30
3	25	35	0	45
4	60	55	50	0

9.8 Suppose in problem 9.1 the changeover to a base involves changeover cost in terms of cost of running the base, and cost of shifting personnel. Determine the location of the base.

	(2,4)	(3,1)	(7,6)	(15,25)	(10,8)
(2,4)	0	50	150	200	−60
(3,1)	50	0	−65	225	30
(7,6)	150	−65	0	120	200
(15,25)	200	225	120	0	220
(10,8)	−60	30	200	220	0

9.9 The demands from the customers change over a time period, as shown in the table. To keep up with the demands the company has to move its sales office to coordinate the sales. The company is planning for next 8 years. Determine the optimum locations if one changeover is allowed. Locations of the customers are also shown in the table.

	x	y	Demand
1	2	4	$3+t$
2	3	6	$18+t$
3	4	5	$8+2t$
4	2	8	$12+t$

9.10 Because of other expansions and developments the company is planning on the changeover of the sales office is only possible at time 4. If so, to which location should the sales office move to incur minimum cost?

10

Simultaneous Facility Location

One common problem in location analysis involves placement of n facilities in n locations. Each of the service facility may be different types and may serve some outside customers, or each facility itself may become a customer for the other facilities. The cost is measured as the sum of the products of flow times and the distance that the flow travels; therefore, the location of each facility in the right position becomes a critical issue.

In this chapter, we will first briefly review a procedure that most readers are familiar with (i.e., assignment algorithm) and then illustrate its application in a location problem. We will illustrate examples mainly in manufacturing because they are easy to visualize, but as with any methods that we have seen so far, the procedures developed are applicable in several other fields. For example, customers requiring the services of certain machines are to be assigned so that each customer is associated with one machine only. The objective is to minimize the total fixed cost of the assignment. This cost may consist of the setup time for a job on a machine, fixed cost of placing a machine in a location, or the cost that is dependent on the degree of the incompatibility between a customer and a location.

Next we study an extension of an assignment problem, specifically used in deciding where to add new machines to an existing plant layout; again, an important problem in minimizing material-handling cost.

The next problem is the variable cost problem. In manufacturing a group of products, each unit may require the services of different machines. There is a flow between the machines as the units are processed, first on one machine and then on another. Because the same facility is used to produce many different items, the sequence in which machines are used are not the same. The objective is to place the machines so that the "flow cost" is minimized. This cost may be defined, for example, as the total distance traveled by units as they move from machine to machine in the sequence of processing. The problem is known as a "quadratic assignment problem," and we will present an algebraic procedure to solve it. The quadratic assignment method is well used in computerized plant layout procedures. We will also study a simple extension in which both fixed and variable costs are present.

In another problem, we cannot place one facility at a time, because placement of one facility affects the relative cost of other facilities. This is because there are exactly n locations for placing n facilities, and placement of one of the facility in any one location influences the placement decisions of other facilities. These types of problems, therefore, are called simultaneous facility location problems.

10.1 FIXED COST ONLY

As the term implies, here, only the fixed cost of assigning a facility to a location is involved. A procedure to solve such a problem is an assignment procedure, and it is briefly reviewed here.

Suppose in a machine shop there are three machines that are to be placed in three available locations (only one machine in each location). Some machines are better suited for a particular location than others, perhaps because of the foundation, the proximity to power outlets and auxiliary equipment, or because of the availability of special tools needed in processing. The estimated "cost" of assigning a given machine to a specific location is given in Table 10.1.

We are required to find the optimum position for each machine. The procedure requires conversion of the present cost table into an equivalent cost

TABLE 10.1 Cost

	Location		
Machine	1	2	3
1	5	10	6
2	3	6	4
3	6	7	5

table consisting only of the positive or zero elements, so that the optimum solution to the modified cost table is also the optimum solution to the original cost table. How can this be done? A property in linear algebra allows us to achieve our goal with a simple algebraic manipulation. If a constant is subtracted from all the elements of a row or of a column, then the optimum solution of the resulting table is different from the optimum solution of the original solution by exactly the amount of that constant. Suppose we perform this step so as to generate a table with at least one zero element within each row and each column, and if the zero elements are in strategic positions so that a machine can be assigned to a location with a zero cost, then the resulting solution is obviously optimal with "zero" cost.

Apply the foregoing rule to the problem under consideration by subtracting the lowest-cost element in each row from all of the elements in that row. The result is the equivalent cost table presented as Table 10.2.

It is possible to assign either machine 1 or 2 to location 1, and machine 3 to location 3. However, there is no zero in column 2; therefore, an assignment for location 2 is not quite so obvious. To create a zero in this column, we subtract 2, the minimum element in the column, from all the elements of the column. The results are shown in Table 10.3.

Now machine 3 may be assigned to either location 2 or 3. However, we still cannot assign all machines uniquely. Once machine 3 is assigned to location 2, for example, we have no other zero available in location 3, and thus no other machine may be assigned to location 3. In problems of this type, we can use a

TABLE 10.2 Equivalent Cost I

	Location		
Machine	1	2	3
1	0	5	1
2	0	3	1
3	1	2	0

TABLE 10.3 Equivalent Cost II

	Location		
Machine	1	2	3
1	0	3	1
2	0	1	1
3	1	0	0

very simple rule to determine how many assignments are possible. The maximum number of possible strategic locations, each consisting of a zero element (therefore, the maximum number of possible assignments of machines to locations), can be determined by drawing the minimum number of horizontal or vertical lines required to cover all the zeros. The number of lines drawn is the number of assignments that can be made.

In our problem, Table 10.4 shows that two lines will cover all the zero elements. This indicates that, at the most, only two machines may be assigned to two of the locations. For example, with machine 1 placed in location 1 and machine 3 in location 2, machine 2 may not be placed in location 3 because we do not have a zero element there.

To solve this problem, other zeros in strategic positions must be created. To do this, find the minimum element among those not covered by the lines. In our example, the lowest value is 1, and there are three of them. Arbitrarily select a 1 from column 3 as the minimum. If we subtract 1 from the elements in column 3, the zero element will become −1, as shown in Table 10.5.

However, for the assignments to be obvious, the elements are restricted to positive values (zero included) only. This can be achieved by adding 1 to the elements of the third row. The results are shown in Table 10.6.

But now we have eliminated the zero that previously was in column 2, row 3. To reestablish the zero in column 2, subtract 1 from the elements of column 2 as shown in Table 10.7.

TABLE 10.4 Maximum Possible Assignments I

	Location		
Machine	1	2	3
1	0	3	1
2	0	1	1
3	1	0	0

TABLE 10.5 Maximum Possible Assignment II

	Location		
Machine	1	2	3
1	0	3	0
2	0	1	0
3	1	0	−1

TABLE 10.6 Maximum Possible Assignment III

	Location		
Machine	1	2	3
1	0	3	0
2	0	1	0
3	2	1	0

TABLE 10.7 Equivalent Cost

	Location		
Machine	1	2	3
1	0	2	0
2	0	0	0
3	2	0	0

The process has created zeros in other locations, as well as the one we sought. To cover all the zeros, we need a minimum of three lines, indicating that now the three machines can be assigned to three locations. We assign each machine at a zero-cost location, and see that there are three possible solutions, as in Table 10.8. Referring to the Table 10.1, we obtain the individual costs listed in Table 10.8 and find that each solution has a total cost of 16 units.

A shortcut procedure for creating additional strategic zero elements can now be summarized by the following steps:

1. Select the minimum cost element from among the elements that are not covered by the lines (see Table 10.4).
2. Subtract this element from all the elements that are not covered by the lines, and add it to the element(s) that are on the intersection(s) of the lines.

TABLE 10.8 Alternative Solutions

	Solution 1			Solution 2			Solution 3		
Machine	1	2	3	1	2	3	1	2	3
Location	1	2	3	3	1	2	1	3	2
Cost	5	6	5	6	3	7	5	4	7

This procedure will readily convert Table 10.4 into Table 10.7. In some cases, it may be necessary to repeat the process several times to reach the final solution.

10.2 PLACEMENT OF NEW MACHINES IN AN EXISTING PLANT

We next consider as a sample problem the case in which we want to add a number of new machines in an existing plant. Suppose we need three new machines (A, B, and C) in our plant, which already contains five other machines. The traffic flow (in number of units) between the new machines and the existing machines is shown in Table 10.9.

Three possible locations (X, Y, and Z) are chosen for the new machines. The distances between these choices and the existing machines are listed in Table 10.10.

It is possible for us to calculate how effective a machine would be if it were placed in a given location. For example, if machine A is placed in location X, we will have 10 items from machine A to machine 1 transported for 10 units of distance; 15 items to machine 2, transported for 20 units of distance, and so on. If we add the products of the various flows multiplied by the corresponding distances that result from placing a new machine in a new location, then that quantity will measure the effectiveness of that placement. The smaller the value, the less the total transportation is, and hence, the more effective the machine placement is.

TABLE 10.9 Traffic Flow

	Existing machines				
New machines	1	2	3	4	5
A	10	15	5	8	6
B	3	4	7	9	10
C	3	8	5	6	2

TABLE 10.10 Distance Data

	Location		
Existing machines	X	Y	Z
1	10	20	50
2	20	30	60
3	40	20	20
4	30	40	40
5	20	50	10

Note the difference between this problem and one we saw in Chapter 3. In Chapter 3 we selected appropriate locations from a set of locations to place the new facilities. In addition there was flow between one location and only one new facility unlike in this problem where the flow exists between each location and each of new facility.

Here, placing machine A in location X results in an effectiveness measure of $(10 * 10) + (15 * 20) + (5 * 40) + (8 * 30) + (6 * 20) = 960$. Similarly, if machine A is placed in location Y, then the effectiveness measure is $(10 * 20) + (15 * 30) + (5 * 20) + (8 * 40) + (6 * 50) = 1{,}370$. Naturally, placing machine A in location X is better than placing it in location Y.

The effectiveness measure of installing each machine in each location can be constructed in a tabular form by taking the product (matrix multiplication) of the flow table and the distance table. In our example, we obtain:

$$\begin{bmatrix} 10 & 15 & 5 & 8 & 6 \\ 3 & 4 & 7 & 9 & 10 \\ 3 & 8 & 5 & 6 & 2 \end{bmatrix} \begin{bmatrix} 10 & 20 & 50 \\ 20 & 30 & 60 \\ 40 & 20 & 20 \\ 30 & 40 & 40 \\ 20 & 50 & 10 \end{bmatrix} = \begin{bmatrix} 960 & 1370 & 1880 \\ 860 & 1180 & 990 \\ 610 & 740 & 990 \end{bmatrix}$$

Table 10.11 is simply the result, with designations assigned to indicate the machines and the locations.

We can now apply the assignment procedure to determine where to locate the new machines such that the sum of the individual measures of effectiveness will be as small as possible. In our example, the final solution is to place machine A in location X, machine B in location Z, and machine C in location Y, yielding a total effectiveness measure of 2,690.

We may find it necessary to analyze the problem in greater detail. For instance, it is not too difficult to envision a machine or facility that produces small bolts and another that manufactures diesel engine crankshafts. If each were the same distance from a third facility, and each had the same flow to that facility, our procedure just presented would give an erroneous indication of the effectiveness if only the distance were to be considered. The proper way to take into account the differences in transporting bolts and crankshafts, is to select appropriate "weighting factors" by which the elements of the flow table are multiplied.

TABLE 10.11 Measure of Effectiveness

	Location		
Machine	X	Y	Z
A	960	1370	1880
B	860	1180	990
C	610	740	990

10.3 QUADRATIC ASSIGNMENT PROBLEM

Problems involving products that are processed by multiple facilities, and in which there is a cost associated with moving the units from facility to facility, fall under the variable-cost category. There are numerous examples that follow this cost structure. Consider a machine shop that is designed to produce various products. The machines are all general-purpose, such as lathes, grinders, drilling machines, planers, and milling machines. The production process calls for using these machines at different times in the manufacturing sequences. Knowing the annual output of the products, we can develop a flow matrix that indicates the number of times that machine i is to be used after machine j. If k locations are available for placing the k machines, the problem is to determine which machine should be placed in which location to minimize the transportation cost, defined as the sum of flow between machines times distance between their locations.

Another example is that of development of a panel board of gauges. The gauges that are more interdependent should be placed as close together as possible. A change in the setting of one gauge parameter, such as pressure, may cause other gauges, such as the one recording temperature or the one measuring volume, to vary. If these three parameter are to be maintained within acceptable operating ranges, we need to monitor the appropriate gauges continuously. Therefore, it is more convenient if they are placed as near each other as possible.

In a storeroom, placement of the storage bins is a similar type of problem. To minimize customer waiting time, those parts that are most often ordered together should be stored close to each other. The procedure that is also used in computerized plant layout methods in developing efficient departmental arrangements also follows the same cost structure.

The mathematical formulation of the foregoing problem, called *quadratic assignment problem* (QAP) is as follows:

$$\text{Minimize} \sum_{i=1}^{n} \sum_{k=1}^{n} \sum_{j=1}^{n} \sum_{l=1}^{n} f_{ik} d_{jl} x_{ij} x_{kl}$$

subject to:

$$\sum_{i=1}^{n} x_{ij} = 1 \quad \text{for all } j$$

$$\sum_{j=1}^{n} x_{ij} = 1 \quad \text{for all } i.$$

$$x_{ij} = 1 \text{ or } 0 \quad \text{for all } j, i.$$

where,

f_{ik} is the nonnegative flow that multiplies the distance between facilities i and k.

d_{jl} the distance between the locations j and l.

$$x_{ij} = \begin{cases} 1 & \text{if facility } i \text{ is assigned to location } j \\ 0 & \text{otherwise} \end{cases}$$

n = total number of machines to place

= total number of available locations

10.4 BRANCH AND BOUND METHOD

Now suppose we want to install n facilities in n locations so that these facilities have flow only among themselves. In this case, there are $n!$ different combinations that can be formed. For example if $n = 3$ and the facilities are designated as A, B, and C, and locations are designated as 1, 2, and 3, then the combinations are

1. A–1 (A to location 1), B–2, and C–3
2. A–1, B–3, C–2
3. A–2, B–1, C–3
4. A–2, B–3, C–1
5. A–3, B–2, C–1
6. A–3, B–1, C–2

To select the optimum, we will have to examine the cost of each and select the combination that has minimum cost. For n greater than 15, this becomes an impossible task as the number of combinations increase in almost exponential manner. One alternative to solve such problems then is use of branch and bound method. This method reduces the number of combinations to examine however the actual number depends on the data of the problem.

Suppose there are four facilities to be placed in four locations and the total to-and-from flow between facilities is given in Table 10.12. Because the table is symmetrical, only the upper half is shown.

The four available sites are marked as A, B, C, and D and the distance configuration is as shown in Figure 10.1.

Traveled distance between each successive location—the distances between locations—are calculated and tabulated in Table 10.13. Here we assume that the travel between nodes is direct (no transshipment); therefore, even though the shortest distance between nodes 1 and 2 is 2, the direct distance is 5.

The first step is to construct number of pairs for facilities and for locations. For example, A–B, A–C, ..., and 1–2, 1–3, ... If there are n facilities then there are $n(n-1)/2$ such pairs that can be formed. Construct a matrix in which the facility pairs are listed in columns in decreasing order of flow, and distance pairs are listed in a row in increasing order of distances, as shown in Table 10.14. Taking the product of distance times the flow indicates the cost of assigning a facility pair into the location pair. For example cost of assigning C–D into location 1–3 is $5 \times 1 = 5$. Within 1–3, facility C may be assigned to location 1 or 3, whereas facility D is assigned to the remaining location 3 or 1.

Note the cost structure in Table 10.14. It is increasing from left to right, and for each column it has the highest values in the first row then in second and so on. This will always be true because distances are ranked in ascending order and flows are ranked in descending order in the construction of the table.

Next, we state a well-known theorem in matrix algebra. Given two vectors $f = (f_1, f_2, f_3 \ldots)$ and $d = (d_1, d_2, d_3 \ldots)$, the inner product $\boldsymbol{fd} = \sum f_i d_i$ is minimized when either $\boldsymbol{f}$ or $\boldsymbol{d}$ elements are arranged in descending order and other is arranged in the ascending order. Thus this inner product also gives the lower bound of the solution, although the associated solution may or may not be feasible.

In our case, the lower bound of a solution, when the flows are arranged in descending order, and distances are arranged in ascending order is 31, calculated as follows:

Flow	5	4	3	2	1	1
Distance	1	1	2	3	5	5

$$= 5 \times 1 + 4 \times 1 + 3 \times 2 + 2 \times 3 + 1 \times 5 + 1 \times 5 = 31$$

Note that these elements are also the diagonal elements of the matrix in Table 10.14. We could have obtained the lower-bound solution to the problem (using assignment algorithm principle) by reducing the diagonal elements to zero and making assignments at these zero positions. The best way to do this is as follows:

1. Subtract the diagonal entries from all the elements of that column (call it column operations) for each column. Add these numbers together to obtain the solution value S.

TABLE 10.12 Flow Matrix

	A	B	C	D
A	–	2	1	1
B	–	–	3	4
C	–	–	–	5
D	–	–	–	–

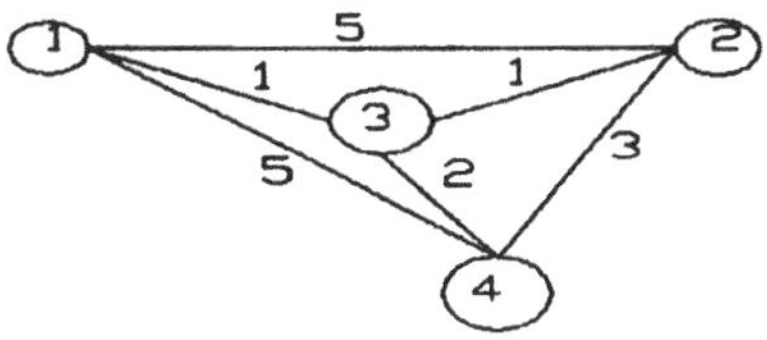

FIGURE 10.1 Distances between locations.

2. Perform the row operation. Subtract from each row the minimum entry in that row, which could be negative.
3. Add these numbers in the solution value S.
4. Continue steps 1 through 3 until all the elements are zero or positive and the diagonals are zero.
5. The associated S value is the solution value.

To illustrate, consider the data in Table 10.14. Subtracting the diagonal values from each column leads to the matrix in Table 10.15.

The present value of the sum of the cost elements added or subtracted so far from the original table is, $S = 5 + 4 + 6 + 6 + 5 + 5 = 31$. As there are still negative elements in the table, we must continue the row–column operations until all the elements are positive and assignments can be made at the zero cost elements. Perform the row operation. Subtract the least value from among the row elements from every element of that row giving Table 10.16.

The value of the S now is $S = 31 - (-1 - 2 - 3 - 4 - 4) = 45$.

Because there are still elements in the diagonal that are not zero, subtract the diagonal elements from every element of that column, which results in Table 10.17.

The S value is $45 - (1 + 2 + 3 + 4 + 4) = 31$.

As there are still negative elements in Table 10.17 subtract the least-cost element from each row, which results in Table 10.18.

TABLE 10.13 Distance Matrix

	1	2	3	4
1	–	5	1	5
2	5	–	1	3
3	1	1	–	2
4	5	3	2	–

The S after the second row operation is $31 - (-5) = 36$. As diagonals are not zero, subtract the least element in each column from the respective column, which results in Table 10.19.

$$S = 36 - (5) = 31.$$

Because the procedure is now fairly clear, from here on we will just show the operations and the associated results.

Performing the row operations leads to the matrix shown in Table 10.20.

$S = 31 - (-2) = 33$. Performing the column operations leads to the matrix shown in Table 10.21.

$$S = 33 - (2) = 31$$

Now in Table 10.21 we can make the assignments in zero cost cells, because all the cells are either positive or zero. If all the assignments are feasible then we have an optimum solution. Let us start by assigning C–D to 1–3, facility C can be in location at 1 or 3, whereas facility D is in the remaining location. Assigning C–D to 1–3 blocks the first row and first column from any further assignments. That is, no zero-cost cells from the first row or column can be used to make any other assignments. We will indicate that by X-ing the corresponding cells. It also makes some other zero-cost assignments unfeasible. For example A–B cannot be assigned to 3–4, because C has to be either in 1 or 3 and not A or B. Such nonassignability can be determined by the following rule: Determine the number of common letters between the facility pair and the present assignment and also determine the number of common numbers between the associated locations. If they have the same value, assignment may be made, if not such

TABLE 10.14 Matrix with Flow–Distance Arranged Data

	Between pair	1–3	2–3	3–4	2–4	1–2	1–4
	Distance Flow	1	1	2	3	5	5
C–D	5	5	5	10	15	25	25
B–D	4	4	4	8	12	20	20
B–C	3	3	3	6	9	15	15
A–B	2	2	2	4	6	10	10
A–C	1	1	1	2	3	5	5
A–D	1	1	1	2	3	5	5

TABLE 10.15 Diagonal Operations (Column Operation) I

		1–3	2–3	3–4	2–4	1–2	1–4
		1	1	2	3	5	5
C–D	5	0	1	4	9	20	20
B–D	4	−1	0	2	6	15	15
B–C	3	−2	−1	0	3	10	10
A–B	2	−3	−2	−2	0	5	5
A–C	1	−4	−3	−4	−3	0	0
A–D	1	−4	−3	−4	−3	0	0

assignment is not feasible. Again, 3–4 has one number common to 1–3 whereas A–B has no letter common to C–D; therefore, assignment is not feasible.

C–D and B–D have one common letter, whereas 1–3 and 2–3 have one common number; therefore, such assignment may be feasible and, hence, the associated 0 is maintained. Following similar logic and marking out unfeasible locations with zero cost by X, the resultant matrix is displayed in Table 10.22.

To make the other assignments we will follow the usual assignment algorithm. We know that the minimum number of lines that can be drawn through the number of zeros are the number of assignments that are possible. In our case we need six lines to cover all the zeros. So let us see what assignments are possible. C–D is fixed at 1–3. The only possible assignment for B–D is 2–3.

TABLE 10.16 Row Operations I

		1–3	2–3	3–4	2–4	1–2	1–4
		1	1	2	3	5	5
C–D	5	0	1	4	9	20	20
B–D	4	0	1	3	7	16	16
B–C	3	0	1	2	5	12	12
A–B	2	0	1	1	3	8	8
A–C	1	0	1	0	1	4	4
A–D	1	0	1	0	1	4	4

TABLE 10.17 Column Operations II

		1–3	2–3	3–4	2–4	1–2	1–4
		1	1	2	3	5	5
C–D	5	0	0	2	6	16	16
B–D	4	0	0	1	4	12	12
B–C	3	0	0	0	2	8	8
A–B	2	0	0	−1	0	4	4
A–C	1	0	0	−2	−2	0	0
A–D	1	0	0	−2	−2	0	0

Hence, D has to be location 3, and it follows that C is in 1, B is in 2, and by default, A is in 4. Next assignment B–C must be in 2–1 but the B–C 2–1 (or 1–2) cell does not have a zero cost; hence, this solution is not feasible.

Because C–D at 1–3, leads to no solution, make that assignment impossible by assigning a large cost (∞). The next least cost for assigning C–D is the next entry in the first row 2–3. Note again that because of our initial arrangement of facilities and location pairs, the first row has the largest (or equal to the largest) cost in each column and the costs are increasing from left to right. The next least cost solution would, therefore, be the next sequential assignment for C–D.

TABLE 10.18 Row Operation II

		1–3	2–3	3–4	2–4	1–2	1–4
		1	1	2	3	5	5
C–D	5	0	0	2	6	16	16
B–D	4	0	0	1	4	12	12
B–C	3	0	0	0	2	8	8
A–B	2	1	1	0	1	5	5
A–C	1	2	2	0	0	2	2
A–D	1	2	2	0	0	2	2

TABLE 10.19 Column Operations III

		1–3	2–3	3–4	2–4	1–2	1–4
		1	1	2	3	5	5
C–D	5	0	0	2	5	14	14
B–D	4	0	0	1	3	10	10
B–C	3	0	0	0	1	6	6
A–B	2	1	1	0	0	3	3
A–C	1	2	2	0	−1	0	0
A–D	1	2	2	0	−1	0	0

Assigning C–D to 2–3, we apply the same logic as before to determine the remaining feasible zero-cost elements. They are displayed in Table 10.23. By applying the assignment algorithm to this table, the minimum number of lines that can be drawn through the available zeros are 5 (5 vertical lines through columns 1, 2, 3, 4, and 5), and the number of assignments to be made are 6. So such assignment is not possible and we must create an additional zero in some other location(s) where the assignment is not possible. To create an additional zero, in column 6 subtract the minimum value, which is 2 from the column values. The resultant table is shown in Table 10.24.

This adds an additional cost to the solution, which now is $31 + 2 = 33$. Assigning C–D to 2–3, the only possible assignment for B–D is 1–3. Hence, D

TABLE 10.20 Row Operations III

		1–3	2–3	3–4	2–4	1–2	1–4
		1	1	2	3	5	5
C–D	5	0	0	2	5	14	14
B–D	4	0	0	1	3	10	10
B–C	3	0	0	0	1	6	6
A–B	2	1	1	0	0	3	3
A–C	1	3	3	1	0	1	1
A–D	1	3	3	1	0	1	1

TABLE 10.21 Column Operation IV

		1–3	2–3	3–4	2–4	1–2	1–4
		1	1	2	3	5	5
C–D	5	0	0	2	5	13	13
B–D	4	0	0	1	3	9	9
B–C	3	0	0	0	1	5	5
A–B	2	1	1	0	0	2	2
A–C	1	3	3	1	0	0	0
A–D	1	3	3	1	0	0	0

has to be in location 3, with C in location 2 and B in location 1, and the remaining A in location 4. For the next assignment B–C must be in 3–4, which is not possible; hence, the solution is unfeasible.

The next possible assignment for C–D, in increasing order of cost, is 3–4. In the original cost table (see Table 10.21), there is a cost of 2 in the C–D, 3–4 position. To make an assignment, we want to create a zero in the C–D, 3–4 location. This is achieved by subtracting 2 from the first row, as shown in Table 10.25. As a result, the cost of the solution is now $31 + 2 = 33$. Table 10.25 includes ∞ cost for C–D in the positions already checked.

Five vertical lines through columns 1, 2, 3, 4, and 6 cover all zeros, to create a zero in column 5, subtract 2 from the column, resulting in Table 10.26. This adds 2 units of additional cost to the solution ($33 + 2 = 35$), if there is one.

TABLE 10.22 Assign C–D to 1–3

		1–3	2–3	3–4	2–4	1–2	1–4
		1	1	2	3	5	5
C–D	5	0	X	2	5	13	13
B–D	4	X	0	1	3	9	9
B–C	3	X	0	0	1	5	5
A–B	2	1	1	X	0	2	2
A–C	1	3	3	1	X	0	0
A–D	1	3	3	1	X	0	0

No feasible solution can be found for the cost matrix; hence, the assignment C–D to 3–4 is not feasible.

Similarly proceeding to check the next possible assignment C–D to 2–4, results in Table 10.27.

Again, note that we are back to Table 10.21 for cost data, except the positions for which the C–D assignment was checked and found to produce no feasible solution. These positions have cost of ∞. Taking 5 units of cost off the first row creates a zero entry in the C–D, 2–4 cell. This also adds 5 units of cost to the solution $31 + 5 = 36$). Place X in all assignments that are not possible results in Table 10.27.

Follow the procedure of the assignment problem to check whether the minimum number of lines that can be drawn through the zeros is equal to the number of possible assignments. We see that this is not so. Hence, we apply the next step by subtracting the minimum value from among the uncovered cells from all the uncovered cells and further adding this minimum value to the cells on the intersection of the lines. In our problem the minimum value is 1. The resulting table is Table 10.28.

This adds 1 unit cost to the solution, which is now $36 + 1 = 37$.

The assignments are now feasible and the final assignments are shown in Table 10.28. The assignments are:

1	2	3	4
A	D	B	C

TABLE 10.23 Assign C–D to 2–3

		1–3	2–3	3–4	2–4	1–2	1–4
		1	1	2	3	5	5
C–D	5	∞	0	2	5	13	13
B–D	4	0	X	1	3	9	9
B–C	3	0	X	0	1	5	5
A–B	2	1	1	X	X	2	2
A–C	1	3	3	1	0	0	X
A–D	1	3	3	1	0	0	X

TABLE 10.24 Creation of Additional Zero

		1–3	2–3	3–4	2–4	1–2	1–4
		1	1	2	3	5	5
C–D	5	∞	0	2	5	13	11
B–D	4	0	X	1	3	9	7
B–C	3	0	X	0	1	5	3
A–B	2	1	1	X	X	2	0
A–C	1	3	3	1	0	0	X
A–D	1	3	3	1	0	0	X

The cost of the solution is the sum of cost between A–D + A–B + A–C + D–B + D–C + B–C: that is, $(1 * 5) + (2 * 1) + (1 * 5) + (4 * 1) + (5 * 3) + (3 * 2) = 37$.

Because there is no other branch open (Figure 10.2), this is the final solution.

10.5 ELIMINATION PROCEDURE

An interesting procedure that gives almost optimum, if not the optimum, solution in most cases is suggested by Das and Gunn (Industrial Engineering Research Conference, 1998). The procedure is based on the fact that some assignments are

TABLE 10.25 Create Zero in C–D, 3–4 Location

		1–3	2–3	3–4	2–4	1–2	1–4
		1	1	2	3	5	5
C–D	5	∞	∞	0	3	11	9
B–D	4	0	0	1	3	9	9
B–C	3	0	0	0	1	5	5
A–B	2	1	1	0	X	2	2
A–C	1	3	3	1	0	X	0
A–D	1	3	3	1	0	X	0

TABLE 10.26 Creation of Additional Zero

		1–3	2–3	3–4	2–4	1–2	1–4
		1	1	2	3	5	5
C–D	5	∞	∞	0	3	9	9
B–D	4	0	0	1	3	7	9
B–C	3	0	0	X	1	3	5
A–B	2	1	1	X	X	0	2
A–C	1	3	3	1	0	X	0
A–D	1	3	3	1	0	X	0

definitely worse than others. For example, in Table 10.14, C–D assignment to 1–4 is the most expensive and, therefore, should not be made unless no other option is available. We present here Das and Gunn's procedure, which has been slightly modified. To this end the following steps are suggested.

Develop a table consisting of machines in the column and locations in rows, in the same order as in Table 10.14 and shown in Table 10.29-A. Add a column for machine combination, location number, and elimination number (EN). The elimination number is initially defined as $n - 1$ (i.e., the number of possible locations minus 1), and is subsequently reduced by 1, every time the same machine appears in the machine combination column. For example, in Table 10.29-A, the combination **C–D** is recorded by **C** followed by **D** in the machine combination column. **D** has an EN value of 3. The next combination to appear in

TABLE 10.27 Assign C–D to 2–4

		1–3	2–3	3–4	2–4	1–2	1–4
		1	1	2	3	5	5
C–D	5	∞	∞	∞	0	6	6
B–D	4	X	0	1	3	9	9
B–C	3	X	0	0	1	5	5
A–B	2	1	1	X	X	2	2
A–C	1	3	3	1	X	0	0
A–D	1	3	3	1	X	0	0

TABLE 10.28 Final Assignments

		1–3	2–3	3–4	2–4	1–2	1–4
		1	1	2	3	5	5
C–D	5	∞	∞	∞	0	6	6
B–D	4	X	0	1	3	9	9
B–C	3	X	0	0	1	5	5
A–B	2	0	1	X	0	2	2
A–C	1	3	3	1	0	0	0
A–D	1	3	3	1	0	0	0

the column is **B–D**. Since **D** is appearing again in a different combination, its EN value is the previous value of **D** minus 1 or in this case, it is equal to 2.

Start with most expensive machine–location combination in Table 10.29-A and record it in the machine combination column. Mark the associated locations in the new location columns (see Table 10.29-B). Continue this procedure, successively selecting the remaining largest-cost machine combination element from Table 10.29-B. When the number of tally marks in a location are equal to EN number for a machine in its present position in the machine combination column, then the associated machine cannot be assigned to this location. Continue the procedure until there is only one location to which the machine can be assigned and make the associated assignment.

Once the machine has been assigned, eliminate corresponding combinations involving the selected machine and the selected location. With the remaining combinations of machines and location reapply the procedure. Continue this process until the last two machines and locations are remaining. Then check the combinations that results by alternatively placing each machine in each location.

Again consider Table 10.29-B for applying the elimination method. EN indicates the elimination number that has the initial value of 3 $(4-1)$ for each machine. The most expensive combination is C–D and 1–4, with the cost of 25. So, C and D are marked by x in locations 1 and 4. The next largest cost element is 25, still with combination C–D, but with locations 1–2. Locations 1 and 2 are marked with x's.

Continue this process with B–D 1–4; B–D 1–2. At this point D is eliminated from assignment to location 1 because its tally counts, 2, match with the associated elimination number for D also 2. Continuing we have B–C 1–4; and B–C 1–2. At this point, the tally marks for B and C are in locations 1 and

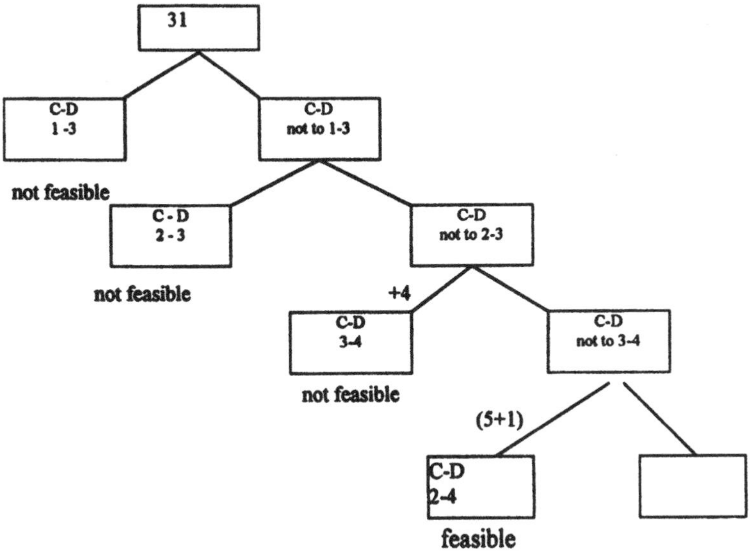

FIGURE 10.2 Branch and bound tree.

2, which correspond to the elimination numbers for those machines. So neither B nor C will be placed in location 1. This leaves only A, which can be assigned to location 1. Make such assignment. This is also shown in allocation Table 10.30.

Develop the elimination Table 2 (i.e., Table 10.31) by eliminating combinations involving machine A, and location 1 from Table 10.29 and apply the procedure again. This time the elimination number is 2 to start and every time a machine appears again in the machine combination listing, the value is reduced by 1 from the previous value.

Now D is eliminated from being considered for locations 2 and 4. hence, D has to be located in location 3 as shown in Table 10.32.

Only B and C are competing for locations 2 and 4, and each assignment must be checked. The possible sequences are ABDC and ACDB. We find the costs for both these combinations and choose the least.

Locations	1	2	3	4
Departments	A	B	D	C

From Tables 10.12 and 10.13 we obtain the cost for ABDC = $(2 \times 5) + (1 \times 1) + (1 \times 5) + (4 \times 1) + (3 \times 3) + (5 \times 2) = 39$

TABLE 10.29-A Initial Table

							Machine combination	Locations				
	1–3	2–3	3–4	2–4	1–2	1–4		1	2	3	4	EN
C–D	5	5	10	15	25	25	**C**					3
							D					3
B–D	4	4	8	12	20	20	**B**					3
							D					2
B–C	3	3	6	9	15	15	**B**					2
							C					2
A–B	2	2	4	6	10	10	**A**					3
							B					1
A–C	1	1	2	3	5	5	**A**					2
							C					1
A–D	1	1	2	3	5	5	**A**					1
							D					1

TABLE 10.29-B Elimination Table 1

							Machine combination	Locations				
	1–3	2–3	3–4	2–4	1–2	1–4		1	2	3	4	EN
C–D	5	5	10	15	25	25	**C**	xx	x		x	3
							D	xx	x		x	3
B–D	4	4	8	12	20	20	**B**	xx	x		x	3
							D	xx	x		x	2
B–C	3	3	6	9	15	15	**B**	xx	x		x	2
							C	xx	x		x	2
A–B	2	2	4	6	10	10	**A**					3
							B					1
A–C	1	1	2	3	5	5	**A**					2
							C					1
A–D	1	1	2	3	5	5	**A**					1
							D					1

Locations	1	2	3	4
Departments	A	C	D	B

The cost for ACDB = (1 × 5) + (1 × 1) + (2 × 5) + (5 × 1) + (3 × 3) + (4 × 2) = 38. We choose ACDB, because it gives a lower cost.

The solution could be checked against the lower bound, and if there is room for further improvement we might apply a two-way interchange procedure described in phase II of the next heuristic.

10.6 ANOTHER HEURISTIC PROCEDURE

We continue our discussion on the quadratic assignment problem by describing another heuristic procedure. The procedure starts with a good initial solution and tries to improve on it by checking all pairwise exchanges (i.e., interchanging placement of facilities from two locations). As with the elimination method, this is not a procedure that can guarantee the optimum, nor is there any good way of knowing which exchange of a pair may lead to an improvement. The procedure is similar to the one used in a computerized plant layout algorithm called CRAFT (computerized relative allocation of facilities technique), which has been a very popular approach in industry. This is because it can provide a good solution for a large value of facilities—locations combinations. Thus, having a value of n greater than 15 is not a problem. The procedure is divided into two phases. To begin with, a best approximate solution is found (we may use the elimination method solution as a starting point if desired), and then in the second phase, that initial solution is improved on by pairwise exchanges, to obtain an improved solution, if possible.

TABLE 10.30 First Allocation

	1	2	3	4
A	Yes			
B	No			
C	No			
D	No			

TABLE 10.31 Elimination Table 2

				Machine combination	Locations			
	2–3	3–4	2–4		2	3	4	EN
C–D	5	10	15	C D	x x		x x	2 2
B–D	4	8	12	B D	x x		x x	2 1
B–C	3	6	9	B C				1 1

TABLE 10.32 Second Allocation

	2	3	4
B			
C			
D	No	Yes	No

10.6.1 Phase I: Approximate Solution

Construct a Facility Chain

The purpose here is to connect all the facilities in a chain such that those facilities with the maximum flow between them are connected together. Each node (facility) of the chain may have at the most two other nodes (facilities) connected to it. The detailed steps are given as follows:

1. Construct a total flow table by adding the flow from facility i to facility j, to that from facility j to machine i for each pair of the facilities. As this table is symmetrical about the diagonal, only one half (top or bottom) need be constructed.
2. From the total flow table, select the maximum flow element and connect the facilities associated with this element. Mark this as step 1 connection.
3. Cross off this element from any further consideration in phase I. This results in the modified total flow table.

4. From the modified total flow table, determine the new maximum flow element in the remaining data.
5. Connect the facilities associated with this maximum flow element. Remember that a node may be connected to a maximum of only two other nodes. If either of the two facilities designated for connection already has two other nodes attached to it, then this new connection may not be made; hence, return to step 3. If the connection can be made, then do so and mark it as the next sequential step number. Continue with the procedure. The chain thus obtained may or may not be unique, but at this point we are only interested in obtaining a good starting solution. If the solution needs improvements, they will be made in phase II.
6. Check to see if all the facilities have been connected in the chain. If they are, proceed to the following section; if they have not, return to step 3.

Construct Location Chain

The steps followed in the development of a location chain are similar to those involved in constructing the facility chain. However, here we will work with the distance table and select the shortest distance in determining which locations to connect. The details are as follows:

1. From the distance table, select the minimum distance element and connect the locations associated with this element together. Note this as a step 1 connection.
2. Cross off this element from any further consideration in phase I. We will designate the resulting table the modified distance table.
3. From the modified distance table, determine the new minimum distance element in the remaining data.
4. Connect the locations associated with this element. As in constructing the facility chain, the location (node) may be connected to a maximum of only two other locations (nodes). If either of the two locations designated for connection are already attached to two other locations, then this new connection may not be made; hence, return to step 2. If the connection may be made, then do so and mark this as next sequential step connection. Continue with the procedure.
5. Check to see whether all the locations have been connected in the chain. If they have, proceed to the following section; if they have not, return to step 2.

Assignment

Finally, sequentially assign each facility in the facility chain to the corresponding location in the location chain; that is, the first facility in the facility chain is assigned to the first location in the location chain, the second facility to the second location, and so forth. The resulting pairing is the initial solution.

Another way of making initial assignments is to pair a facility node with a location node, to the extent possible, in the order (steps) in which the nodes were selected from their respective tables. It is not critical which method we use in making initial assignments, because the initial solution will be checked and improved on as appropriate. The number of calculations required in the second phase, however, may be influenced by the initial selected solution. One good indicator of a better initial arrangement is to check the cost associated with each and, then, select the one with the minimum cost.

10.6.2 Phase II: Improvement Routine

After obtaining the initial solution, it is necessary to determine whether improvement in the solution is possible. This is accomplished by interchanging the locations of two facilities while keeping all other facilities in their respective locations. If any improvement in the solution is obtained, this becomes the new interim solution, and the process continues. For K facilities, exchanging two at a time will require, $K!/(K-2)!2! = K(K-1)/2$ combinations. For example, for $K = 4$, we will have to check $4(3)/2 = 6$ combinations. The detailed solution procedure is developed in the following sections. Initially, it may appear complicated, but as we will observe later, such is not actually so. The procedure is divided into two steps.

Initial Solution

1. Construct a facility-to-facility distance table based on the present placement of the facilities. The order (sequence) of the facilities in this table must be the same as the order of the facilities in the flow table.
2. Multiply each element of the facility-to-facility distance table by the corresponding element of the flow table. Designate this as the solution table.
3. Total the elements in each row and in each column.
4. The value of this solution can be obtained in two ways: either by taking the grand total of all the columns or by taking the grand total of all the rows.

Improvement Check

We now check to see if any improvement is possible if facilities *i* and *j* are interchanged. For the purposes of this procedure, the interchange is called a two-way exchange, and the improvement may be checked in either of two ways. The first method is to calculate only the incremental value in the solution owing to the exchange, and the second is to calculate the solution value of the exchange. The first method may be preferable when small numbers of locations and facilities are to be investigated, whereas the incremental method results in considerable savings in time and effort when large numbers of facilities and locations are involved.

1. Begin with the solution table.
2. Interchange facilities *i* and *j* in the facility-to-facility distance table by interchanging the columns associated with the *i*th and *j*th facilities, and also interchanging the row associated with the *i*th and *j*th facilities. The order is immaterial; either the columns or the rows may be switched first. The net effect is to change the elements in the rows and columns of *i*th and *j*th facilities except for the intersecting elements; that is, elements *Cij*, *Cii*, *Cji*, and *Cjj*. Because the *i*th and *j*th rows and columns are the only ones of interest, we shall ignore all other elements. We will refer to the resulting matrix as an exchange table.
3. The next step is to take the product, element by element, of the interchanged rows and columns from the exchange table and the corresponding elements from the flow table. Denote this as the product table.
4. Take the sums of all the elements in columns *i* and *j*, and the sums of all the elements in rows *i* and *j* from the product table. Calculate the grand total by adding all four totals together.
5. Add the four totals associated with the rows and columns *i* and *j* in the solution table.
6. If the grand total in step 4 is less than the grand total in step 5, the exchange of facilities *i* and *j* is profitable, and we proceed to step 8. On the other hand, if the grand total in step 4 is more than the grand total in step 5, then locations of facilities *i* and *j* should not be interchanged, and we proceed to step 7.
7. See if all possible combinations have been checked. If they have, go directly to step 10; if not, continue with the next combination by assigning appropriate values for *i* and *j*. Return to step 2.
8. Form the new facility-to-facility distance table using the exchanged elements and the remaining elements of the previous facility-to-facility distance table.

9. Construct the new solution table by repeating steps 2 through 4 in the Initial Solution section, using the new facility-to-facility distance table. Return to step 7 in the improvement check section.
10. We have the optimum solution.

Sample Problem 1

Consider placing four machines in four locations. The expected flows between the machines are given in Table 10.33 and the distances between the locations are provided in Table 10.34. Note, unlike in the previous problem, in this problem letters A, B, C, D are the locations and numbers 1, 2, 3, 4 are the machines. Practically, it makes no difference in the notation we use to denote machines and locations.

In general, there is no reason to expect that flow from machine i to machine j, Cij, would be the same as the flow from machine j to machine i, Cji. The distance matrix, though, is generally symmetrical on the diagonal, indicating that the distance from location i to location j is the same as from j to i. This is not a requirement of the solution procedure discussed earlier.

Initial Solution. From Table 10.33 we develop the total flow table, by adding flow from i to j plus flow from j to i as in Table 10.35.

To build the facility chain, observe that the largest element is 13, associated with machines 1 and 2; therefore, we connect these machines together.

(1)–(2)

Remove element 13 from further consideration. The next largest element in the table is 12, the flow between machines 3 and 4, indicating these facilities should be connected together.

(3)–(4)

Remove the element 12 from further consideration; also note that the next largest element is 11, which will mean that facilities 3 and 2 should be connected together. Inasmuch as both 3 and 2 presently have only one node connected to each, 3 and 2 can be connected together. Thus we have:

(1)–(2)–(3)–(4)

as the facility chain. Because all the facilities are now connected, the chain is complete.

The next step is to develop the distance chain. In the distance table (see Table 10.34), the elements are symmetrical about the diagonal; therefore, only the elements above or below diagonal need be considered. The smallest element above the diagonal is 2, associated with locations A and C, and locations B and D. We may break the tie arbitrarily by selecting locations A and C and joining them together.

(A)–(C)

TABLE 10.33 Traffic Flow

Origin (machine)	Destination (machine)			
	1	2	3	4
1	0	10	5	6
2	3	0	3	6
3	5	8	0	9
4	1	4	3	0

TABLE 10.34 Distance Data

Origin (location)	Destination (location)			
	A	B	C	D
A	0	6	2	4
B	6	0	5	2
C	2	5	0	3
D	4	2	3	0

The next smallest element is the 2 associated with the distance between locations B and D. We will connect B and D together.

(B)–(D)

The third smallest distance element is 3 associated with C–D. Connecting locations C and D gives:

(A)–(C)–(D)–(B)

as the completed chain.

Pairing sequentially from the facility and location chains, we obtain the following assignment:

TABLE 10.35 Total Flow

	Machine			
Machine	1	2	3	4
1	0	13	10	7
2		0	11	10
3			0	12
4				0

Facility	1	2	3	4
Location	A	C	D	B

or

Location	A	B	C	D
Facility	1	4	2	3

The next step is to develop the facility-to-facility distance table. First, we merge the initial solution with the distance matrix (see Table 10.34) as shown in Table 10.36.

We will now arrange the elements in the preliminary distance table so that the facilities are in the same order as the facilities in the flow table (i.e., in the order 1, 2, 3, 4). This arrangement leads to the facility-to-facility distance table (Table 10.37). This table will be the basis for calculating the solution table as described in step 2 of the Initial Solution section.

Multiplying each element of the facility-to-facility distance table (see Table 10.37) by the corresponding element of the flow table (see Table 10.33), we obtain the initial solution, Table 10.38.

The sum of the row or column totals is the grand total, 215. This is the value of our initial solution.

Improvement Check. There are four facilities, and if we interchange two at a time, there will be ${}^{n}C_{m} = 4!/2!(2!) = 6$ combinations to check. These combinations are $1 \rightarrow 2$ (i.e., facility 1 exchanges with facility 2), $1 \rightarrow 3$, $1 \rightarrow 4$, $2 \rightarrow 3$, $2 \rightarrow 4$, and $3 \rightarrow 4$. Investigate the effects of all six interchanges, one at a time, to see whether any one or more will improve the solution.

Incidentally, in this example, the minimum cross-product theorem (lower bond) gives us the following listing:

Flow	13	12	11	10	10	7
Distance	2	2	3	4	5	6

With the cross product of $13 \times 2 + 12 \times 2 + 11 \times 3 + 10 \times 4 + 10 \times 5 + 7 \times 6 = 215$.

As our solution has the same cost, the solution is optimum, and there is no need to check any further.

Sample Problem 2

Initial Solution. In the previous analysis, no new solution table had to be formed because the initial assignment proved to be the best assignment. This,

TABLE 10.36 Preliminary Facility-to-Facility Distance

Location		A	B	C	D
	Facility	1	4	2	3
A	1	0	6	2	4
B	4	6	0	5	2
C	2	2	5	0	3
D	3	4	2	3	0

however, may not always be true. To demonstrate the method of further exchange when the initial solution is not the optimum, we will select the initial following assignment, which is deliberately different from that previously chosen.

Facility	1	2	3	4
Location	C	D	B	A

Table 10.39 identifies the corresponding facility-to-facility distance table derived from the foregoing assignments and the distance data given in Table 10.34. Multiplication of each element of Table 10.39 by its respective flow element from Table 10.33 produces Table 10.40, the initial solution with a total cost of 237 units.

This is greater than the lower bond established by the cross-product theorem of 215; hence, we continue with an improvement check.

Improvement Check. This phase of the procedure is to check for improvement of the solution by interchanging pairs of facilities. Recall that there are six such pairs.

TABLE 10.37 Facility-to-Facility Distance

Location		A	C	D	B
	Facility	1	2	3	4
A	1	0	2	4	6
C	2	2	0	3	5
D	3	4	3	0	2
B	4	6	5	2	0

TABLE 10.38 Initial Solution Values

	Facility				
Facility	1	2	3	4	Total
1	0	20	20	36	76
2	6	0	9	30	45
3	20	24	0	18	62
4	6	20	6	0	32
Total	32	64	35	84	215

1. Interchanging facilities 1 and 2 results in the following updated facility assignments:

Facility	1	2	3	4
Location	D	C	B	A

The associated facility-to-facility distance table is Table 10.41. By multiplying the elements of Table 10.41 by their corresponding flow elements from Table 10.35 results in the modified solution found in Table 10.42. Because the total cost shown in Table 10.42 is less than the cost in Table 10.40, the new arrangement becomes the current solution.

2. To check if the next exchange of facilities 1 and 3 is profitable, we should start with Table 10.36 inasmuch as it is the current facility-to-facility distance table. Exchanging facilities 1 and 3 will result in the following assignments:

TABLE 10.39 Facility-to-Facility Distance

		C	D	B	A
Location	Facility	1	2	3	4
C	1	0	3	5	2
D	2	3	0	2	4
B	3	5	2	0	6
A	4	2	4	6	0

Facility	1	2	3	4
Location	B	C	D	A

The new facility-to-facility distance table, Table 10.43, is derived simply by switching rows 1 and 3, and then columns 1 and 3 (we could have used the same method as when converting Table 10.41 from Table 10.39 earlier). To demonstrate the method, we have reproduced Table 10.41 as the original matrix, then switched rows 1 and 3 and finished by interchanging columns 1 and 3.

In the same manner as for earlier solutions, the elements of Table 10.43 are multiplied by their corresponding flow elements of Table 10.35 to give us Table 10.44, the second modified solution value table.

The total cost, 228, is less than the 234 from Table 9.42; therefore, the exchange should be made. The distance matrix (see Table 10.43) becomes the current facility-to-facility distance table, and it is used in developing the next exchange table.

3. Checking the remaining pairs of facilities (1 and 4, 2 and 3, 2 and 4, and 3 and 4) is left as an exercise for the reader. The final solution obtained should show a total cost 215 with the assignments listed as follows. It is, as one might anticipate, the same as that calculated in earlier section.

Facility	1	2	3	4
Location	A	C	D	B

10.7 THREE-WAY EXCHANGE

The exchange of positions of three facilities at a time is aptly called the three-way exchange. For example, the exchange of $i \rightarrow j \rightarrow k$, meaning that facility i is to be moved to where facility j is, facility j is to be located where facility k is, and

TABLE 10.40 Initial Solution Values

	Facility				
Facility	1	2	3	4	Total
1	0	30	25	12	67
2	9	0	6	24	39
3	25	16	0	54	95
4	2	16	18	0	36
Total	36	62	49	90	237

TABLE 10.41 Facility-to-Facility distance (1 → 2)

		D	C	B	A
Location	Facility	1	2	3	4
D	1	0	3	2	4
C	2	3	0	5	2
B	3	2	5	0	6
A	4	4	2	6	0

facility k will be placed where facility i is, is a three-way exchange. The procedure to evaluate the effectiveness of the move is similar to that for the two-way exchange. The new distance table is constructed by first performing the i to j exchange and then the j to k exchange. For example, consider the initial facility assignment as follows:

Facility	1	2	3	4
Location	C	D	B	A

Table 10.45 lists the facility-to-facility distance data when we merge the foregoing initial solution with the distance matrix (see Table 10.36). Now, suppose we

0	3	2	4	2	5	0	6	0	5	2	6
3	0	5	2	3	0	5	2	5	0	3	2
2	5	0	6	0	3	2	4	2	3	0	4
4	2	6	0	4	2	6	0	6	2	4	0
a. Original matrix				b. Switch rows 1 and 3 of a.				c. Switch columns 1 and 3 of b.			

wish to check the 1 → 2 → 3 exchange. The new facility assignments are as shown. Facility 4 is not changed.

Facility	1	2	3	4
Location	D	B	C	A

TABLE 10.42 Modified Solution Values (1 → 2)

	Facility				
Facility	1	2	3	4	Total
1	0	30	10	24	64
2	9	0	15	12	36
3	10	40	0	54	104
4	4	8	18	0	30
Total	23	78	43	90	234

TABLE 10.43 Facility-to-Facility Distance (1 → 3)

		B	C	D	A
Location	Facility	1	2	3	4
B	1	0	5	2	6
C	2	5	0	3	2
D	3	2	3	0	4
A	4	6	2	4	0

TABLE 10.44 Modified Solution Values (1 → 3)

	Facility				
Facility	1	2	3	4	Total
1	0	50	10	36	96
2	15	0	9	12	36
3	10	24	0	36	70
4	6	8	12	0	26
Total	31	82	31	84	228

To develop the new facility-to-facility distance table, we will first exchange facilities 1 and 2, and then facilities 2 and 3. Our original matrix is taken from Table 10.45.

The final matrix resulting when columns 2 and 3 are exchanged is the body of Table 10.46, the new facility-to-facility distance table when facilities 1, 2, and 3 are interchanged.

With the same method as before, the solution is obtained by multiplying the elements of Table 10.46 by the corresponding elements in the flow Table 10.34.

When the multiplication is performed, Table 10.47 shows the costs between the various facilities and the total of 223 for this modification to the initial solution.

Because this three-way solution of 223 is more expensive than 215, the optimum solution obtained by the two-way exchange (by the way, we know that this is the optimum solution), this exchange is rejected and we may continue by making other three-way exchanges.

In this example, with four facilities, we will have the following three-way exchanges: $1 \to 2 \to 3$, $1 \to 2 \to 4$, $1 \to 3 \to 4$, $2 \to 3 \to 4$, call these primary arrangements along with their complimentary arrangements of $1 \to 3 \to 2$, $1 \to 4 \to 2$, $1 \to 4 \to 3$, and $2 \to 4 \to 3$. Suppose we had six facilities; the three-way exchanges involved then would be as follows:

$$1 \to 2 \to 3, 1 \to 2 \to 4, 1 \to 2 \to 5, 1 \to 2 \to 6, 1 \to 3 \to 4,$$
$$1 \to 3 \to 5, 1 \to 3 \to 6, 1 \to 4 \to 5, 1 \to 5 \to 6, 2 \to 3 \to 4,$$
$$2 \to 4 \to 5, 2 \to 5 \to 6, 3 \to 4 \to 5, 3 \to 5 \to 6, \text{ and } 4 \to 5 \to 6,$$

along with their 15 compliments we would have a total of 30 combinations to examine.

This gives us a generalized rule in forming the combinations. If there are M facilities, the possible primary three-way exchanges are

$$i \to j \to k$$
$$i = 1, 2, 3, \ldots, (M-2)$$

For each i, $j = (i+1),\ (i+2), \ldots, (M-1)$
For each j, $k = (j+1),\ (j+2), \ldots, (M)$

TABLE 10.45 Preliminary Facility-to-Facility Distance

Location		C	D	B	A
Facility		1	2	3	4
C	1	0	3	5	2
D	2	3	0	2	4
B	3	5	2	0	6
A	4	2	4	6	0

0	3	5	2	3	0	2	4	0	3	2	4
3	0	2	4	0	3	5	2	3	0	5	2
5	2	0	6	5	2	0	6	2	5	0	6
2	4	6	0	2	4	6	0	4	2	6	0
a. Original matrix				b. Switch rows 1 and 2 of a.				c. Switch columns 1 and 2 of b.			

0	3	2	4	0	3	2	4	0	2	3	4
3	0	5	2	2	5	0	6	2	0	5	6
2	5	0	6	3	0	5	2	3	5	0	2
4	2	6	0	4	2	6	0	4	6	2	0
c. New matrix				d. Switch rows 2 and 3 of c.				e. Switch columns 2 and 3 of d.			

Knowing this primary arrangement, we can easily develop an associated equal number of complementary arrangements.

10.8 MULTIPLE EXCHANGE

If we perform interchanges with more than three facilities at a time, the procedure involved is similar to that previously described. For example, if the exchange involves $a \rightarrow b \rightarrow c \rightarrow d \rightarrow e$, then the modified facility-to-facility distance table is constructed by making transfers in the sequence: $a \rightarrow b$, $b \rightarrow c$, $c \rightarrow d$, and $d \rightarrow e$. It is not necessary to construct a transfer table for $e \rightarrow a$ because the

TABLE 10.46 Facility-to-Facility Distance ($1 \rightarrow 2 \rightarrow 3$)

		D	B	C	A
Location	Facility	1	2	3	4
D	1	0	2	3	4
B	2	2	0	5	6
C	3	3	5	0	2
A	4	4	6	2	0

TABLE 10.47 Modified Solution Values (1 → 2 → 3)

Facility	1	2	3	4	Total
1	0	20	15	24	59
2	6	0	15	36	57
3	15	40	0	18	73
4	4	24	6	0	34
Total	25	84	36	78	223

procedure will result in a distance table that will incorporate that transfer automatically. Once the modified distance table is obtained, the solution can be obtained by multiplying the elements of the new distance table by the corresponding elements of the flow table, and then taking the grand total.

One can try all possible exchanges; however, usually, the two-way exchange, or at the most, the three-way exchange, will prove sufficient to obtain a solution that is either the optimum or very close to it.

10.9 FIXED AND VARIABLE COST

When both fixed and flow costs are present, the problem is a combination of the two problems we have studied before. The exact solution procedure is quite complex mathematically. It requires using integer 0–1 programming. Here, however, we will present a procedure that will give an approximation that, at least, is very close to the optimum solution, and often, it will even provide the exact solution.

The procedure involves using a combination of the steps we have studied in the previous two sections. Consider the same problem that we have already examined; however, now let us add the fixed cost matrix to the problem, as shown in Table 10.48.

The matrix indicates that if machine 1 were placed in location A, it would cost \$5; in location B, it would cost \$3 and so on.

10.9.1 Optimizing with Relative to Variable Cost

We will proceed by first ignoring the fixed cost matrix and working only on the flow cost matrix. As shown in the previous section, the approximate first solution to the problem is as follows:

Machine	1	2	3	4
Location	A	C	D	B

The variable cost (flow cost) corresponding to this solution is 215. The associated fixed cost from Table 10.48 is 18 units, and the total cost is 233 units.

Now, let us observe the change in each segment of the cost as we examine the effect of exchanging facilities. If the change will decrease the total cost, not just the variable cost, then that assignment will become the new solution matrix, and the procedure will continue.

Exchanging $1 \rightarrow 2$ increases the variable cost by $221-217=4$ units. The corresponding fixed cost is $4+3+5+3=15$, a decrease of $18-15=3$ units. The net change, however, is an increase of $4-3=1$ unit. Similar calculations for other exchanges are summarized in Table 10.49. Based on that data, the best solution is the minimum variable cost solution with a total cost equal to $233.

10.9.2 Optimizing Relative to Fixed Cost

We will repeat the search procedure, except that this time our starting point will be the minimum fixed cost solution instead of the minimum variable cost solution. Start by applying the assignment procedure.

One possible solution of assignment algorithm is as follows:

Machine	1	2	3	4
Location	C	A	D	B

The fixed cost for this assignment is $13. We then must check the corresponding variable costs of this solution. The facility-to-facility distance matrix is presented as Table 10.50. When each of the distance elements of Table 10.50 is multiplied by its corresponding flow element from flow Table 10.33, the results are presented in Table 10.51 as the initial variable cost solution. This shows the variable cost to be equal to 219. Hence, the total cost for this solution is $219+13=232$. We next must investigate the effects of interchanging the various facilities, and Table 10.52 displays the modified distances when the first interchanges, rows 1 and 2 and columns 1 and 2 of Table 10.50 have been made.

TABLE 10.48 Fixed Costs

	Location			
Facility	A	B	C	D
1	5	3	2	4
2	3	8	5	6
3	3	5	5	6
4	4	2	3	4

TABLE 10.49 Facility Interchange Cost Summary

Exchange	Change in variable cost	Change in fixed cost	Net change	Exchange (yes/no)
1 → 3	227 − 205 = 22	15 − 18 = −3	+19	No
1 → 4	237 − 224 = 13	17 − 18 = −1	+12	No
2 → 3	218 − 206 = 12	18 − 18 = 0	+12	No
2 → 4	240 − 225 = 15	> 0	> 0	No
3 → 4	221 − 213 = 8	> 0	> 0	No

TABLE 10.50 Facility-to-Facility Distance

Location		C	A	D	B
	Facility	1	2	3	4
C	1	0	2	3	5
A	2	2	0	4	6
D	3	3	4	0	2
B	4	5	6	2	0

TABLE 10.51 Initial Solution (Variable Cost)

	Facility	1	2	3	4	Total
3	1	0	20	15	30	65
1	2	6	0	12	36	54
4	3	15	32	0	18	65
2	4	5	24	6	0	35
	Total	26	76	33	84	219

TABLE 10.52 Modified Facility-to-Facility distance (1 → 2)

Location		A	C	D	B
	Facility	1	2	3	4
A	1	0	2	4	6
C	2	2	0	3	5
D	3	4	3		
B	4	6	5		

TABLE 10.53 Modified Variable Cost (1 → 2)

Facility	1	2	3	4	Total
1	0	20	20	36	76
2	6	0	9	30	45
3	20	24			
4	6	20			
Total	32	64			

TABLE 10.54 Facility Interchange Cost Summary

Exchange	Change in variable cost	Change in fixed cost	Net change	Exchange (yes/no)
1 → 3	208 − 189 = 19	Not required	+	No
1 → 4	239 − 210 = 29	Not required	+	No
2 → 3	239 − 228 = 11	Not required	+	No
2 → 4	263 − 249 = 14	Not required	+	No
3 → 4	225 − 217 = 8	Not required	+	No

Table 10.53 is obtained by multiplying the distance elements of Table 10.52 by the corresponding flow elements in Table 10.33.

The change in variable cost is $(76 + 45 + 32 + 64) - (65 + 54 + 26 + 76) = -4$; the change in fixed cost is $(5 + 5 + 6 + 2) - 13 = 5$; and the net change is $-4 + 5 = 1$.

The exchange would increase the cost for the solution by 1 unit. Consequently, we will not implement this transformation.

Similar analyses of other exchanges leads to the summary in Table 10.54. We should realize that we started out with the lowest fixed cost solution, and any other solution will lead to an increase in the fixed cost. The only way the total cost will decrease is if there is a decrease in the variable cost. If, in an exchange, there is no such decrease, then there is no need to check the change in the fixed cost.

Given the foregoing summary, we conclude the initial solution of \$232, is the best in this fixed-cost sequence.

10.9.3 Selection of the Best Alternative

It is now necessary to compare the solutions from both sequences. Starting with the minimum variable cost led to a solution with a total cost of \$233, and as we just saw, beginning with the minimum fixed cost led us to a solution of \$232. Therefore, we select the solution with the lesser cost (i.e., \$232).

Exercises

10.1 Location analysis plays a very important role in ergonomics. A machine control panel is to be designed, taking into consideration the ergonomics. Five such keys are to be placed on the panel. A scale from 1 to 10 is calibrated in which the lower the value in the scale for a particular location, the more ergonomical it is to place the key at that location. The table shows such values for all the keys for each of the locations. Determine the optimum location for each key so that they are ergonomically friendly.

	Location				
Key	1	2	3	4	5
R	6	5	4	2	1
A	2	1	2	3	4
D	2	5	3	1	4
N	4	3	2	2	1
P	10	6	4	2	1

10.2 Flow between two new machines A and B and four existing machines 1, 2, 3, and 4 are given in the following table

	Existing machines			
New machines	1	2	3	4
A	10	15	12	5
B	8	10	15	7

Machines A and B may be placed in either locations X, Y, or Z. The distance between these locations and the existing machines are given in the next table. Determine the optimum location for A and B.

	Existing machine locations			
New locations	1	2	3	4
X	10	6	8	9
Y	7	12	9	5
Z	8	6	6	11

10.3 If in 10.2, owing to technical reasons machine A cannot be placed in location X what would be the optimum machine locations?

10.4 A hospital wishes to place three new file cabinets to maintain the patient records. The file cabinets will be used by four offices of different doctors in the hospital. File cabinets are of three types, depending on the size and height of the shelves. Each office needs to use the three types of cabinets. Depending on the type of hospital patients the flow from each office to each of the cabinets is given in the flow matrix. The hospital has identified four possible locations—W, X, Y, and Z—for these cabinets. The distance required to travel from the doctors, office to each of these locations is represented in the following distance matrix. Determine the optimum location of the file cabinets.

Flow Matrix

	Existing offices			
File cabinets	1	2	3	4
A	15	18	12	15
B	10	14	21	11
C	14	9	12	19

Distance Matrix

	Locations			
Existing offices	W	X	Y	Z
1	15	8	9	12
2	7	14	22	15
3	18	12	20	16
4	25	5	17	18

10.5 Why is placing facilities simultaneously important but at the same time, difficult? How is this problem different from the problem is question 10.2? By using the following flow and distance matrices develop the initial assignment using the heuristic explained in this book. Does the solution equal the lower-bound value of the problem? If not perform two iterations in two-way exchanges toward optimization.

Flow Matrix

	A	B	C	D
A	0	15	6	8
B	15	0	25	13
C	8	7	0	4
D	11	9	5	0

Distance Matrix

	1	2	3	4
1	–	7	4	2
2		–	8	3
3			–	9
4				–

10.6 Given the following flow and distance matrix determine

1. The optimum placement of departments A, B, C, and D and the associated cost using branch and bound method.
2. Determine the placements by the elimination procedure and compare your results with your solution in part 1.

Flow Matrix

	A	B	C	D
A	–	16	20	4
B	6	–	18	9
C	21	8	–	7
D	14	9	7	–

Distance Matrix

	1	2	3	4
1	–	6	2	4
2	6	–	8	9
3	2	8	–	7
4	4	9	7	–

10.7 There are five departments with material flow as shown in the flow matrix. The departments can be placed in any one of the five sheds 1, 2, 3, 4, 5. The distances between the sheds are shown in the distance matrix.

Flow Matrix

	A	B	C	D	E
A	–	2	1	3	5
B	5	–	3	6	4
C	0	4	–	–	7
D	2	7	4	–	8
E	5	3	7	4	–

Distance Matrix

	1	2	3	4	5
1	–	15	16	11	12
2		–	6	12	8
3			–	19	10
4				–	8
5					–

1. Determine the lower bound for the solution.
2. Determine the best location of the departments and the associated cost by initial phase of the heuristic method.
3. Solve the problem by the elimination procedure and compare the results with part 2.
4. Choosing best solution from between 2 and 3 and apply one iteration of the two-way exchange.

10.8 A large manufacturer of castings has recently decided to invest in four foundries. Because of the large investment, the company has decided to buy old machines along with the new ones to be placed in these facilities. Owing to the unequal technology in these plants, there will always be a flow between these facilities as shown in the flow matrix. The company has also narrowed the choice down to four locations, with the distance between them as shown in the distance matrix. Help decide the locations of the four foundries in these four locations to incur the minimum cost

Flow Matrix

	A	B	C	D
A	–	15	20	5
B		–	10	25
C			–	10
D				–

Distance Matrix

	1	2	3	4
1	–	6	3	5
2		–	2	6
3			–	4
4				–

10.9 How does the allocation change in problem 10.8 if there is a fixed cost associated with placing a facility in a particular location. The cost is shown in the cost matrix for each facility, for each location.

Cost Matrix

	1	2	3	4
A	5	4	6	8
B	8	6	9	10
C	3	4	7	5
D	8	2	6	8

10.10 A carpenter wishes to place five new machines in five possible locations in his workshop. Given the average expected flow between machines in a day in the flow matrix and the distances between locations in the distance matrix, determine the optimum locations for the five machines and the associated cost.

Flow Matrix

From (machine)	To (machine)				
	A	B	C	D	E
A	0	5	10	5	10
B	5	0	9	7	8
C	2	2	0	1	1
D	3	3	6	0	7
E	8	4	4	9	0

Distance Matrix

From locations	To locations				
	1	2	3	4	5
1	0	25	40	70	15
2	25	0	80	50	45
3	40	80	0	55	75
4	70	50	55	0	65
5	15	45	75	65	0

10.11 What is a three-way exchange? Supposing in problem 10.8 we start with initial solution of

Machines	A	D	C	B
Locations	1	2	3	4

Develop initial machine-to-machine distance table. Perform the following three-way interchange:

A → D
D → C
C → A

What is the associated cost?

10.12 A research laboratory purchases three pieces of equipment that perform various tests. During an experiment tests are required to be performed on this equipment. There is flow between the equipment depending on the type of experiments conducted. The laboratory has three places to locate the equipment. Given the fixed cost, traffic flow, and distance data determine the best location for each piece of equipment using the branch and bound method. Compare the solution using the heuristic method.

Fixed Cost

	Location		
Machines	1	2	3
A	3	11	8
B	5	13	10
C	9	4	12

Traffic Flow

	Destination (machine)		
Origin (machine)	A	B	C
A	0	100	150
B	200	0	75
C	80	175	0

Distance Data

	Locations		
Locations	1	2	3
1	0	10	8
2	10	0	7
3	8	7	0

10.13 A retail manufacturer has to locate four warehouses at four sites in the country. These warehouses are located in different parts of the country. Owing to seasonal changes, the demand for a particular product may be higher from one warehouse than from another. Hence, there is flow between these warehouses to keep up with the demands. The flow between the warehouses change once a year as shown in the flow matrices. The distance between these locations is shown in the distance matrix. Find how allocation of these warehouses to the locations should be made. Assume that the warehouses once assigned to these locations cannot be relocated. (Assume average flow to determine location.)

Flow Matrix 1

	W1	W2	W3	W4
W1	–	12	23	9
W2		–	15	10
W3			–	8
W4				–

Flow Matrix 2

	W1	W2	W3	W4
W1	–	16	19	7
W2		–	13	15
W3			–	14
W4				–

Distance Matrix

	1	2	3	4
1	–	12	18	20
2		–	15	21
3			–	8
4				–

11

Transportation Network Problems

So far in a network analysis (outlined in Chapter 7), a node had represented a customer and service was provided by a new facility that was placed either on a node or on a branch. The objective was to develop (or select) the optimum location for the new facility. Consider now a set of problems for which some nodes represent customers, whereas others represent suppliers. We want to provide service from the existing facilities (suppliers) to existing customers to satisfy some objective. Thus, the problem now is to allocate the customers to the existing facilities, and the only paths we are concerned about are the paths between customers and suppliers. When the objective is to minimize the cost of transportation by selecting specific paths from supplier to customers, the problem is called transportation problem. There are several good books in operations research that illustrate the procedures for solving such problems. Most readers are familiar with transportation, for a brief description was provided in Section 9.2.

11.1 APPLICATION OF THE TRANSPORTATION ALGORITHM

Here is one example of a simple allocation problem. Flowers that are cut fresh in a garden need to be transported to a flower market within 12 h. There are seven flower gardens in the estate and the product can be sold in three different markets. A different profit is associated with each market because of its location. Each

TABLE 11.1 Flower Gardens: Market Cost Data

	A	B	C
1	70	58	M
2	73	60	84
3	M	M	80
4	68	56	M
5	M	63	81
6	72	60	M
7	67	M	82

garden has a limit on amount it can supply and, similarly, each market site has a limit on how much can be sold in that market. Table 11.1 shows different markets accessible to the flower gardens within 12 h, and if so the profit associated with it. Those not accessible are marked "M". The accessibility is determined by the road conditions and time and distance involved in making the trip for which a road network is important. The problem is to develop optimum strategy of distribution of flowers to maximize the profit.

We can transform the problem into an equivalent transportation problem for which the objective is to minimize the total "transportation cost." This is done by converting profit into cost by subtracting each cost element (not M element) from a large number, say 100. Knowing the amount available to sell at each garden and amount that can be sold in each market we obtain Table 11.2.

The total demand is for 250 units, whereas the total supply is 270 units. A dummy column, representing a dummy demand of 20 units, as shown in Table 11.3, must be added to make the total demand equal to the total supply. And

TABLE 11.2 Tea Gardens: Market Profit Matrix

	A	B	C	Supply
1	30	42	M	45
2	28	40	16	35
3	M	M	20	30
4	32	44	M	45
5	M	37	19	50
6	28	40	M	35
7	33	M	18	30
Demand	100	80	70	

TABLE 11.3 Data for Application of Transportation Algorithm

	A	B	C	D	Supply
1	30	42	M	0	45
2	28	40	16	0	35
3	M	M	20	0	30
4	32	44	M	0	45
5	M	37	19	0	50
6	28	40	M	0	35
7	33	M	18	0	30
Demand	100	80	70	20	170

TABLE 11.4 Solution to Transportation Problem

	A		B		C		D		Supply
1	45	30		42		M		0	45
2	20	28	5	40	10	16		0	35
3		M		M	30	20		0	30
4		32	25	44		M	20	0	45
5		M	50	37		19		0	50
6	35	28		40		M		0	35
7		33		M	30	18		0	30
Demand	100		80		70		20		

because the dummy column represents a supply that is never shipped associated cost is marked with zero values.

Solving the algorithm results in the final assignments shown in Table 11.4. The profit associated with this solution is

$$(45*70) + (20*73) + (5*60) + (10*84) + (30*80) + (25*56) + (20*0)$$
$$+ (50*63) + (35*72) + (30*82) = 17{,}680.$$

11.2 MINIMIZATION OF THE MAXIMUM TRANSPORTATION TIME

Frequently, minimizing the maximum time takes the form of a standard transportation problem. We have different customers, each with a specific

TABLE 11.5 Cost–Demand–Supply Data

Customer	Warehouse 1	2	3	4	Demand in units
1	20	30	20	30	39
2	50	25	20	40	20
3	40	20	42	45	32
Supply	20	30	25	15	

demand, being supplied from different warehouses (production facilities) with limited supplies. It is necessary to ship these items, which may include perishable goods, military equipment, critical parts in production, or even human transplant organs, so that maximum time of transportation between any "warehouse" to any "customer" is as small as possible. For example, consider the data given in Table 11.5. The "cost" of transportation here represents the time to transport (truckload, carload) between a warehouse and a customer, and we wish to develop a shipping schedule in which the maximum time of shipment between a warehouse and customer is as small as possible.

The procedure starts by seeking an initial solution. As we have seen before, an assignment in a cell indicates the number of units transported from a source to a destination. We will use the "least-cost rule" for obtaining an initial solution. The process involves assigning maximum possible value to the lowest-possible "cost" cell. In Table 11.5, the least cost is 20, found in cells (1,1), (1,3), (2,3), and (3,2). Choose any one of these and assign it as much as possible. For example, select cell (1,1). The maximum amount that can be assigned here is 20 units, limited by the supply from garden (warehouse) 1. The corresponding column is stricken from any future consideration because the supply has been exhausted. Continuing in a similar manner we obtain a least-cost solution as shown in Table 11.6. To reach this particular solution, the cells were assigned values in the sequence (1,1), (3,2), (1,3), (2,3), and (2,4).

Because this is a transportation problem, we must have $m + n - 1$ variables with values (basic variables), in which m is the number of rows and n is the number of columns. In our problem we should have $3 + 4 - 1 = 6$ assignments, which we do. If that had not been so, we would have created new (dummy) assignments. The maximum time associated with this solution is (cell 3, 4). We will now follow a procedure that may reduce this time. The first step is to convert each time element into 0 or a 1 cost element (modified cost) based on the

TABLE 11.6 Initial Solution

Customer	Gardens 1	2	3	4	Demand
1	20 20	30	20 19	30	39
2	50	25	20 6	40 14	20
3	40	20 30	42	45 1	31
Supply	20	30	25	15	90

following rules (for convenience we place these costs in the corner beside the time values as shown in Table 11.7):

1. If a cell has a time that is less than the present maximum, assign that cell 0 cost.
2. If a cell has a time that is equal to or greater than the present maximum, and does not have an entry in it, drop that cell from further consideration in this phase, mark by X.
3. If the time associated with a filled cell is equal to the maximum time, or if there are also some other filled cells with the same maximum time; assign the cost for each of those cells as 1.

Application of these rules to Table 11.7 results in Table 11.8.

TABLE 11.7 Cost Modification I

Customer	Gardens 1	2	3	4	Demand
1	20 0 20	30 0	20 0 19	30 0	39
2	50 X	25 0	20 0 6	40 0 14	20
3	40 0	20 0 30	42 0	45 1 1	31
Supply	20	30	25	15	90

TABLE 11.8 U–V Method Application, Cycle I

	Gardens					
Customer	1	2	3	4	Demand	U_i
1	20 0 20 $-\theta$	30 0	20 0 19 $+\theta$	30 0	39	0
2	50 X	25 0	20 0 6 $-\theta$	40 0 14 $+\theta$	20	0
3	40 0 −1 $+\theta$	20 0 30	42 0 −1	45 1 1 $-\theta$	31	1
Supply	20	30	25	15		
V_j	0	−1	0	0		

Now we will use the U–V method on the modified cost data in Table 11.7. Starting with $U_1 = 0$, the other values of U_i and V_j are developed in Table 11.8. The table also shows the relative cost values for empty cells, $(C_{ij} - U_i - V_j)$, which are negative. (All other values are zero or positive.)

Any one of the negative cells can be used to bring new variables into the solution. Let us select the cell with the minimum time among these, (3,1), to be a part of the new solution. Table 11.8 also shows the quantity that can be moved, as well as the corresponding θ loop. The maximum amount that can be moved in this loop is 1 unit. Moving this reduces the assignment in the cell with a "1" cost; that is, the present maximum time cell (cell 3, 4) to 0.

Table 11.9 shows the present assignments. The maximum time is now 40 associated with cells (3,1) and (2,4). All the unfilled cells with costs equal to or greater than 40 are scratched as shown in Table 11.9.

Continuing with the procedure, we will convert the original time values in Table 11.9 to a 0–1 cost value and mark them next to the time values. A cost of 1 is assigned to cells (3,1) and (2,4), the maximum time cell, and 0 to all other admissible cells. The values of U_i and V_j are calculated using a starting value of $U_1 = 0$ as shown in Table 11.10. The table also shows the negative cell (1,4) in which θ units are to be moved.

The maximum value of θ is 15. Moving this many units still maintains the present cell (cell 1, 3) as the maximum cell; therefore, there is no change in the 0–1 cost data, except that cell (2,4), with the time of 40 released in the previous stage, now can be scratched. The Table 11.11 shows the results.

New U_i and V_j values are shown on the same table, starting with $U_1 = 0$. No admissible table has a negative relative cost; therefore, the present solution is

TABLE 11.9 Cycle II

Customer	Gardens 1	2	3	4	Demand
1	20 0 19	30 0	20 0 20	30 0	39
2	50 X	25 0	20 0 5	40 1 15	20
3	40 1 1	20 0 30	42 X	45 X	31
Supply	20	30	25	15	

TABLE 11.10 U–V Calculations Cycle II

Customer	Gardens 1	2	3	4	Demand	U_i
1	20 0 19	30 0	20 0 20 $-\theta$	30 0 −1 $+\theta$	39	0
2	50 X	25 0	20 0 5 $+\theta$	40 1 15 $-\theta$	20	0
3	40 1 1	20 0 30	42 X	45 X	31	1
Supply	20	30	25	15		
V_j	0	−1	0	1		

TABLE 11.11 Cycle III

Customer	Gardens 1	2	3	4	Demand	U_i
1	20 0 19	30 0	20 0 5	30 0 15	39	0
2	50 X	25 0	20 0 20	40 X	20	0
3	40 0 1	20 0 30	42 X	45 X	31	1
Supply	20	30	25	15		
V_j	0	−1	0	0		

optimal. The corresponding maximum value of time is 40 associated with cell (3,1), which is the minimum possible value of the maximum transportation time.

11.3 NONLINEAR TRANSPORTATION TIME

Frequently, the time required to ship items from warehouse to a customer consists of two components: a direct component, which depends on only the locations of a warehouse and its customer; and an indirect component, which depends on the quantity shipped. For example, if all of the quantity that is demanded by a customer could be shipped in one truckload from the warehouse, a small variation in this shipment will not affect the time of transportation because the truck must still make the trip. This is called the direct, or fixed, time of the shipment. The indirect time, however, is directly proportional to quantity delivered, and may be due to factors such as modifying each unit to meet the customer's specifications, loading and unloading each unit, and even perhaps attaching a pamphlet containing operating instructions or a parts list to each unit.

The time required for both the direct and indirect components may vary with the customer–warehouse combination. For instance, food items, such as shrimp, may be packed differently depending on the geographic location of the customer and how long the shrimp will have to remain fresh after the delivery. The packaging variation may include using either plain or dry ice, the possibility of either requiring or not needing treating the shrimp before packaging, and using small, cellophane-wrapped packages or large bulk container storage. The available facilities in each boat terminal or processing plant (warehouse) may cause the indirect time to vary for each type of packaging. The direct time depends on the distance involved between the customer and the processing plant. The objective in this type of problem is to devise a shipping schedule that will minimize the maximum supply time.

11.3.1 Solution Procedure

Here, we will assume, as we did earlier in the transportation problems, that the total demand equals the total supply. If this is not actually true, however, an appropriate dummy customer or warehouse will have to be created, with a 0 transportation costs, to take up the slack.

Because the cost in this case is not linear, the problem is no longer a true transportation problem. The number of variables forming a solution, therefore, are not restricted to $m + n - 1$ (number of warehouses plus number of customers minus 1). Indeed, the number of variables in the solution is restricted only by the objective of the problem, that of minimization of maximum time.

The steps involved in solving the problem are listed in the following (refer to the solved example).

Initial Assignments

1. Determine the minimum fixed time component for each customer.
2. Select the customer with the maximum fixed time from among the times in step 1. Set the present value of the maximum time as the value of the next higher fixed cost associated with the selected customer.
3. Form inequalities using the cells that have fixed time components less than or equal to the present maximum time.
   a. In a manner similar to that for the transportation problems, we will define X_{ij} as the variable indicating the quantity shipped from warehouse j to customer i. Each inequality consists of the sum of the fixed time component and the variable time component multiplied by the corresponding variable X_{ij}^k the kth component of X_{ij}. The lower limit of the variable is 0, whereas the upper limit is set by the maximum time for the step.
   b. The value of the left-hand side of each inequality must always remain less than or equal to the right-hand side. To achieve this, assign values to each variable in each inequality, increasing them in a manner such that the total time associated with each inequality is approximately the same as for the others. Stop assigning any further values to a variable if:

      1. The unassigned demand from a customer has been reduced to zero.
      2. Any further assignment will make the total time of the left-hand side of the equation greater than the right-hand side in step 3a.

   c. For each customer, reduce the present value of the unassigned demand (initially the entire demand is unassigned demand) by the assignments made in step 3b. This gives new values for the unassigned demands, and if for any customer the value if not 0, proceed directly to step 3e; if they are all equal to 0, go to step 3d.
   d. Take the sum of the values of the variable components in each step (x_{ij}^k). This gives the net value of the variable. Enter this in the appropriate cell of the tabular data format. Go to to step 1 of "Making the Solution Feasible."
   e. Using the present value for each variable, determine the total transportation time for each inequality. For each variable, formulate a new inequality consisting of the left-hand side of the total transportation time so far plus the variable time, that sum multiplied by the value for that variable, as identified in this step. The right-hand side is the next higher fixed time (higher than present maximum value) if one exists. If all of the fixed time values are

less than or equal to the present maximum, the right- hand side is expressed symbolically as U, where the value of U is to be determined by an iterative process. Return to step 3b.

Making the Solution Feasible

The initial solution obtained by the previous steps may or may not be feasible. The following steps will obtain the feasibility.

1. Take the sum of each column and list it in the "Assign" row (see the example problem). Subtract the amount supplied for each column from that assigned, and if the difference is positive, list it in the "Excess" row (excess assignment). Similarly, if the difference is negative, there is some slack; that is, not all the available supply from the warehouse has been distributed. Indicate this slack in the "Shortage" row.
2. Select the column with the greatest shortage and perform the following steps.
   a. Formulate inequalities from the cells of the selected column having fixed times less than the present maximum.
   b. Set the right-hand side equal to the present maximum value.
   c. Assign the values to the variables so that the total time of any inequality does not exceed the current maximum time, and simultaneously reduce the shortage by the total amount that is assigned to the variables. If the shortage goes to zero, proceed to step 3. If no additional value can be assigned, go to step 2d.
   d. Take the next higher fixed cost in the column greater than the present maximum and designate it as the new maximum time. If there is no cell in the column with a fixed cost greater than the present maximum, then denote the maximum time as U. The value of U will be determined so that all of the shortage is eliminated using the least possible value for U. Continue on to the next step.
   e. Using the present assignments, calculate the fixed time for each inequality. Form a new set of inequalities to include all the variables that have a fixed cost less than or equal to the present maximum value. Return to step 3b, from the initial assignments.
3. Follow steps 2a through 2e for each of the shortage columns. When all of the shortage columns have been examined, proceed to step 4.
4. Using the present maximum value, determine the maximum amount to be added to each of the variables in each of the shortage columns. Enter these values in the appropriate cells.
5. Make moves from the excess columns to the shortage columns, starting with the column that determined the limiting value of the maximum

time; the details are similar to the θ procedure, in transportation algorithm. If the θ move is not possible, go to step 6, otherwise, in this step we should reach a feasible schedule.

6. Perform moves to transfer the present assignments and to create cell availability for θ moves (details are shown in the examples). If the cells can be created from θ moves, go to step 5, otherwise, increase the present maximum time by one time and return back to step 2.

11.3.2 Illustrative Example

We will demonstrate the procedure by applying it to a sample problem. Suppose a sausage-producing plant has three facilities (warehouses) where the processing can be performed. There are three grocery chains who buy sausage from this manufacturer, but each has its own recipe (specification) and location to ship, varying the delivery time and the time for processing demand from a customer. In addition, the processing times are also influenced by the equipment and manpower available in the plants. Because the manufacturer wants to fill his orders as quickly as possible so that he can maintain the quality of the product for each customer, his main objective is to minimize the maximum time of delivery.

The cost coefficient in each cell indicates the fixed transportation time component, which is independent of the quantity shipped, and a variable unit time component, which is proportional to the quantity processed and shipped. For example, if 10 units were shipped from warehouse 1 to customer A, the total time for shipment would be 35 time units ($25 + 1 \times 10$). The problem is to develop a schedule that will minimize the maximum transportation time.

Application of the Solution Procedure

For the data in Table 11.12, Table 11.13 displays the fixed minimum time for each customer. The notation used in Table 11.14 in the solution of the inequalities, is of the form $a + bX_{ij}^k$, where X_{ij}^k is the amount shipped in the kth step, i and j (both numbers) indicate the customer and the warehouse, respectively, the constant a is

TABLE 11.12 Shipping and Preparation Times

	Warehouse			
Customer	1	2	3	Demand
A	25, 1	30, 1	20, 2	10
B	40, 1	25, 2	20, 5	18
C	40, 1	20, 1	30, 3	22
Supply	15	20	15	

TABLE 11.13 Minimum Fixed Times

Customer (number)	Time
A (1)	20
B (2)	20
C (3)	20

TABLE 11.14 Customer C Calculation

		Variable				
Step	Inequality	x^k_{31}	x^k_{32}	x^k_{33}	Unfilled demand	Maximum time in step k
0					22	20
1	$20 + 1x^1_{32} <= 30$		10		12	30
2	$30 + 1x^2_{32} <= 40$		9		3	39
	$30 + 3x^2_{33} <= 40$			3	0	
Total		0	19	3		

the fixed shipping time, and the coefficient b is the preparation time per unit shipped. (Note that k is not an exponent for the variable X_{ij}.) Because i is a number, the customer designations of A, B, and C should be considered interchangeable with 1, 2, and 3, respectively.

In Table 11.13, we see that the maximum of the minimum times is 20 for all customers. We choose one of these, customer C, at random and begin our procedure there. Basically we will be determining for each customer what portion of its demand should be allotted to each warehouse to minimize the maximum of the transportation times (fixed plus variable components). Obtaining an initial solution requires the assumption that each warehouse has an infinite supply. Because this is obviously an invalid assumption, we must use an iterative procedure to bring the initial solution to the optimum. The details are discussed in the following paragraphs and summarized in the tables.

Initial Assignments. Step 0 begins by indicating the initial unfilled demand of customer C. In step 1, the new maximum time, 30, is the value of the fixed time that is next higher than 20, for customer C. There is only one cell 3.2 or 32 that has a fixed cost less than the present maximum; hence, the one inequality $20 + 1X^1_{32} <= 30$, is formed: X^1_{32} can be made 10, reducing the unfilled demand to 12.

TABLE 11.15 Customer A Calculation

Step	Inequality	Variable x_{11}^k	x_{12}^k	x_{13}^k	Unfilled demand	Maximum time in step k
0					10	39
1	$25 + 1x_{11}^1 <= 39$	10			0	39
	$30 + 1x_{12}^1 <= 39$		0			
	$20 + 2x_{13}^1 <= 39$			0		
	Total	10	0	0		

The next step is to increase the maximum time value to the next higher fixed cost and form the inequalities as shown in Table 11.14 for step 2. Only two inequalities are possible because of limitations on maximum time.

Step 2 may be of some special interest because it clearly illustrates simultaneous value assignments in each inequality. The new maximum time is fixed at 40 (cell 31) in this step. However, the use of 40 will make the unfilled demand go to 0, with partial assignments to X_{33}^2 ($X_{32}^2 = 10, X_{33}^2 = 3$). It may, therefore, be possible to reduce the maximum time and still be able to reduce unfilled demand to 0. With some manipulation, we observe that the maximum time of 39 is sufficient to achieve this objective. The corresponding variable values are shown in Table 11.14.

The total value of each variable is obtained by adding the assignments in each step for the particular variable. With a maximum time of 39 as the beginning value, we move on to customer A. The details of the calculations are similar to those described for customer C, and the results are shown in Table 11.15.

The maximum time is still 39. Continue the analysis with customer B. The results are shown in Table 11.16. Step 3 is of interest. U in the right-hand side of the inequality stands for open upper limit. The value of which should be adjusted to a minimum value, which permits the unfilled demand to 0. This is a cut-and-try method, and in this step the least value of U has to be 45 to reduce the unfilled demand to 0. Again note that we could have obtained different assignments, but we chose the values given in the table.

All of the customers have now been checked, and we enter the values for the appropriate variables in Table 11.17. These are the initial assignments.

Taking the sum of the assignments in a particular column gives the present assigned total for that warehouse. The total assignment in each column is not equal to what is available for that column. The difference is the excess or the shortage and is shown as such in the appropriate row. Table 11.17 summarizes the initial assignments.

TABLE 11.16 Customer B Calculations

		Variable				
Step	Inequality	x^k_{21}	x^k_{22}	x^k_{32}	Unfilled demand	Maximum time in step k
0					18	39
1	$25 + 2x^1_{22} <= 39$		7		11	
	$20 + 5x^1_{23} <= 39$			3	8	
2	$40 + 1x^2_{21} <= 40$					40
	$39 + 2x^2_{22} <= 40$					
	$35 + 5x^2_{23} <= 40$			1	7	
3	$40 + 1x^3_{21} <= U$	5			2	45
	$39 + 2x^3_{22} <= U$		2		0	
	$40 + 5x^3_{23} <= U$					
	Total	5	9	4		

TABLE 11.17 Initial Assignments

	Warehouse			
Customer	1	2	3	Demand
A	25, 1	30, 1	20, 2	
	10			10
B	40, 1	25, 2	20, 5	
	5	9	4	18
C	40, 1	20, 1	30, 3	
		19	3	22
Supply	15	20	15	
Assign	15	28	7	
Excess	0	8		
Short	0		8	

Making the Solution Feasible

The next few steps are to convert the initial assignment table to a feasible solution, if the solution is infeasible.

Shortage Column Analysis. The shortage (or maximum shortage) of 8 is in column 3; therefore, we will begin with that column. Using the present

TABLE 11.18 Column 3 Calculations

Step	Inequality	Variable x^k_{13}	x^k_{23}	x^k_{33}	Shortage in step K	Maximum time in step k
0					8	45
1	$20 + 2x^1_{13} <= 45$	8				
	$20 + 5x^1_{23} <= 45$				0	
	$30 + 3x^1_{33} <= 45$					
	Total	8	0	0		

maximum time of 45 and the associated inequalities, we proceed in much the same manner as before, but this time we are reducing a shortage instead of filling a demand (working on column rather than row). Because all of the shortage could not be eliminated with a maximum time of 45, and there is no other shortage column to examine, the maximum time remains 45.

The purpose for checking the shortage columns was to determine the limit for the maximum time; namely, does the problem appear to be feasible with the maximum time determined in the previous calculations, or to accommodate for the shortage, will this time have to be increased at this stage? In this case, we have found that the maximum time of 45 is sufficiently large to take care of all of the shortage. The "assignments" in Table 11.18 were actually assignments only in the sense that they were temporarily made to be able to make this determination. Now that we have the maximum time, the assignments revert back to those of Table 11.17, and we will continue with the procedure to reconcile the excesses and shortages. The next step is to calculate the maximum possible value for each of the variables in the shortage column(s) based on the present maximum time. Table 11.19 lists the values so determined.

The next calculations are made to determine the maximum slack available in each cell of the shortage columns. This is done by subtracting the present assigmnent in the cell from the maximum possible value for that cell. The values are entered in the upper right-hand corner of each of the appropriate cells in each

TABLE 11.19 Maximum Possible Values in the Shortage Columns

Inequality	Maximum possible variable value
$20 + 2x_{13} <= 45$	$x_{13} = 12$
$20 + 5x_{23} <= 45$	$x_{23} = 5$
$30 + 3x_{33} <= 45$	$x_{33} = 5$

TABLE 11.20 Initial Assignment and Slack

	Warehouse			
Customer	1	2	3	Demand
A			12	
	10			10
B			1	
	5	9	4	18
C			2	
		19	3	22
Supply	15	20	15	
Assigned	15	28	7	
Excess	0	8		
Shortage	0		8	

of the shortage columns as shown in Table 11.20. The table also shows the entire present assignment. Note that for convenience the cost data is now eliminated from the table, for it plays very little part in the calculations, and when necessary, one can always refer back to the original cost table (see Table 11.12).

Theta (θ) Procedure. We now develop a chain in which θ represents the amount to be moved such that all of the excess assignments are transferred to the shortage column(s) and reduce that shortage(s) there to 0. Note that when an assignment is moved to a cell in a shortage column, that cell must have a slack sufficient to accommodate the transfer, otherwise minimum of the maximum time will increase.

Begin by selecting a cell having slack available and that is located in a shortage column. Increase the assignment in that cell by θ, increase the amount in the assigned row for that column by θ, and reduce the amount in the shortage row for that column. For the same customer, subtract θ from the assignment in an excess column, and likewise subtract from both the assigned and excess rows for that column. The value of θ can be no more than the smallest of the following: (1) the slack available in the cell selected, (2) the amount in the shortage row for the column in which the cell is located, (3) the amount already assigned for the customer in the excess column, or (4) the amount listed in the excess row for the column that is providing the assignment for transfer. Repeat the process for different cells until all shortages and excesses have been eliminated.

We continue in Table 11.21 with the solution of our sample problem. Cell B–3, with an assignment of 4 and a slack of 1, is chosen first. The assignment becomes $4 + \theta$, the slack is $1 - \theta$, and the assigned and shortage rows for column 3 become $7 + \theta$ and $8 - \theta$, respectively. Cell B–2 is chosen to supply the excess, and the assignment there changes to $9 - \theta$. The assigned and excess amounts for column 2 becomes $28 - \theta$ and $8 - \theta$, respectively. Investigation to determine the value for θ finds that the limiting factor is the slack available in cell B–3.

Many different θ-chains could have been formed in Table 11.21. The selection of a specific theta-chain is immaterial as long as both the shortage and the excess are reduced by the indicated move.

When we make the moves indicated, Table 11.22 shows the result. The procedure followed in simultaneously reducing the excesses and shortages to zero is a series of such iterations.

TABLE 11.21 θ-Chain

	Warehouse			
Customer	1	2	3	Demand
A			12	
	10			10
B			$1 - \theta$	
	5	$9 - \theta$	$4 + \theta$	18
C			2	
		19	3	22
Supply	15	20	15	
Assigned	15	$28 - \theta$	$7 + \theta$	
Excess		$8 - \theta$		
Shortage			$8 - \theta$	

TABLE 11.22 First θ-Moves

	Warehouse			
Customer	1	2	3	Demand
A			12	
	10			10
B			0	
	5	8	5	18
C			2	
		19	3	22
Supply	15	20	15	
Assigned	15	27	8	
Excess	0	7		
Shortage	0		7	

TABLE 11.23 Second θ-Chain

Customer	Warehouse 1	2	3	Demand
A	10		12	10
B	5	8	5	18
C		$19-\theta$	$2-\theta$ $3+\theta$	22
Supply	15	20	15	
Assigned	15	$27-\theta$	$8+\theta$	
Excess		$7-\theta$		
Shortage			$7-\theta$	

The second θ chain is shown in Table 11.23. Slack is available in shortage column 3 (cell C–3), and there is an assignment in excess colunm 2 in that same row (cell C–2). Therefore, θ is added to the assignment for C–3, subtracted from the slack for that same cell, added to the assigned row for column 3, subtracted from the assigned row for that same column, subtracted from the assignment for cell C–2, subtracted from the assigned row for column 2, and subtracted from the excess row for that same column. The maximum value of 0 is now 2, and the resulting table is shown as Table 11.24.

TABLE 11.24 Third θ-Chain

Customer	Warehouse 1	2	3	Demand
A	$10-\gamma$		12 $+\gamma$	10
B	5	8	0 5	18
C	$+\gamma$	17	0 $5-\gamma$	22
Supply	15	20	15	
Assigned	15	25	10	
Excess		5		
Shortage			5	

It is not possible to find a θ-chain very conveniently in Table 11.24. To form a chain, we need at least one entry for a customer in both the excess and the shortage columns, and there must also be slack available in the shortage column. (An excess column is one with excess total assignments, and a shortage column is one with less than total required assignments.) Because we cannot find such a customer, a θ-chain is not readily available. We now apply a γ-chain procedure to shift some assignments as follows.

We must shift the present assignments to develop a cell, which can accommodate all or part of the shortage. As cell A–3 has slack capacity of 12 units, we can move γ units in that cell. But that means taking off γ units from cell A–1. We can not add any units in B–1 without exceeding the maximum time of 45; therefore, go to the next cell C–1. Up to 5 units can be added in cell C–1 without going over the maximum time. γ units must be taken off from C–3. The loop is thus complete. The maximum permissible value of γ is 5, defined by both cell C–1 and C–3. Moving 5 units or making $\gamma = 5$ results in Table 11.25.

Now we can apply new θ-loop consisting of excess in column 2, and shortage in column 3. The result is also shown in Table 11.25.

Because only 4 units of shortage are present, we will move 5 units in this θ move to reduce the excess and shortage for columns 2 and 3, respectively. The final solution is presented in Table 11.26.

In this section, we have seen a procedure to minimize the maximum time when the cost (time) of delivery is not linear. The method is lengthy, but that is generally characteristic of any iterative procedure. With practice, one can run through it quickly and accurately, or one may use a computer to handle the details. Even though the preparation time (variable time) is assumed in this chapter to be linearly proportional to the quantity shipped, the same procedure may be used if that function is also nonlinear.

TABLE 11.25 Resultant of a γ-Move and New θ-Loop

	Warehouse			
Customer	1	2	3	Demand
A	5		5	10
B	5	8	5	18
C	5	$17 - \theta$	$0 + \theta$	22
Supply	15	20	15	
Assigned	15	$25 - \theta$	$10 + \theta$	
Excess		$5 - \theta$		
Shortage			$5 - \theta$	

TABLE 11.26 Final Solution

	Warehouse			
Customer	1	2	3	Demand
A	5		5	10
B	5	8	5	18
C	5	12	5	22
Supply	15	20	15	
Assigned	15	20	15	
Excess	0	0	0	
Shortage	0	0	0	

Exercises

11.1 There are four warehouses that store frozen fish. These warehouses serve three customers in different parts of the country. The capacity of each warehouse and the distances of the warehouses from the customers are shown in the following table. Find the minimum value of the maximum distance necessary from any warehouse to serve these customers, so that the fish can be efficiently distributed.

	Warehouse				
Customer	1	2	3	4	Demand
1	29	36	43	18	34
2	47	24	42	34	31
3	32	50	21	13	45
Supply	35	40	20	15	

11.2 A local Pizza King has three pizzerias from which he can supply the parties in town. To prepare each batch requires a large oven and each location has a different type oven (some old, some new) that restricts the maximum number of fresh pizzas that can be made in a reasonable time, and are noted as supply limit.

Given below is a matrix representing the fixed time of travel and bake, plus the variable time for making each pizza for three parties for which he has just received the orders. Determine how pizzas should be supplied to minimize the maximum time to serve all customers.

Parties	Locations 1	2	3	Order quantity
A	15, 2	20, 3	35, 1	25
B	10, 3	40, 2	25, 3	18
C	30, 3	10, 1	20, 2	12
Supply limit	20	25	10	

11.3 Three customers are to be supplied with their demands from three warehouses 1, 2, and 3. The fixed setup times and the variable times for each warehouse corresponding to each customer are shown in the table following. Make the assignments in such a way that the maximum time taken to serve all the customers is minimized.

Customer	Warehouse 1	2	3	Demand
A	20, 1	30, 2	35, 2	16
B	30, 2	20, 1	40, 1	22
C	25, 4	30, 3	20, 5	15
Supply	20	14	18	

11.4 Shipping and preparation times for customers A, B, and C from each warehouse, 1 and 2, are given in the following table. The total time for shipping follows the equation $Y = a + bX$, where a is the fixed time and b is the variable time. Each quantity is displayed in the table within each cell, the first number being a and the second number being b. Determine the optimum schedule to minimize the maximum time in shipment.

Customer	Warehouse 1	2	Demand
A	20, 2	15, 3	10
B	10, 1	3, 2	15
C	8, 2	9, 1	10
Supply	18	17	

11.5 In the manufacture of silicon chips in an semiconductor industry, it is required that the chips be transported between stations with minimum transportation time. There are four existing stations that supply the chips to the next stage. Each station has different capacity, owing to the different type of machines at these stations. The next stage has three new machines that can process these chips according to their capacity. Because to the large capital invested for these new machines, each should be fully utilized to its capacity. The matrix shows the transportation time between the stations to the new machines of the next stage. Determine how the chips should be supplied to the new machines to minimize the maximum transportation time.

	Old machine				
New machine	1	2	3	4	Demand
1	30	40	45	60	30
2	70	45	50	30	35
3	20	60	40	50	40
Supply	20	15	30	40	

11.6 An automobile manufacturer recently introduced a new SUV into the market. Because of its success it has demand from three regions that have huge markets for this SUV. The SUV is manufactured at four different facilities in the country, with each production facility having a limited supply. The table shows the time required to satisfy unit demand from production facility to the customer (region). So that the customer is not lost the company wants to minimize the maximum time of transportation between the production facility and the customer. If so determine the optimum assignments.

Customer	Production facility				Demand
1	30	23	15	30	42
2	40	75	55	20	20
3	60	40	60	35	28
Supply	20	35	15	20	

11.7 A frozen food industry wants to ship its product, in minimum time possible, from its warehouse. The fixed setup times and the variable times for each warehouse corresponding to each customer are shown here in the table.

Make the assignments in such a way that the maximum time taken to serve all the customers is minimized.

	Warehouse			
Customer	1	2	3	Demand
A	40, 1	20, 2	30, 2	25
B	35, 2	40, 1	30, 1	20
C	25, 4	15, 3	15, 5	15
Supply	25	15	20	

12

Allocation–Selection Models in the Production Environment

Basic premises in location allocation problems have been to select appropriate locations and assign customers to these location to minimize cost and provide the necessary service. So far we have seen the examples of facilities location models as they are applied in transportation, social, and economic topics. In this chapter we will study some additional variations of the cost structures that make the problems somewhat complex to analyze. Because these cost structures are easily understood in manufacturing terms, the models are called models in the production environment.

12.1 LINEAR PRODUCTION COST PROBLEM

Consider a variation of site selection problem within the context of production planning in a global environment. We have seen international companies periodically evaluating their production options, opening plants at some locations while closing plants at other locations. The underlying reason is economics, but how does it work? We will take a somewhat simple example to illustrate the concepts. Suppose the production cost, G_j, at a location j (plant) can be expressed as $G_j = A_j + b_j F_j$, where A_j is the lump sum component (setup or fixed cost), b_j is the variable cost of production per unit, and F_j is the total production (demand) assigned to a location. Depending on the local economic and labor conditions, the fixed cost and variable cost are different in different locations. The customer demand originates from various locations, and the cost of transportation from production location j to customer i is noted as C_{ij}. The problem is to decide how much production to have in each location (some locations may not have any production), and how the demands from the customers should be satisfied. There are a number of examples in which such a cost structure exists even within a small geographic area. For instance, in the warehouse example in Chapter 3, the total cost of building a warehouse at a given location may consist of the cost of land and cost of construction. The construction cost depends on how large the warehouse must be, which in turn, depends on the demands (customers) that the warehouse is assigned to serve. There is no reason to expect that the cost of the land or the cost of construction, A_j and b_j would be identical in any available location. Other similar examples may include the placement of health care centers, laundromats, and even tool rooms within a manufacturing facility.

This problem can be formulated as a 0–1 linear-programming problem, or alternatively, it can be solved by a heuristic solution method explained in Section 12.2.

12.1.1 Mathematical Formulation

The problem could be stated as follows:

$$\text{Min} \sum_{j=1}^{m} A_j I_j + \sum_{j=1}^{m} b_j F_j + \sum_{j=1}^{m} \sum_{i=1}^{m} C_{ij} d_{ij}$$

Subject to constraints

$$\sum_{j=1} d_{ij} = D_i \qquad \text{for} \qquad i = 1, 2, \ldots, n$$

$$\sum_{i=1}^{N} d_{ij} = F_j \qquad \text{for} \qquad j = 1, 2, \ldots, n$$

$$I_j M - F_j \geq 0 \qquad \text{for} \qquad j = 1, 2, \ldots, m$$

when $I_j = 0$, M is a large positive number; d_{ij} is the demand from customer i to location j.

The first constraint assures us that all the demand from source i, D_i, is assigned to a facility location. The second constraint measures the total demand assigned to the location j, given by F_j. The third constraint assigns the value to I_j. If $F_j > 0$ (i.e., a demand is assigned to location j) then $I_j = 1$. On the other hand if $F_j = 0$, then the objective function will force I_j to be equal to 0. The solution to the problem is obtained by application of the 0–1 programming technique.

12.1.2 Alternative Solution Procedure

To solve this problem, we can use the fixed cost procedure of Chapter 3, with one simple modification. The transportation cost element, from each customer i to each location j, is now modified to include the unit production cost associated with location j (i.e., the new cij equals the old $C_{ij} + b_j$ for $i = 1, \ldots, n$ and $j = 1, \ldots, m$. The fixed cost location j is given by the setup cost (fixed cost) A_j. With this data, we can apply the procedure detailed in Section 3.2 and find the desired solution.

Illustrative Example

A pharmaceutical manufacturer whose products are sold in all parts of the country is in the process of establishing two regional headquarters that would serve the customers more efficiently, and at the same time, would perhaps also decrease the cost of production and distribution. Six cities are under consideration: (1) New York, (2) Atlanta, (3) New Orleans, (4) Dallas, (5) San Francisco, and (6) Chicago. Costs of operation in the cities vary and depend on how large a territory the facility is assigned to serve. The variation occurs as a result of differing labor costs, rents, taxes, and insurance premiums applicable in the different cities. Table 12.1 gives the cost data for the various cities.

For convenience, the country is divided into five territories. The annual demand from each region and the unit distribution (transportation) cost from each city under consideration to each region are given in Table 12.2.

TABLE 12.1 Production Cost Data

	City					
Type of cost	1	2	3	4	5	6
Fixed cost per year	10,000	8500	6500	7500	9500	9000
Variable cost per year per unit of production	10	9	12	13	15	6

TABLE 12.2 Distribution Cost and Regional Demand

	Cities						
Regions	1	2	3	4	5	6	Demand
A	2	3	4	6	8	0	500
B	5	3	1	2	5	14	600
C	4	4	5	3	5	6	800
D	10	8	15	12	2	8	400
E	8	7	10	10	5	8	900

TABLE 12.3 Distribution Cost and Regional Demand

	Cities						
Regions	1	2	3	4	5	6	Demand
A	12	12	16	19	23	6	500
B	15	12	13	15	20	20	600
C	14	13	17	16	20	12	800
D	20	17	27	25	17	14	400
E	18	16	22	23	20	14	900
Fixed cost	10,000	8,500	4,000	7,500	9,500	9,000	

TABLE 12.4 Optimum Solution

	Selected locations	
Customers	3	6
A		X
B	X	
C		X
D		X
E		X

To decide where to locate the new regional offices and assign the territories they will serve, simply add the per unit distribution and service cost and then apply the fixed cost algorithm to give the solution. The associated cost table is shown in Table 12.3 and the solution by applying procedure from Section 3.2 is displayed in Table 12.4.

The minimum cost is $6 \times 500 + 13 \times 600 + 12 \times 800 + 14 \times 400 + 14 \times 900 + 4000 + 9000 = 51{,}600$.

12.2 QUANTITY DISCOUNT PROBLEM

Now we study more realistic cost structure, the one for which economy of large-scale production can be accommodated. For example, if production quantity is large, more efficient production machines and processes, more effective material handling equipment, better storage and warehousing facilities, and productive plant layout, may be employed. These alternatives although high in first cost may be justified because of large production quantity may result in lowering the unit cost. We call this a quantity discount problem.

Another example is when the discount applies to transportation cost based on units ordered and shipped. A large order quantity may mean a truckload of supply with less unit price for shipment. A smaller quantity may have to be shipped by other means, such as parcel postal or a messinger service (e.g., UPS). We may, for example, have one or more suppliers who could provide various parts to our plant (customers). The price quoted for a product by each supplier shows a fixed charge for each unit, a minimum fixed transportation cost per order, and a variable cost of production and transportation based on the total quantity ordered. There is a quantity discount on this cost, the larger the order the less the unit cost.

Example

Consider an example: We have plants in locations A, B, C, ..., requiring units that can be supplied by manufacturers in locations 1, 2, 3, ... The transportation cost varies based on the distances; however, unit cost of the product depends on the total quantity ordered from the supplier. As the quantity ordered is increased the unit production cost decreases, and the supplier passes part of the savings on to the customer by means of a quantity discount. The discount given by each supplier depends on his cost of production. Thus, the transportation cost, the handling cost, and the quantity discount associated with a purchase (a step function) vary from supplier to supplier. Let $Z_i(tj)$ describe the total purchase cost under the quantity discount structure associated with location i when tj units are ordered. The decision is to choose the supplier and decide the quantity to order from that supplier to minimize the total cost of operation for our plants. In this example, we will also assume that a customer's order cannot be broken and supplied by more than one source, even though the procedure can be modified to accommodate the relaxation of this assumption.

Production or Purchase Cost

$$Z1(t1) = \begin{cases} 1.2t1 & 0 < t1 < 200 \\ 1.2*200 + 1.1(t1 - 200) & 200 < t1 < 700 \\ 240 + 550 + 0.9(t1 - 700) & t1 > 700 \end{cases}$$

$$Z2(t2) = \begin{cases} 1.3t2 & 0 < t2 < 300 \\ 390 + (1.1)(t2 - 300) & 300 < t2 < 800 \\ 390 + 550 + 0.8(t2 - 800) & t2 > 800 \end{cases}$$

$$Z3(t3) = \begin{cases} 1.4t3 & 0 < t3 < 450 \\ 630 + 1.0(t3 - 450) & 450 < t3 < 1000 \\ 630 + 550 + 0.9(t3 - 1000) & t3 > 1000 \end{cases}$$

$$Z4(t4) = \begin{cases} 1.1t4 & 0 < t4 < 100 \\ 110 + 1.0(t4 - 100) & 100 < t4 < 600 \\ 110 + 500 + 0.9(t4 - 600) & t4 > 600 \end{cases}$$

Table 12.5 provides various data. For example, it costs \$5/unit for transportation between supplier 1 and plant A, and in addition, the processing cost is \$330. The total demand cost is dependent on the quantity discount function. For the first 200 units ordered from supplier 1, the cost is \$1.2/unit; if between 200 and 700 units are ordered the cost is 200*1.2 = \$240 for the first 200 units and from there on, \$1.1/unit. If more than 700 units are ordered, the first 200 units cost \$1.2 each, the next 500 units cost \$1.1 each, and from there on the cost is \$0.9/unit.

Solution Procedure. As in the previous chapters, the solution procedure consists of determining which suppliers to select (where to place the facilities), and how many units to order from each (assignment of each customer's demands). Because of the cost of components involved, however, we need a procedure that is slightly different from the ones we have seen so far. The flow diagram is illustrated in Figure 12.1, and the detailed steps of the new procedure are as follows:

1. Determine for each location (supplier), the cost of bringing all the customers to that location. This cost consists of the transportation costs, fixed costs, and the production cost.
2. Select the location with the minimum cost as the location for the first facility. As before, a location with a facility is called an assigned location, and one without a facility is called an unassigned location. Similarly, when a customer is assigned to a location, that location becomes the source for the customer.

TABLE 12.5 Shipping and Purchase Cost from Location and Customer Demand

	Locations (Suppliers)				
Customer	1	2	3	4	Demand
A	5	4	3	5	100
B	3	4	2	1	50
C	2	3	1	4	80
D	6	5	2	3	120
E	2	4	5	3	70
F	1	2	3	4	160
G	4	2	3	3	220
H	3	4	5	3	300
FC	330	500	350	400	

Check each unchecked, unassigned location by forming steps 3 through 10. When all the locations have been checked, the current solution is optimal. Initially, all unassigned locations are unchecked.

3. Select an unassigned location as a candidate to become assigned and denote it as the destination location. Calculate the savings that could be obtained by moving the presently assigned customers to the destination location. Note, all the customers must be checked, even though they

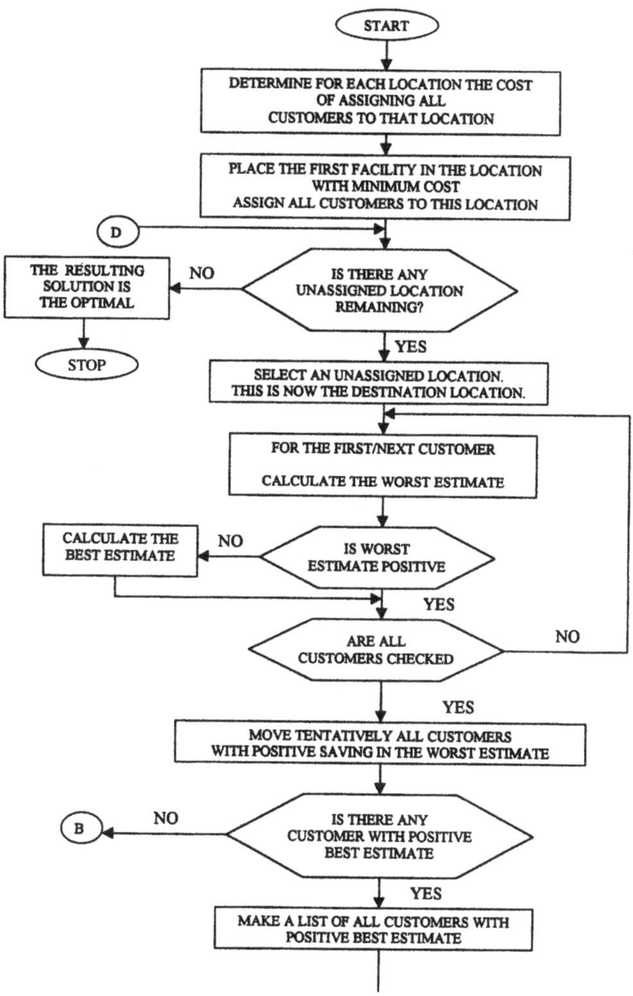

FIGURE 12.1 Quantity discount problem.

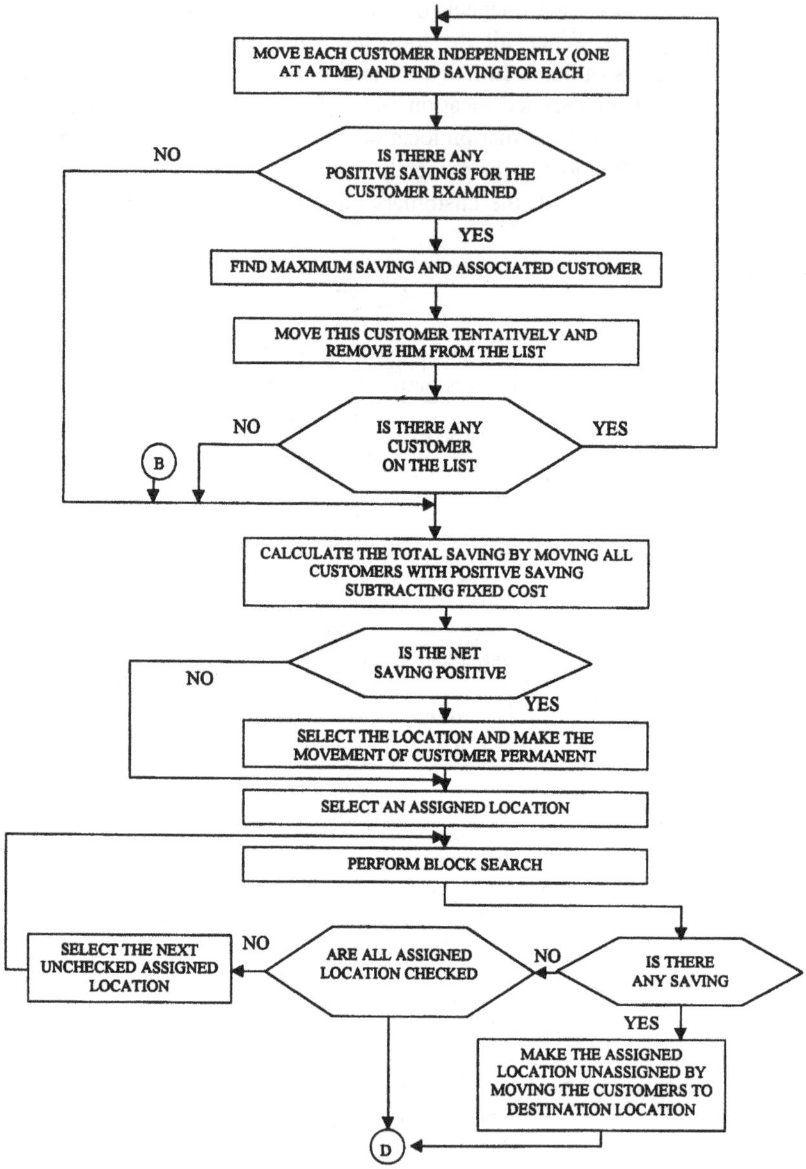

FIGURE 12.1 *(continued)*

may have been assigned to different locations. Savings for each customer consist of three components:

a. Savings in transportation cost: This equals the transportation cost for the customer at the source location less the transportation cost for the customer at the destination location.
b. Savings at the source: This equals the cost of production at the source station with the present assignment less the cost of production at that same location when the customer is moved.
c. Savings (negative cost) at the destination: This is the cost of production at the destination location with the present assignment less the cost of production at that destination when the customer is moved.

4. Sequence of assigmnent: Components b and c are dependent on the sequence in which the customers are moved, because the demand of the customer that is moved affects the total quantity assignment; therefore, the cost, both at the source and the destination locations. For example, if the source location 1 is presently assigned customers A, B, and C, with 100, 50, and 30 units of demand, respcctively, and if a check is being made for destination location 2, moving customer A first followed by customer C will result in 50 units at location 1 and 130 units in location 2. On the other hand, moving customer B first, followed by customer C will result in 100 units at location 1 and 80 units at location 2. The cost of moving customer C in each case is different. This being so, a procedure must be developed that indicates the sequence in which the customers that are under consideration for reassignment are to be moved. To develop such a sequence the following method is utilized

   a. As a first step, we calculate the least-possible savings (LPS) to be realized if the customer under consideration is moved. The three components in the LPS estimate are the following:

      (1) Transportation cost savings: This equals the transportation cost for the customer at the source location less the transportation cost for that same customer at the same destination.
      (2) Savings at the source: This is the cost of production for the customer at the lowest possible unit rate at the source.
      (3) Savings (negative cost) at the destination: This is the cost of production for the customer at the highest possible unit rate at the destination.

The sum of the foregoing components gives the least-possible savings for the customer if he is moved from the presently assigned location to the destination location. If these savings are positive, then, even under the worst conditions, it is profitable to move the customer. If the savings are negative, the customer does not show promise under the worst conditions, but may still be profitable to move if the existing conditions are not as bad as the worst possible conditions.

b. In the second step, for the customers with negative net savings in the LPS estimate, we calculate what the savings would be under the "best" conditions. The best conditions are those in which a customer is allowed to make a move in the manner that will result in the maximum savings. If savings under the best conditions are also negative, then the customer is no longer a candidate for the move. The best-case estimate is calculated as the sum of the following three savings:

(1) Transportation savings: This figure is calculated in the same manner as the LPS estimate.
(2) Savings at the source: This equals the cost of production for the customer at the source at the highest possible applicable unit rate.
(3) Savings (negative cost) at the destination: This cost of production for the customer at the destination at the lowest possible rate. If for a customer, the LPS estimate is negative, and the best estimate is positive, it might be possible to obtain savings if he is moved in a proper sequence. Such a customer is examined further.

5. Temporarily move all the customers with positive savings in the LPS table to the destination location.
6. If there are customers having both negative and positive savings in the LPS table and positive savings in the best-case table, list these customers and continue to step 7.
7. For each customer on the list, determine the net savings (sum of transportation, source, and destination savings) based on the present assignments. The customer with the greatest positive savings is selected to be moved first. If there is no customer with positive savings, go to step 8. Delete this customer from consideration and repeat step 7 for the remaining customers on the list. When all the customers have been moved, or if there is no customer with positive savings proceed to step 8.

8. Subtract the fixed cost of the destination location from the net savings obtained in step 6 or step 7, if that step was performed. If the result is positive, net savings are to be obtained by choosing the destination location under consideration and assigning the customers as in steps 6 and 7. Regardless of whether the location is selected, proceed to step 9.
9. Perform a block search (or block move) for each assigned location. *The block search* is defined as moving all of the demand from the previously assigned location to the destination location. In addition, to be included are the customers in the move that showed positive net savings in the LPS table and those that are assigned to locations other than the assigned location that is presently under consideration for a block move. Calculate the savings for all of the customers that have moved to the new destination. The customers with positive net savings in the LPS estimates are next checked relative to their present assignments following the procedure outlined in step 7. Determine the change in the fixed cost as the destination location becomes the assigned location, and the assigned location under the consideration becomes an unassigned location.
10. Determine if there are any net savings in steps 8 and 9. If there are any, choose the largest savings and permanently move the associated customers to the destination location. If there are no savings in either step, the destination location is not profitable; therefore, terminate it from any further consideration. This completes the check for the next assigned location.

A shortcut in the block move is possible if the destination location can be selected in step 8. If so we should select the location and move the customers contributing to the savings to that location. The block move now consists of independently checking for each assigned location, to determine whether savings can be realized by moving all the remaining customers from that location to the destination location. If additional savings can be made, this location is made an unassigned location and dropped from any further consideration.

Solution to the Quantity Discount Problem. We can now apply the solution procedure to the data for Table 12.6. Total demand from all customers is 1100 units. The first facility should be placed in a location where the cost of assignment of all the demand is a minimum. For example, cost for location 1 is $3610 + 330 + 240 + 500 + 0.9(1100 - 700) = 5090$. Table 12.6 shows the cost for each location.

Because the first location has the minimum total cost of \$5090, we begin by assigning all customers to that location. The next step is to check the unassigned location, starting with the location 2, for improvement (reduction in total cost) in the solution.

TABLE 12.6 Cost of Assigning all Demand to One Location

	Locations			
Customer	1	2	3	4
A	500	400	300	500
B	150	200	100	50
C	160	240	80	320
D	720	600	240	360
E	140	280	350	210
F	160	320	480	640
G	880	440	660	660
H	900	1200	1500	900
Total transportation cost	3610	3680	3710	3640
Fixed cost	330	500	350	400
Production cost (purchase)	1150	1180	1270	1060
Total cost	5090	5360	5330	5100

Location 2 Analysis. A savings table is constructed that shows the least possible savings (LPS) for each customer when moved to location 2. For example customer A, with a demand of 100, is presently assigned to location 1. If he were moved from this to location 2, the smallest possible savings in the production (in purchase), source savings (SS), at the location 1 is $0.9 * 100 = \$90$. The maximum cost at the destination for location 2 is $1.3 * 100 = \$130$ or in terms of savings at destination (SD), it is $-\$130$. The savings in transportation (ST) cost are $500 - 400 = \$100$. These numbers are first entered in the first row of Table 12.7. The net savings for the customers are $100 + 90 - 130 = \$60$, which shows that even under the worst conditions, it is possible or beneficial to move customer A to location 2.

TABLE 12.7 Least Possible Savings (LPS) to Location 2

Customer	Demand	ST	SS	SD	Net savings
A	100	100	90	−130	60
B	50	−50	45	−65	−70
C	80	−80	72	−104	−112
D	120	120	108	−156	72
E	70	−140	63	−91	−168
F	160	−160	144	−208	−224
G	220	440	198	−286	352
H	300	−300	270	−390	−420

Similar analyses are performed for all other customers. Positive net savings with the customers A, D, and G indicate that these customers can be moved profitably even under the worst conditions and the sequence of the move is immaterial.

Because the customers A, D, and G have shown positive savings in LPS, they need not be checked again in the maximum savings (MPS) calculations. Table 12.8 is a listing of the maximum possible savings (i.e., the savings at the source are calculated at the maximum rate and the costs are determined at the minimum rate). For example, for customer B, savings at the source are $1.2 * 50 = \$60$, and the net cost at the destination is $0.8 * 50 = \$40$. Negative net savings in this table indicate that the associated customer should not be moved even under the best of conditions.

The customers who have negative net savings in the LPS table, but positive net savings in the MPS table, are candidates for the move if the sequence is favorable. But as shown by the negative savings for all the customers in the Table 12.8 there is no such candidate.

We next check the effect of moving the customers A, D, and G. Table 12.9 provides the details. It should be noted that because the rates used in calculating SS and SD depend on the amount of demand, three customers must be treated as a group, not as individuals.

The total demand moved from location 1 is $100 + 120 + 220 = 440$ units. The savings at location 1, to which 1100 units are presently assigned by moving 440 units, is the difference in the cost of producing 1100 units and that of

TABLE 12.8 Maximum Possible Savings (MPS) for Location 2

Customer	Demand	ST	SS	SD	Net savings
B	50	−50	60	−40	−30
C	80	−80	96	−64	−48
E	70	−140	84	−56	−112
F	160	−160	192	−128	−96
H	300	−300	360	−240	−180

TABLE 12.9 Moving Positive Savings Customer to Location 2

		Location					
Customer	Demand	1		2			NS
		SS	SF	ST	SD	SF	
A, D, and G	440	404	0	660	−544	−500	20

producing $1100 - 440 = 660$ units. It can be calculated as $[1.2 * 200 + 1.1 * 500 + 0.9 * 400] - [1.2 * 200 + 1.1 * (660 - 200)] = \$404$. The cost at the location 2, at which 440 units are now assigned, is $1.3 * 300 + 1.1 * (440 - 300) = \$544$, and the fixed cost is \$500 (SF, the savings in the fixed cost is \$500). The savings in transportation is $100 + 120 + 440 = \$660$.

The net savings (NS), therefore, is \$20(660 + 404 − 544 − 500). Because they are positive, location 2 can be chosen, and customers A, D, and G are assigned there, as shown in the demand assignment Table 12.10.

The next step is to check the effect of a block move; namely, to determine the consequences if all the customers (B, C, E, F, and H) that are still assigned to location 1 are moved to location 2; however, it is not necessary to perform any calculations because we already know the result. We previously had calculated for Table 12.6 that the cost of assignments of all customers to location 2 is \$5360. The cost of the original solution was \$5090; therefore, the block move would not be profitable. Based on the foregoing discussion, location 2 is chosen only for customers A, D, and G, and the other customers remain at location 1. Table 12.10 shows the present assignment for all customers. The cost associated with this solution is $5090 - 20 = \$5070$.

Location 3 Analysis. Continuing the process, LPS and MLS calculations are made for location 3 and presented in Tables 12.11 and 12.12, respectively. Note that the customers A, D, and G, are moved from location 2 and customers B, C, E, F, and H are moved from location 1.

Because even under the best conditions, none of customers E, F, G, and H show net savings, they are not the candidates for a move.

TABLE 12.10 Present Assignments with Locations 1 and 2 as Assigned Locations

Customer	Locations				
	1	2	3	4	D_i
A		100			100
B	50				50
C	80				80
D		120			120
E	70				70
F	160				160
G		220			220
H	300				300
Total	660	440			1100

TABLE 12.11 Least Possible Savings (for Location 3)

Customer	Demand	ST	SS	SD	NS
A	100	100	110	−140	70
B	50	50	55	−70	35
C	80	80	88	−112	56
D	120	360	132	−168	324
E	70	−210	77	−98	−231
F	160	−320	176	−224	−368
G	220	−220	258	−308	−270
H	300	−600	330	−420	−690

TABLE 12.12 Maximum Possible Savings (for Location 3)

Customer	Demand	ST	SS	SD	NS
E	70	−210	84	−63	−189
F	160	−320	192	−144	−272
G	220	−220	286	−198	−132
H	300	−600	350	−270	−520

Table 12.13 shows the savings derived as customers A, B, C, and D are moved to location 3. Savings from the source location are divided between the sources presently utilized: two in this case. Savings are calculated relative to the total demand associated with the customers in question as they are moved. For example, from location 1 the total demand that is designated to move is $50 + 80 = 130$ units associated with the customers B and C. The savings then are the cost for production of the current assignment of 660 units minus the cost of production of 530 units (i.e., \$143 for the 130 units).

TABLE 12.13 Savings by Moving to Location 3

Customers	Demand	Location							NS
		1		2		3			
		SS	SF	SS	SF	ST	SD	SF	
A and D	220			258	0				258
B and C	130	143	0						143
A, B, C, D	350	350				590	−490	−350	−250
Total		143	0	258	0	590	−490	−350	151

The cost at the destination is the cost of assigning 350 units to the location 3 is ($1.4 * 350 = \$490$). The fixed cost incurred by selecting the location is $350, and the net savings are $151. Because the savings are positive, the location is selected. The present assignment is shown in Table 12.14, and the cost for this solution is $5070 - 151 = \$4919$.

The next step is to perform a block move. The only customer presently assigned to location 2 is customer G. If he is moved to location 3 the savings in the transportation cost will be −220 and the source savings will be $1.3 * 220 = \$286$. The total demand assigned to location 3 would change from 350 to 570 units. Thus the cost of moving 220 units is $630 + 1(570 - 450) - 1.4(350) = 750 - 490 = 260$ (or the savings of −260). In addition, because location 2 is made an unassigned location, there is a fixed saving (SF) of $500, giving the net saving of $306. The individual and the net savings are shown in Table 12.15.

Therefore, customer G is moved to location 3. The cost of this latest solution is $4919 - 306 = \$4613$.

A block move for location 1 would require moving all customers presently assigned to location 1 to location 3. They are customers E, F, and G. But moving these three to location 3 would result in assigning all of the customers to location

TABLE 12.14 Present Assignment with 1, 2, and 3 as Assigned Locations

Customer	Location				D_i
	1	2	3	4	
A			100		100
B			50		50
C			80		80
D			120		120
E	70				70
F	160				160
G		220			220
H	300				300
Total	530	220	350		1100

TABLE 12.15 Block Move for Location 3

Customer	Amt	ST	SS	SD	SF	NS
G	220	− 220	286	− 260	500	306

TABLE 12.16 Final Demand Assignment with 1, 2, and 3 as Assigned Locations

	Locations				
Customer	1	2	3	4	Demand
A			100		100
B			50		50
C			80		80
D			120		120
E	70				70
F	160				160
G			220		220
H	300				300
Total	530		570		1100

3. The cost of such a solution from Table 12.4 is \$5330. That figure is higher than the cost of the present solution; therefore, the block move for the location is not profitable and is rejected. Table 12.16 presents the current assignments.

Location 4 Analysis. Similar analyses with location 4 produce no savings; therefore, the solution in Table 12.16 is the best solution. Location 1 and 3 are selected with 530 and 570 units of total order at each location, respectively.

12.3 GROUP ITEMS WITH LARGE SETUP COSTS

We can extend the analysis farther, even to include nonlinear production cost. Consider an example of a large, diversified manufacturing firm, who has production facilities in various parts of the country and who receives orders from various customers also distributed throughout the country. All of these plants are either capable, or can readily be made capable, of producing any or all the items that the firm supplies to the customers. With the large setup costs involved in production, however, it is impracticable to produce every item in every plant. The problem is to develop a production schedule, or select the plants and associated customer assignments, such that the costs of production and transportation for all customers are minimized.

Here, the production cost consists of three primary parts: a comparatively large setup cost, a nonlinear variable cost, and a changeover cost based on the number of different customers a production facility is assigned to serve. This latter cost is incurred when it is necessary to alter the production setup that is suitable for one customer, but that does not meet the requirements of another customer who has ordered the same basic item, except with slight modifications.

We can understand this problem by studying an example of a glass manufacturer having production facilities in different locations and receives all its orders from various customers in its headquarters. Here the orders are grouped in terms of product type, each of which has the same major characteristics except for the minor variations based on individual customer preference. For example, several slightly different drinking glasses of a particular style can be made from a common mold. A customer's variation may occur as a result of a change in the glass color or the design differences of the decals used to decorate the glasses. It is also possible to make glasses of different capacities and stem lengths from the same mold.

The planner at the home office is responsible for distributing the workload among the different production facilities. His or her objective is to assign product types to each plant to minimize the total cost of production and distribution. The major cost of production for the glass manufacturer is the fixed cost of the mold, ranging anywhere from 20,000 to \$50,000. Because such a large investment is required, it is not economical to have identical molds in all plants. The customers' orders, which may consist of many different items, therefore, are broken down into types of items. The responsibilities for the production of various item types are distributed among the available plants. By doing so, only selected plants will need the mold for a particular product type.

Individual customer variations may also cost additional dollars as the setup on the mold is modified. Generally, these are very minor modifications, such as a change in the glass color, or a change in decoration. The production cost structure will incorporate these changeover costs.

12.3.1 Cost Structure of the Problem

While breaking down the total cost structure, we find that it consists of four major components: (1) transportation cost; (2) fixed cost, such as the cost of the mold; (3) changeover cost, based on the number of customers assigned to a production facility; and (4) variable production cost. We are already familiar with all the cost components, except for changeover cost, which is briefly explained next.

Changeover cost results when production facilities change from one customer to another, increasing as the number of customers assigned to a plant increases. The increment may be either linear or nonlinear. As an illustration, let us again examine the example of the glass manufacturer. The changeover cost is a result of three main factors: (1) downtime cost for the molding machine when it is adjusted for the change in capacity or stem length and color of glass, (2) downtime cost for the decorating machine when the design (screen) is changed and the paints are replaced to accommodate color changes, and (3) clerical cost associated with the development of production schedules and billing.

As the solution procedure is very similar to the one we studied in the previous section, we will just illustrate it by means of an example.

Sample Problem

The data for an illustrative problem having four customers and three production facilities are given in Tables 12.17–12.19. The data for the transportation cost are given in Table 12.17. An asterisk (*) indicates that the customer cannot be supplied by the associated facility (for reasons such as inaccessibility, political differences between countries, trade embargo, or others).

Table 12.18 shows the lump sum and incremental portions of the fixed cost. The number of customers assigned to a location determines the incremental cost. Again, a minus sign (–) indicates that the number assigned cannot exceed a

TABLE 12.17 Transportation Cost Data

	Location			
Customer	1	2	3	Demand
A	4	7	*	20
B	*	10	7	60
C	6	5	4	35
D	8	9	6	55

TABLE 12.18 Fixed Cost and Incremental Cost per Customer

	Location		
Customer	1	2	3
A	12	10	9
B	10	6	8
C	8	5	4
D	–	4	–
Lump sum	60	40	50

TABLE 12.19 Locations and Associated Production Cost Functions

Location	1	2	3
Production cost function	$2q_1^{0.85}$	$4q_2^{0.75}$	$5q_3^{0.70}$

limiting value: for example, neither location 1 nor location 3 can handle more than three customers. This could be because of a limitation on available free production time at those locations. Thus, the total fixed cost of assigning three customers to location 1 would be $60 + 12 + 10 + 8 = 90$.

If we were to attempt to assign the fourth customer to location 1, the cost would approach infinity because that location does not have the facilities or manpower necessary to process four customers. The cost of production at each location is given in Table 12.19. The problem is to assign the customers to production facilities to fill all demands and to minimize the total cost of setup, production, and transportation.

Our first step in solving the problem is to construct Table 12.20, the total transportation cost table, the cost of assigning the entire demand to particular location.

Initial Assignment. For first assignment, select the least lump sum cost location from Table 12.18, location 2, and assign all of the customers there. The resulting solution may or may not be feasible. If it proves to be, infeasible, it will have infinite cost. Location 2 has now become an assigned location.

Now arrange the remaining locations in ascending order of their lump sum costs. Here the order is 3–1, and we will seek improvement in the solution by checking each unassigned location in that sequence.

Check for Unassigned Location 3. If each customer is moved independently to location 3, the transportation and production cost savings would be calculated as follows (Table 12.21). Because the transportation savings for customer A is $140 - \infty = -\infty$, there is no need to calculate production cost savings because the net sum is $-\infty$. For all other customers, both transportation and production cost savings are calculated whether each quantity is negative or positive. For example, for customer B the transportation cost savings are $600 - 420 = 180$. To calculate the production cost savings, we observe that the source, location 2, has 170 units of demand assigned. If customer B with its demand of 60 units is moved to location 3, it will reduce the assignment at the source location to

TABLE 12.20 Total Transportation Cost

	Location		
Customer	1	2	3
A	80	140	–
B	–	600	420
C	210	175	140
D	440	495	330

$170 - 60 = 110$ units. The cost of production location 3 for these 60 units is $5(60)^{0.7}$, giving a net savings in production cost of -36. Similar calculations are performed for the other customers, and they are then ranked based on their total savings. There is no need to include customer A in the ranking because, with its high negative savings, moving that customer could never be profitable.

We will now observe the effect of moving the customers sequentially according to their rankings. Moving customer B first, we find production savings of (MPB stands for moving production of customer B),

$$\text{MPB} = 4(170)^{0.75} - 4(170 - 60)^{0.75} - 5(60)^{0.7} = -36$$

and transportation savings of B, $\text{TB} = 600 - 420 = 180$, giving total savings of SB

$$\text{SB} = \text{MPB} + \text{TB} = 144.$$

Because the savings are positive, customer B could be moved. Given this tentative move, we will next calculate the effect of moving customer D. Present demand at source location 2 is $170 - 60$, and the assignment at the destination, location 3, is 60 units. Moving customer D will reduce the assigmnent at location 2 to $170 - 60 - 55 = 55$ units, whereas it will increase the assigmnent at location 3 to $60 + 55 = 115$. This provides savings in production cost of

$$\begin{aligned}\text{MPD} = {} & 4(170 - 60)^{0.75} - 4(170 - 60 - 55)^{0.75} \\ & - [5(60 + 50)^{0.7} - 5(60))^{0.7}] = 4\end{aligned}$$

transportation cost savings $\text{TD} = 495 - 330 = 165$, and the total savings of

$$\text{SD} = \text{MPD} + \text{TD} = 169$$

Again, positive net savings indicate that customer D could be moved. Next, based on the assumption that along with customer B, customer D has also been moved, we will examine customer C.

TABLE 12.21 Independent Analysis (for Location 3)

Customer	Transportation savings	Production savings	Total savings	Rankings
A	–	–	–	–
B	180	$4(170)^{0.75} - 4(170 - 60)^{0.75} - 5(60)^{0.7} = -36$	144	1
C	35	$4(170)^{0.75} - 4(170 - 35)^{0.75} - 5(35)^{0.7} = -30$	5	3
D	165	$4(170)^{0.75} - 4(170 - 55)^{0.75} - 5(55)^{0.7} = -36$	130	2

The savings in production cost for customer C, if moved to location 3, are

$$\text{MPC} = 4(170 - 60 - 55)^{0.75} - 4(170 - 60 - 55 - 35)^{0.75}$$
$$- [5(60 + 70 + 35)^{0.7} - 5(60 + 55)^{0.7}] = 14.65$$

and the savings in the transportation cost are

$$\text{TC} = 175 - 140 = 35.$$

Giving a total saving of

$$\text{SC} = \text{MPC} + \text{TC} = 49.65$$

As this figure is positive, customer C can also be moved. The total savings by moving all three customers is $144 + 169 + 49.65 = 362.65$.

We must next consider the change in the fixed cost. Selection of location 3 and the assignment of three customers there would cost $50 + 9 + 8 + 4 = 71$; however, there is a savings of $4 + 5 + 6 = 15$ at location 2 because the three customers are moved from that location. This gives a net increase in fixed cost of $71 - 15 = 56$.

Because the net fixed cost increase is less than the total savings ($56 < 362.65$), location 3 is selected and customers B, C, and D are moved there.

Let us now investigate the effect of a block move, that of moving the remaining customer, customer A, from location 2 to location 3. The cost of transportation for the customer at location 3 is infinite, because he cannot be assigned to that location. That being the case, this block move would not be profitable. This leaves the present assignment as shown in Table 12.22 below.

Check for Unassigned Location 1. Moving the customers independently to location 1 from their present assignments will lead to savings Table 12.23.

Moving customer A first results in a net savings of $60 + 12.3 = 72.30$. As the savings are positive, we tentatively move him to location 1.

Next, check customer C. If he is moved, the production cost savings, based on the present tentative assignment, are $5(150)^{0.7} - 5(150 - 55)^{0.7} - 2(55)^{0.85} -$

TABLE 12.22 Present Assignments After Evaluation of Locations 2 and 3

	Location		
Customer	1	2	3
A		X	
B			X
C			X
D			X

TABLE 12.23 Independent Analysis (for Location 1)

Customer	Transportation savings	Production savings	Total savings	Rankings
A	60	$4(20)^{0.75} - 0 - 2(20)^{0.85} = 12.3$	72.3	1
B	–	Not necessary	–	–
C	−70	$5(150)^{0.7} - 4(150 - 35)^{0.7} - 2(35)^{0.8}$ $= -12.6$	−82.6	2
D	−110	$5(150)^{0.7} - 5(150 - 55)^{0.7} - 2(55)^{0.85}$ $= -14.3$	−124.3	3

$2(35)^{0.8} = 9.06$. The transportation cost savings are −70, giving a net savings of −60.94. Because this figure is negative, the customer will not be moved, and there is no need to check the remaining customers with lower rankings.

Customer A, therefore, is the only one that can be moved. Investigating the change in fixed cost, we find that the cost will increase by $(60 + 12) - (40 + 10) = 22$. As the fixed cost increment is less than the savings $(22 < 72.3)$, customer A is moved to location 1.

Only one block move needs to be checked, that of making location 3 an unassigned location. The production cost savings would be the production costs of customers B, C, and D at location 3, less the increase in the production cost at location 1 resulting when those same customers are moved to location 1. As customer A is already there, this latter cost is that for all customers at location 1 less that cost of customer A at location 1; that is,

$$\text{MP(block)} = g_4 q_4^{a4} - (g_1 q_1^{a1} - g_1 d_1^{a1})$$

where

$$q_1 = d_1 + d_2 + d_3 + d_4 \qquad \text{and} \qquad q_4 = d_2 + d_3 + d_4$$

Substituting numerical values,

$$\text{MP(block)} = 5(150)^{0.7} - [2(170)^{0.85} - 2(20)^{0.85}] = 34.97$$

Fixed cost savings are obtained from Table 12.18, and transportation cost savings are found in Table 12.20. The various savings are presented in Table 12.24.

As can be seen, both the transportation and the fixed cost savings approach a negatively infinite amount, or from a different view, the added cost for making the block move from location 3 to location 1 is infinite. Obviously, the block move is rejected.

We could have saved ourselves the trouble of performing the calculations by first checking Tables 12.18 and 12.20. Table 12.18 tells us that location 1 can serve no more than three customers, and Table 11.16 specifically informs us that

TABLE 12.24 Block Move (3 → 1)

Customer	Transportation savings	Production savings	FC savings	Total
B	–			
C	−70			
D	−110			
B, C, and D	–	34.97	–	–

TABLE 12.25 Final Customer Assignments

	Location		
Customer	1	2	3
A	X		
B			X
C			X
D			X

customer B cannot be assigned to location 1. Both tables advise us that the block move to location 1 is impossible. With the block move rejected, the present assignments for the customers are as shown in Table 12.25.

All locations have now been examined; therefore, the foregoing selection is the best possible. The cost of this solution is $1269, which is calculated based on the present assignments.

It should be realized that in problems with large dimensions, solution by hand will require a great deal of time. A computer program can, however, solve these problems very easily.

12.4 DISCRETE FACILITIES WITH COSTS DEPENDENT ON SERVICE CAPABILITIES

We now extend the facility location–allocation concept to include a machine selection problem. Basic features of the problem are similar to the ones we have seen in the previous sections: There are customers from whom demands originate, and there are many available locations, from which we are required to select a specific number at which our facilities (machines) will be placed. We now, however, must also choose the capacity of each facility, the cost of which is dependent on the capacity selected. For example, there may be four machines, each with the capacity, and the amortized cost follow, from which a selection may be made.

Machine	Capacity	Cost
1	150	200
2	300	400
3	600	750
4	1000	1200

In addition to transportation cost and the price of the machine, the other cost considered here is that of the site preparation, which may depend on the capacity of the machine used and the location selected. We can also include, with only a minor modification to the solution procedure, the production cost at each location. The problem is to select the location and the respective machine capacities, and then to assign the customers to minimize the total cost.

The examples are numerous: In almost every field there are alternatives with different costs. For instance, copy machines, with different speeds, capacities, and costs, are made by different manufacturers. The planner for the office complex could select from many machines and suppliers in making this decision.

In the problem we are presently discussing, the capacity of each machine is important. Generally, the larger the capacity, the more expensive the unit is. That being true, it would not be economical to buy units such that the total available capacity would significantly exceed that required. Because the machines are available in discrete size increments, it may not be possible to have the total available capacity exactly equal to the total needed, but our goal should be to keep the excess as low as possible. To achieve this objective, it may often be necessary to divide a customer's demand so that a portion of the demand is assigned to one facility in a different location. This is an important departure from the principle that we have seen in the previous sections; previously, it was not considered feasible to divide a customer's demand.

As has been our practice, the mathematical formulation is furnished next; however, the reader may go directly to the heuristic procedure if so desired.

12.4.1 Mathematical Formulation

The notations used in the formulation are as follows:

X_{ij} = the amount of demand of customer i is assigned to location j, $i = 1, 2, \ldots, n; j = 1, 2, \ldots, m$

$Y_{jk} = 0$ if facility k is not placed in location j; $j = 1, 2, \ldots, m$; $k = 1, 2, \ldots, h$

d_i = the demand from customer i
D = the total demand on the system
TC_{ij} = the transportation cost per unit between the customer i and location j; $i = 1, 2, \ldots, n$; $j = 1, 2, \ldots, m$
MC_{jk} = the cost of a k-type facility installed in location j; $j = 1, 2, \ldots, m$; $k = 1, 2, \ldots, h$
SC_{jk} = the site cost for location j, if facility k is chosen; $j = 1, 2, \ldots, m$; $k = 1, 2, \ldots, h$
M/C_k = the upper limit on capacity of facility k; $k = 1, 2, \ldots, h$ when the facilities are arranged in ascending order of their capacities.

We wish to minimize Z (total cost), where,

$$z = \sum_{i=1}^{n} \sum_{j=1}^{m} X_{ij} TC_i + \sum_{j=1}^{m} \sum_{k=1}^{h} (MC_{jk} + SC_{jk}) Y_{jk}$$

Subject to

$$\sum_{j=1}^{m} X_{ij} = d_i \qquad i = 1, 2, \ldots, n$$

$$\sum_{k=1}^{h} Y_{jk} = 1 \qquad j = 1, 2, \ldots, m$$

$$\sum_{1}^{n} d_i = D \qquad i = 1, 2, \ldots, n$$

where

$$Y_{jk} = \begin{cases} 1 & \text{if } k\text{th facility is chosen for location } j \\ 0 & \text{otherwise} \end{cases}$$

Since Y_{jk} is the integer variable indicating the selection of facility type, we, therefore, define

$$Y_{jk} = 1, \qquad \text{if} \qquad M/C_{k-1}, \sum_{i=1}^{n} X_{ij} < M/C_k \quad \text{for each location.}$$

$$Y_{jk} = 0, \qquad \text{if} \qquad \sum_{i=1}^{n} X_{ij} = 0.$$

Alternative Method

Before describing the steps involved in the heuristic procedure, we must define some of the terms that will be used in the discussion. Unless modified herein, those terms previously defined in earlier chapters will retain those meanings.

1. *Block move*: The move that drops a facility from a higher capacity. (Note, this is a refinement of the definition used in the previous section).
2. *Positive move*: The demand in which only the customers showing positive or zero savings in transportation cost are moved.
3. *Excess capacity*: The difference between the amount assigned to a location and the capacity of the facility installed at that location.
4. *Dropping amount table*: A table that shows the amount of demand assigned to a location, the quantity of demand that must be moved to drop to a lower capacity facility, and the associated savings in site and facility costs.
5. *Savings table*: This table shows the location(s) to which each customer is assigned, the savings in transportation cost if a unit of demand is moved from where it is presently assigned to the location under consideration, and the maximum demand that can be moved.

The algorithm is presented in the following steps. It might be easier to follow the steps if one simultaneously refers to Figure 12.2 under the solution example

1. Construct a total transportation cost table as usual by calculating the transportation cost if all the demands from all of the customers were assigned to each location.
2. Select the minimum cost facility that is at least capable of producing the total demand (D) for each location.
3. Add the site, machine, and total transportation costs for each location.
4. Select the location that has the minimum total cost as the initial location; place the facility there, and assign it all of the demand.
5. Based on the present allocation, develop the dropping amount table for the assigned location(s) (e.g., Table 12.22).
6. Select an unassigned location j for examination; refer to it as the destination location.
7. Generate the savings table for the destination location (e.g., Table 12.23).
8. For each facility that can be installed at the destination location, follow steps 9–12.
9. Perform a positive move check. Calculate the savings that could be achieved by moving demands from the presently assigned location(s)

to the destination location. The customers' demands are moved in the order of their unit savings in transportation cost, the customer with the maximum savings is moved first, then the customer with the next highest savings, and so on. Only the customers with positive or zero savings need be considered. In addition, it should be noted that the capacity of the facility installed in the destination location may also restrict the amount of the of demand that can be moved there. The total assigned demand may not exceed the available capacity. (Note: Do not make any actual moves at this time; see step 11) (refer to Table 12.24).

10. Perform block move checks. The objective of the block move check is to see whether if by moving all or a portion of the demand presently supplied by an assigned location to a destination location, it is possible to reduce the capacity of the machine in the assigned location. If this can be done, there might be savings associated with such a move. To this end, the dropping amount table is important. It shows how demand can be moved from an assigned location to cause the capacity requirement to drop to that of the next lower capacity unit and produce the associated savings. The block move is divided into two parts: first, from an assigned location only (each of the assigned locations is individually checked in this phase), and then from combinations of the assigned locations.

    a. To perform a block check on an assigned location, go to the dropping table. Move the exact amount of demand from the assigned location so that the machine unit in that location may be dropped. This move must be made with the least cost, or even with profit if one is possible. From the assigned location under examination, the customer with the maximum savings in transportation cost is moved first; then the one with the next highest is moved, and so on, until the required demand has all been moved. Remember that the objective is to move a necessary demand, which may mean breaking up an individual customer's demand so that only a portion of it is moved to the destination location, or moving a customer, even though it may mean an increase (negative savings) in the transportation costs.

Once the required amount of demand is moved, if some excess capacity remains at the destination location, this capacity may be filled if there are any customers still showing positive or zero savings in transportation cost. These customers may be presently assigned to any of the locations, not just the location on which the block move is performed. Again, the customer with maximum transportation savings is moved first, and so on. One important point must be made clear, and that is: It is not required that the remaining customers be moved to the destination location simply because there is excess capacity; do so only as long as it is profitable. Also, for an assigned location, there may be more than one block search, each associated with dropping the required capacity from that of the existing machine to that of a lower-capacity machine. We even need to check a block move that would completely remove a machine from the assigned location and make the latter an unassigned location.

b. The next step is to perform a combination block move check as shown later in Tables 12.46, 12.47, and 12.48. A combination block move is one in which capacities are dropped simultaneously in more than one assigned location, call them "locations in the block." These drops are obtained by moving the demands to the destination location; however, this is possible only if the capacity at the destination location is sufficient. Again, as in the block move with a single location, the customer with the maximum savings, presently assigned to the location in the block, is moved first; and then of those remaining, the customer with the next highest savings, and so on. The necessary amount of demand must be moved from the locations in the block, whether it results in savings or an increase in the transportation cost. If there is any remaining capacity, it may be filled by moving customers who had been assigned to any of the other locations, as long as they have positive savings. To reiterate, the customers are moved in the order of their unit savings, but it is not necessary to completely utilize all the capacity of the destination location.

Calculate the net savings in each block move, steps 10a and 10b, by considering savings at the source(s), costs at the destination, and transportation cost changes. (Note: Do not make any actual moves at this time, see step 11).

11. From the calculations in steps 9 and 10, determine the maximum savings alternative. If these savings are negative, the destination location is not profitable and must be dropped from any further consideration. On the other hand, if the savings are positive, the destination location is made an assigned location by selecting the facility that caused these savings, and the demands are moved accordingly. (Note: For the purpose of simplifying the discussion in this chapter, "positive" will be used as meaning nonnegative [i.e., it will include zero]).

12. If all assigned locations have been examined, then the present solution is the best; if not, return to step 5.

Sample Problem

Suppose an automobile corporation plans to produce a part that will be used by its four assembly divisions located in different areas of the nation. Management has information about three candidate locations at which the production facilities could be installed. The planning department has determined the demand from each assembly division for these parts. There are now two types of production facilities (machines) that can be installed, each with a specific capacity and a corresponding cost. The cost-accounting department has furnished the pertinent data for each machine: Table 12.26, capacity and purchase cost; Table 12.27, site preparation, installation, and fixed operating costs; and Table 12.28, unit transportation costs (from different production sites to the subassembly plants). For convenience, in this problem we will consider capacity to be coded in terms of units, and costs to be in dollars per year of operation.

To be compatible, the time periods for machine and site costs, and the customer's demands must be the same. That is, if the demands represent the customer's requirement per month, then the machine and site costs must be prorated on a monthly basis. Any convenient period may be chosen as long as it is common to all.

The problem is to determine which site locations are to be selected, what capacity machines are to be installed in each, and how many demand units are to

TABLE 12.26 Machine Capacity and Cost

Machine type	Capacity	Cost/yr
1	150	150
2	450	580

TABLE 12.27 Site Preparation Cost per Year

	Locations		
Machine type	1	2	3
1	400	425	450
2	700	850	800

TABLE 12.28 Transportation Cost (per Unit)

	Locations			
Customers (plants)	1	2	3	Demand/yr
1	1	8	3	150
2	5	3	3	120
3	3	8	6	50
4	7	2	5	100
Total				420

be supplied by a selected location to an assembly plant (customer). And all of this must be accomplished at the least possible cost.

Preliminary Steps. The first step is to determine which location to select if only one can be chosen (Figure 12.2). The total demand in this problem is 420 units. To produce these units, we will need a machine of type 2, with a capacity of 450. The cost for procuring and installing this machine in each location per year is shown in Table 12.29. The transportation cost of shipping the entire demand from a single source location to a customer is shown in Table 12.30. By adding the total transportation, installation, site, and machine costs, we obtain the grand total for each location as shown in Table 12.31.

Table 12.31 indicates that the least expensive is location 1; a total demand of 420 is now assigned to location 1. Because the type 1 machine can process only 150 units of demand, a type 2 machine, with a capacity of 450 is assigned to location 1, and all the customers are supplied by this machine. The next step is to construct a dropping amount table for location 1.

Table 12.32 shows the savings that could be achieved if a unit of smaller capacity could be used in location 1, and it also tells how much demand will have to be removed from location 1 to achieve this. To go from a type 2 to a type 1 machine will require the $420 - 150 = 270$ units of demand removed. The machine and site costs for the type 2 are $580 + 700 = \$1280$; whereas, the machine and site costs for the type 1 are only $150 + 400 = \$550$. Thus, the changeover would save $730. The detailed table is shown as Table 12.32.

Check for Unassigned Location 2. Begin the checks of the unassigned locations by selecting location 2 for the first examination. This choice is arbitrary, and we will proceed in ascending order of location number just for convenience. The savings in the transportation cost per unit (refer to Table 12.28) of demand moved from location 1 to location 2 are shown in the following Table 12.33.

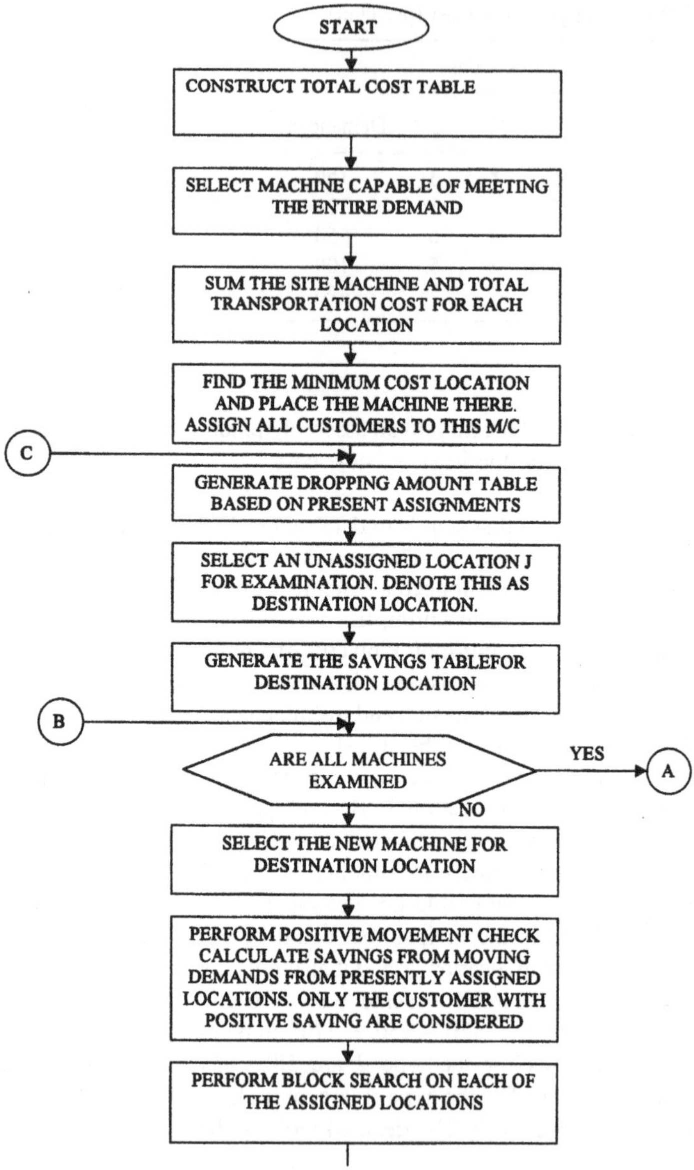

FIGURE 12.2 Discrete facilities with costs depending on the service capabilities.

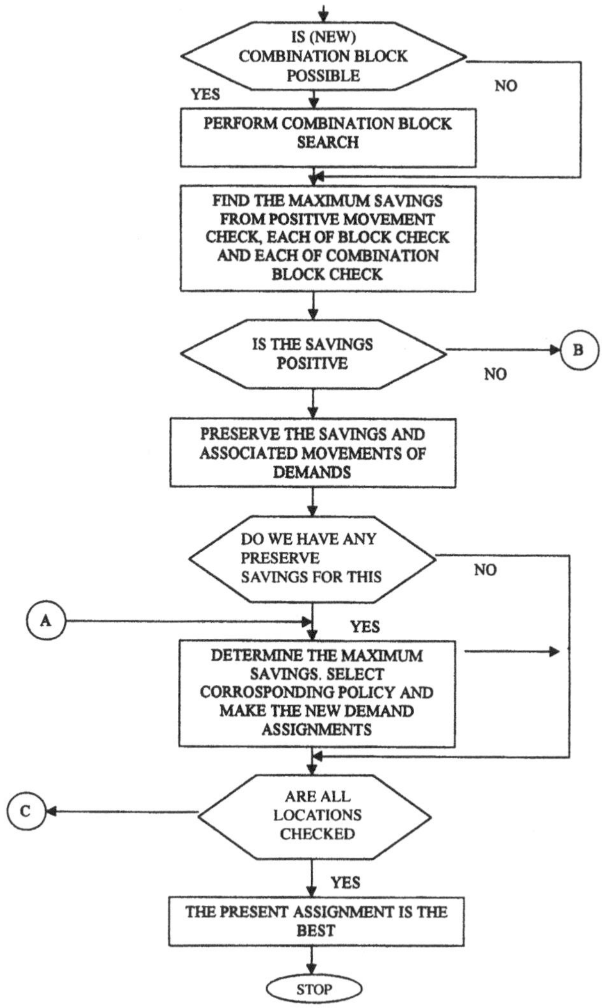

FIGURE 12.2 (*continued*)

TABLE 12.29 Installation Cost (Type 2 Machine)

	Location		
	1	2	3
Machine cost	580	580	580
Site preparation	700	850	800
Total cost/yr	1280	1430	1380

TABLE 12.30 Total Transportation Cost

	Location			
Customer	1	2	3	Demand
1	150	1200	450	150
2	600	360	360	120
3	150	400	300	50
4	700	200	500	100
Total	1600	2160	1610	420

TABLE 12.31 Single Location Total Costs

	Location		
	1	2	3
Machine/site	1280	1430	1380
Transportation	1600	2160	1610
Grand total/yr	2880	3590	2990

TABLE 12.32 Dropping Amount (Location 1)

Dropping to type	Minimum demand to be moved	Savings
1	270	730
—	420	1,280

TABLE 12.33 Unit Savings

Customer	Present location	Savings per unit	Demand
1	1	−7	150
2	1	2	120
3	1	−5	50
4	1	5	100

We must check the effect as each type of machine is placed in location 2, and in doing so, we will perform both a positive move check and a block move check for each one. The details follow:

1. Machine type 1 (capacity 150)

   a. Positive move check: From Table 12.28, we can see that customers 2 and 4, with savings of \$2 and \$5/unit, respectively, could be moved. Select the customer with maximum unit savings—customer 4—move first. The demand from this customer is for 100 units; that leaves $150 - 100 = 50$ unfilled units of capacity at the destination, location 2. The next to move is the customer with savings of \$2/unit. However, the capacity restriction allows only 50 units to be moved. Table 12.34 provides the details.

Note: In the Table 12.34 and similar ones to follow, the following abbreviations are used in the interest of conserving space

L: location
US: Unit savings
EC: excess capacity
TS: transportation savings
SS: source savings
DS: destination savings
NS: net savings.

The cost of the machine and installing it on site at location 2 (from Tables 12.26 and 12.27) is $150 + 425 = 575$. The machine capacity required at location 1 does not decrease enough to permit us to drop from type 2 to a type 1 machine there, because only 150 units of demand are moved. This is reflected by zero savings for the source location.

   b. Block move check: To drop the required capacity of the machine at location 1, it is necessary to move 270 units from that location, according to Table 12.32. This is not possible with a 150-unit

TABLE 12.34 Positive Move (Machine Type 1, Capacity 150)

Customer	L	US	Demand	EC	TS	SS	DS	NS
4	1	5	100	50	500			
2	1	2	50	0	100			
Total			150		600	0	−575	25

capacity machine at location 2; therefore, no block move is attempted.

2 and 3. Machine type 2 (capacity 450) at location 2

a. Positive move check: Again, based on Table 12.28, customer 4 is investigated first, and we find that all of his demand can be moved, leaving 450 − 100 = 350 units of free capacity. Next, customer 2 is moved, reducing the unfilled capacity by 350 − 120 = 230 units. There is no other customer with positive savings; therefore, no other move is made.

Table 12.35 presents the results. The total transportation savings are $740. As 100 + 120 = 220 units of demand are removed from location 1 based on Table 12.32, there are no savings at the source. The cost at location 1 to install a machine with a capacity of 450 units is $1430. After adding all the savings and costs, the net result is a savings of −690.

b. Block move check: For the moment, we ignore the positive move check just completed and determine whether a block would be more profitable. With a capacity of 450 units in location 2, it is possible to move the required 270 units, as well as 420 units, from location 1 to location 2 (refer to Table 12.32); therefore, two block moves are feasible, first moving 270 units and second by moving 420 units. Consider the first block move. This move will drop the machine requirement at location 1 from a type 2 to a type 1. The first task then, is to move 270 units of demand, with profit if possible, or at least-cost otherwise. Table 12.36 shows that moving customers 4 and 2 will result in moving 100 + 120 = 220 units of demand. There are still 270 − 220 = 50 units of demand left in location 1. We may fill that demand by moving any of the remaining customers. There are no other customers with positive savings; therefore, we are left with moving 50 units from customer 3, one with minimum cost. The results are shown in Table 12.36.

TABLE 12.35 Positive Move (Machine Type 2)

Customer	L	US	Demand	EC	TS	SS	DS	NS
4	1	5	100	350	500			
2	1	2	120	230	240			
Total			220		740	0	−1430	−690

TABLE 12.36 Block Move (Location 2, 270 Units)

Customer	L	US	Demand	EC	TS	SS	DS	NS
4	1	5	100	350	500			
2	1	2	120	230	240			
3	1	−5	50	180	−250			
Total			270		490	730	−1430	−210

The second block movement consists of moving enough demand from location 1 to location 2, so that the demand assignment in location 1 drops to next sequential drop unit, 0 in this case, which will require moving 420 units demand from location 1 and assigning it to location 2. But this is equivalent to assigning all demand to location 2, which according to Table 12.31 is not possible; therefore, it is not necessary to perform this block analysis.

c. Combination block move check: As we presently have only one assigned location, it obviously is not possible to arrange a combination block move of assigned locations.

4. Selection of best alternative.

Among the alternatives presented in Tables 12.34 through 12.36, Table 12.34 shows the maximum savings of $25. This alternative is selected, and Table 12.37 shows the corresponding allocations when the indicated customers and demands are moved.

As shown in Table 12.37, a type 2 machine (capacity of 450 units) is used in location 1. Machine type 1 with capacity of 150 units is used in location 2. The dropping amounts and the corresponding savings are presented in Table 12.38 for

TABLE 12.37 Demand Assignment 1

	Location			
Customer	1	2	3	Demand
1	150			150
2	70	50		120
3	50			50
4		100		100
Total	270	150		
Machine type	2	1		
Excess capacity	180	0		

TABLE 12.38 Dropping Amount (Location 1 and 2)

Location 1 (present amount: 270)		
Dropping to type	Dropping amount	Savings
1	120	730
–	270	1280
Location 2 (present amount: 150)		
Dropping to type	Dropping amount	Saving
–	150	575

these two locations (refer to Tables 12.32 and 12.34 for savings data and calculations).

Check for Unassigned Location 3. As in the previous location check, the first step is to develop a unit savings table. The savings resulting from moving a unit of demand for each customer from its presently assigned location to location 3 are listed in Table 12.39.

1. Machine type 1 (capacity 150) at location 2

   a. Positive move check: Table 12.39 shows that only one customer—namely, customer 2—served from location 1, offers positive savings. After moving this customer, there are still $150 - 70 = 80$ units of excess capacity in location 3 (Table 12.40). Customer 2 is also partially served from location 2 without incurring any additional transportation cost. It is also moved to obtain the possible source location savings. Table 12.40 shows the details.

   b. Block move checks: We now have two assigned locations 1 and 2 and a block move must be checked for each. We arbitrarily begin

TABLE 12.39 Unit Savings

Customer	Present location	Savings per unit	Demand
1	1	−2	150
2	1	2	70
2	2	0	50
3	1	−3	50
4	2	−3	50

TABLE 12.40 Positive Move (Machine Type 1)

Customer	L	US	Demand	EC	TS	SS	DS	NS
2	1	2	70	80	140			
2	2	0	50	30	0			
Total			120		140	0	−565	−425

with location 1 and find that we must move 120 (270–150) units of demand from there (see Table 12.38) to obtain the first drop. Table 12.41 presents the details.

After moving customer 2 and a portion of the demand for customer 1 to complete the block, we still have 30 units of excess capacity at the destination location. The most profitable way of filling this is to move a portion of the requirement for customer 2, presently assigned to location 2. The net savings are $205 with machine type 1 in location 3. The second block move consisting of moving 270 units of demand is not possible; therefore, we go to a location 2 block move.

Location 2 is currently assigned 150 units and a drop at that location requires moving all 150 units (see Table 12.38). The details of this block move are provided in Table 12.42. There is no net savings in this move.

TABLE 12.41 Block Move Checks (Location 1, 120 Units)

Customer	L	US	Demand	EC	TS	SS	DS	NS
2	1	2	70	80	140			
1	1	−2	50	30	−100			
2	2	3	30	0	0			
Total			150		40	730	−565	205

TABLE 12.42 Block Move (Location 2, 150 Units)

Customer	L	US	Demand	ES	TS	SS	DS	NS
2	2	0	50	100	0			
4	2	−3	100	0	−30			
Total			150		−30	575	−565	−290

c. Combination block move check: The minimum capacity required at the destination location to form the first combined block is the sum of 120 units from location 1 and 150 units from location 2, giving a total of 270 units. As we are presently examining machines with a capacity of only 150 units, no combination block move can be made.

2 and 3. Machine type 2 (capacity 450) at location 3

a. Positive move check: Table 12.43 shows the check for a positive move with a type 2 at location 3. As with Table 12.40, however, partial demand of customer 2 assigned to location 2 is also moved for possible savings in the source location.
b. Block move check: Because we are investigating the possibility of installing a 450-unit machine in location 2, we must check the effects of three block moves: 120 and 270 units from location 1, and 150 units from location 2. We will arbitrarily begin with moving 120 units from location 1.

The first block of 120 units from location 1 is obtained by moving the 70 units of customer 2 and 50 units of customer 1. There are still 330 units of excess capacity, a portion of which is filled by moving customer(s) from location 2 with positive or zero savings. The details are shown in Table 12.44 and a net savings of $30 results.

The second block checked is that of 270 units from location 1. Table 12.45 provides the summary. Note that customer 2 at location 2

TABLE 12.43 Positive Move (Machine Type 2)

Customer	L	US	Demand	EC	TS	SS	DS	NS
2	1	2	70	380	140			
2	2	0	50	330	0			
Total			120		140	0	−1380	−1240

TABLE 12.44 Block Move (Location 1, 120 Units)

Customer	L	US	Demand	EC	TS	SS	DS	NS
2	1	2	70	380	140			
1	1	−2	50	330	−100			
2	2	0	50	280	0			
Total			170		40	730	−1380	−610

is used in filling the final excess capacity after all 270 units of demand are transferred from location 1. We are allowed to move customer 2 because the capacity exists at location 2, and customer 2 has positive or zero savings. Even so, the net result would be a loss of $410.

The third block move to be checked is that of moving 150 units of demand from location 2. Table 12.46 lists the customers and the savings to be gained. Customer 2 on location 1 is included because doing so provides decrease in transportation costs and might also provide savings at the source. The net result, however, is a loss of $965.

c. Combination block move: The two possible combinations are listed as follows:

	Units to be moved from	
Combination	Location 1	Location 2
1	120	150
2	270	150

The first combined block move consists of 120 units from location 1 and 150 units from location 2. Table 12.47 reveals the revised assignments and their corresponding savings. Again, excess

TABLE 12.45 Block Move (Location 1, 270 Units)

Customer	L	US	Demand	EC	TS	SS	DS	NS
2	1	2	70	380	140			
1	1	−2	150	230	−300			
3	1	−3	50	180	−150			
2	2	0	50	130	0			
Total			320		−310	1280	−1380	−410

TABLE 12.46 Block Move (Location 2, 150 Units)

Customer	L	US	Demand	EC	TS	SS	DS	NS
2	2	0	50	400	0			
4	2	−3	100	300	−300			
2	1	2	70	230	140			
Total			180		−160	575	−1380	−965

TABLE 12.47 Combined Block Move

Customer	L	US	Demand	EC	TS	SS	DS	NS
2	1	2	70	380	140			
1	1	−2	50	330	−100			
2	2	0	50	280	0			
4	2	−3	100	180	−30			
Total			220		−260	1305	−1380	−335

capacity could have been utilized by moving additional demand had we been able to do so without increasing the transportation cost.

Table 12.48 shows the block move for the second combination, location 1 with 270 units and location 2 with 150 units. Again, the source savings are the sum of the savings a locations 1 and 2.

4. Selection of the best alternative

Among the alternatives presented in Tables 12.39 through 12.48, the choice providing the greatest savings is shown in Table 12.42, with a net savings of \$205. This alternative is selected and the corresponding assignments are made. Table 12.49 shows the new solution.

Given the current assignments, Table 12.50 shows the supply schedule that results in the total cost:

$$100 + 150 + 200 + 60 + 150 + 300 + 400 + 425 + 450 + 450 = \$2685$$

This compares with the cost of \$2880 for our original solution, that of installing a type 2 machine at location 1 and assigning all demand there.

TABLE 12.48 Combined Block Move (Location 1, 270 Units; Location 2, 150 Units)

Customer	L	US	Demand	EC	TS	SS	DS	NS
2	1	2	70	380	140			
1	1	−2	150	230	−300			
3	1	−3	50	180	−150			
2	2	0	150	130	0			
4	2	−3	100	30	−300			
Total			320		−610	1855	−1380	−135

TABLE 12.49 Demand Assignment 3

	Location			
Customer	1	2	3	Demand
1	100		50	150
2		20	100	120
3	50			50
4		100		100
Total	150	120	150	420
Machine type	1	1	1	
Excess capacity	0	30	0	

TABLE 12.50 Optimum Solution

Customer	Demand	Transportation cost per unit	Transportation cost
Location 1 supplies			
1	100	1	100
3	50	3	150
Location 2 supplies			
4	100	2	200
2	20	3	60
Location 3 supplies			
1	50	3	150
2	100	3	300

It would be possible to conceive of a solution wherein the optimum solution would require the installation of more than one machine at a given location. These machines might be of the same type, or each might be different. The method to follow in making the various checks would be to identify a dummy machine the capacity of which was equal to the sum of the capacities of the machines it represented. Likewise, the machine and site costs would be the sums of those costs for the individual machines.

Throughout this chapter, we have presented the problem as being that of location of machines to serve customers. The reader should realize that "machine" could represent anything from an electrical outlet to a complete manufacturing plant, and that "locations" could also be broad in definition to include cities, countries, or rooms in a building.

Exercises

12.1 Define production cost and give three examples.

12.2 Distinguish between lump-sum and variable costs.

12.3 What different costs would be involved in constructing the following:

a. A warehouse
b. A hospital
c. An elementary school
d. A power plant

12.4 What is a quantity discount? Why would it be beneficial to a supplier?

12.5 An automobile company is in the process of establishing two dealer outlets in a county. It has identified four cities as probable locations from which to serve the customers. The cost of operating at these cities is represented in the following table. Also, for better service, the county is divided into four regions. The annual demand from each region and the unit transportation cost from each city under

Cost Data for the Four Cities

	Cities			
	1	2	3	4
Fixed cost	10000	6800	15000	8000
Variable cost per year per unit	10	8	12	6

Distribution Cost per Unit

	Cities				
Region	1	2	3	4	Demand
A	2	5	4	8	100
B	3	4	10	15	200
C	5	8	6	8	500
D	6	9	9	8	350

consideration to each region is shown in the next table. Decide the location of the new dealer outlets and the regions they will serve.

12.6 A telephone company wants to distribute its appliances. The company gives discount depending on the quantity ordered. We have four customers, who can be supplied by three suppliers. The shipping cost, and fixed associated costs are given in the following table.

Customer	Location			Demand
A	2	5	1	150
B	3	6	4	280
C	4	3	2	100
D	3	2	3	200
Fixed cost	100	500	350	
Purchase cost	$Z1(t1)$	$Z2(t2)$	$Z3(t3)$	

Purchase cost

$$Z1(t1) = \begin{cases} 1.3t1 & 0 < t1 < 200 \\ 1.3*200 + 1.0(t1 - 200) & 200 < t1 < 400 \\ 1.3*200 + 1*400 + 0.9*(t1 - 400) & t1 > 400 \end{cases}$$

$$Z2(t2) = \begin{cases} 1.4t1 & 0 < t2 < 300 \\ 1.4*300 + 1.2*(t2 - 300) & 300 < t2 < 500 \\ 1.4*300 + 1.2*500 + 0.8*(t2 - 500) & t2 > 500 \end{cases}$$

$$Z3(t3) = \begin{cases} 1.2t3 & 0 < t3 < 350 \\ 1.2*350 + 1.1*(t3 - 350) & 350 < t3 < 450 \\ 1.2*300 + 1.1*500 + 0.8*(t3 - 500) & t3 > 450 \end{cases}$$

Determine the supplier and quantity to order from that supplier to minimize the total cost of operation for the phone company.

12.7 A company wishes to build two plants in a certain state and has four possible locations from which to choose. The state has been divided into five buying regions. The fixed cost per year for each location along with the cost per unit are given in the two following tables. Consider that information as well as the transportation costs from each plant to the buying regions and the demand of each region in finding the best location for the plants.

	Cities			
	1	2	3	4
Fixed cost	14100	12000	15000	16000
Variable cost per year per unit	5	6	10	8

	Location costs				
Region	1	2	3	4	Demand (per month)
A	10	9	8	11	20
B	6	3	15	10	40
C	2	4	6	8	25
D	15	11	7	5	50
E	3	7	8	12	30

12.8 A company wishing to buy yarn to manufacture sweaters has five potential suppliers. The fixed cost for purchasing from each supplier is \$400, \$450, \$500, \$380, and \$450, respectively. The shipping cost from each supplier to each plant, along with the demand of each plant, is given in the next table. If the company has four plants to buy for, order yarn such as to minimize the cost.

Shipping and Purchasing Costs and Customer Demand

	Location					
Customer	1	2	3	4	5	Demand
A	3	3	1	4	2.5	650
B	5	3	1.5	3.5	3	400
C	2	4	4.5	3	1	320
D	4	5	2.5	5	3	200

Quantity discount

$$Z1(t1) = \begin{cases} 2.0t1 & t1 < 100 \\ 200 + 1.9(t1 - 100) & 100 < t1 < 250 \\ 485 + 1.75(t1 - 250) & t1 > 250 \end{cases}$$

$$Z2(t2) = \begin{cases} 1.9t2 & t2 < 200 \\ 380 + 1.8(t2 - 200) & 200 < t2 < 500 \\ 920 + 1.70(t3 - 500) & t2 > 500 \end{cases}$$

$$Z3(t3) = \begin{cases} 2.1t3 & t3 < 150 \\ 315 + 2.0(t3 - 150) & 150 < t3 < 350 \\ 715 + 1.85(t3 - 350) & t3 > 350 \end{cases}$$

$$Z4(t4) = \begin{cases} 1.95t4 & t5 < 250 \\ 487.5 + 1.9(t4 - 250) & 250 < t5 < 400 \\ 772.5 + 1.85(t4 - 400) & t4 > 400 \end{cases}$$

$$Z5(t5) = \begin{cases} 2.1t5 & t5 < 75 \\ 157.50 + 2.0(t5 - 75) & 75 < t5 < 200 \\ 557.50 + 1.9(t5 - 200) & t5 > 200 \end{cases}$$

12.9 Distinguish between the following:
- *a.* Fixed cost
- *b.* Changeover cost
- *c.* Variable cost of production

12.10 A company has just finished a large order that required overtime work to meet the deadline, and it is now cutting back on the overtime shifts. What effect is this likely to have on the variable cost of production?

12.11 The foregoing company supplies an essential ingredient from two plants to three customers. The incremental costs for one to three customers for each location, the lump sum costs for using each location, the production cost at this location, the cost of transporting a unit from each location to each customer, and

Incremental and Lump Sum Costs

	Location	
Number of customers	1	2
1	5	6
2	4	3
3	3	—
Lump sum	15	20

Production Cost at Location

Location	
1	2
$3q_1^{0.45}$	$3q_2^{0.50}$

Unit Transportation Cost and Demand

	Location		
Customer	1	2	Demand
1	—	1	20
2	2	3	25
3	3	2	15

the demand of each customer for the ingredient follows. Determine which plants should supply which customers.

12.12 A company has three semiconductor manufacturing plants that supply three customers. The cost for transporting a unit from each plant to each

Unit Transportation Cost and Demand

	Location			
Customer	1	2	3	Demand
1	10	7	—	50
2	8	9	11	70
3	5	6	4	65

Production Cost

Location		
1	2	3
$1q_1^{0.20}$	$2q_2^{0.15}$	$3q_3^{0.10}$

Incremental and Lump Sum of Costs

	Location		
Number of customers	1	2	3
1	11	9	10
2	7	8	8
3	5	6	—
Lump sum	35	50	47

customer, the number demanded by each customer, the production cost function at each location, the fixed cost at each location based on the number of customer labels to be made, and the lump sum costs at each location are given in the following three tables. Determine which plants should supply which customers.

Bibliography

Aikens CH. 1985. Facility location model for distrubution planning. Eur J Operat Res 22:263–279.

Armour GC, ES Buffa. 1963. A heuristic algorith and simulation approach to relative location of facilities. Manage Sci 9:294–309.

Balinski ML. 1965. Integer programming: methods use computation. Manage Sci 12:253–313.

Baumol WJ, P Wolfe. 1958. A warehouse location problem. Operat Res Q 6:252–263.

Bazaraa MS, AN Elshafei. 1979. An exact branch and bound procedure for quadratic assignment problem. Nav Res Log Q 26:109–121.

Bazaraa MS, O Kirca. 1983. A branch and bound based heuristic for solving quadratic assignment problem. Nav Res Logist Q 30:29–41.

Beckenbach EF, R Bellman. 1965. Inequalities. Berlin: Springer-Verlag.

Belardo S, J Harrald, WA Wallace, JA Ward. 1984. A partial covering approach to sitting response resources for major maritime oil spills. Manage Sci 30:1184–1996.

Bellman R. 1965. An application of dynamic programming to location–allocation problems. SIAM Rev 7:126–128.

Bilde O, J Krarup. 1977. Sharp lower bounds an efficient algorithms for the simple plant location problem. Ann Discrete Math 1:79–97.

Bindschedler AE, JM Moore. 1961. Optimal location of new machines in existing plant layouts. J Ind Eng 12:41–48.

Bos HD. 1965. Special Dispertion of Economic Activity. Rotterdam: University Press.

Brady SD, RE Rosenthal. 1980. Interactive computer graphical solutions for constrant minimax location problems. AIIE Trans 12:241–248.

Brady SD, RE Rosenthal, D Young. 1983a. Interactive graphical minimax location of multiple facility with general constraints. AIIE Trans 15:242–254.

Bukard RE. 1984. Quadratic assignment problems. Eur J Operat Res 15:283–289.

Bukard RE, KH Stratmann. 1978. Numerical investigations on quadratic assignment problems. Nav Res Logist Q 25:129–148.

Burness RC, JA White. 1976. The travelling salesman location problem. Transport Sci 10:348–360.

Calamai PH, AR Conn. 1987. A projected Newton method for l_p norm location problem. Math Programming 38:75–109.

Chalmet LG, RL Francis, A Kolen. 1981. Finding effective solutions for rectilinear distance facility location problem. AIIE Trans 2:132–141.

Chalmet LG, RL Francis, A Kolen. 1981. Finding efficient solutions for rectilinear distance location problems efficiently. Eur J Operat Res 6:117–124.

Charalambous C, 1981. A iterative algorithm of multifacility minimax location problem with euclidean distances. Nav Res Logist Q 28:325–337.

Church RL, CS Revelle. 1974. The maximal covering location problem. Pap Reg Sci Assoc 32:101–118.

Cobot AV, RL Fransis, MA Stary. 1970. A network flow solution to facility location problems. Math Programming 38:75–109.

Converse AO. 1972. Optimum number and location of treatment plans. Water Poll Control Fed 44:1629–1636.

Cooper AO. 1972. Optimum number and location treatment plants. Water Poll Control Fed 44:1629–1636.

Cooper L. 1963. Location–allocation problems. Operat Res 11:331–343.

Cournuejols G, ML Fisher, GL Nemhouser. 1977. Location of bank accounts to optimize float. Manage Sci 23:789–810.

Daskin MS, EH Stern. 1981. A hierarchical objective set covering problem model emergency.

Davis SG, GB Kleindolfer, GA Kochenberger, ET Reutzel, EW Brown. 1986. Strategic planning for bank operations with multiple check-processing locations. Interfaces 16/6:1–12.

Dearing PM, RL Francis. 1974. A network flow solution to multifacility minimax location probem involving rectilinear distances. Transport Sci 15:126–141.

Dohrn PJ, CDT Watson–Gandy. 1973. Depot location with van salesmen—a practical approach. OMEGA 1:321–329.

Drezner Z. 1984. The planar two-center and two-median problems. Transport Sci 18:351–361.

Dutton R, G Hinman, CB Millham. 1974. The optimal location of nuclear-power facilities in the Pacific Northwest. Operat Res 22:478–487.

Economides S, E Fork. 1984. Warehouse relocation and modernization: modeling the managerial dilemma. Interfaces 14/3:62–67.

Eilon S, CDT Watson–Gandy, N Cristofides. 1971. Distribution Management. New York: Hafner.

Elzinga JD, W Hearn, WD Randolph. 1972. The minimum covering sphere problem. Manage Sci 19:96–104.

Francis RL, AV Cobot. 1972. Properties of multifacility location problem involving euclidian distances. Nav Res Logist Q 19:335–353.

Francis RL, LF McGinnis, JA White. 1983. Location analysis. Eur J Operat Res 12:220–252.

Francis RL, LF McGinnis, JA White. 1974. Facility Layout and Location: An Analytical Approach. Englewood Cliffs, NJ: Prentice-Hall.

Gavett JW, NV Plater. 1966. The optimum assignment of facilities to locations by branch and bound. Operat Res 14:210–232.

Gelders LF, LM Printelon, LN Van Wassenhove. 1987. A location allocation problem in large Belgian brewery. Eur J Operat Res 28:196–206.

Geoffrion AM, RF Powers. 1980. Facility design is just the beginning (if you do it right). Interfaces 10/2:22–30.

Geoffrin AM, TJ Van Roy. 1979. Caution common sense planning methods can be hazardous to your corporate health. Sloan Manage Rev 20/4:31–42.

Gillmore PC. 1962. Optimal and suboptimal algorithms for the quadratic assignment problem. SIAM J Appl Math.

Hall KM. 1970. An r-dimensional quadratic placement algorithm. Manage Sci 17:219–229.

Handler GY, PB Mirchandani. 1979. Location on Networks: Theory and Algorithms. Cambridge, MA: MIT Press.

Hansen P, J Perreur, JF Thisse. 1980. Location theory dominance and convexity: some further results. Operat Res 28:1241–1250.

Harvey ME, MS Hung, JR Brown. 1973. The application of P-median algorithm to the identification of nodal hierarchies and growth centers. Econ Geogr 50:187–202.

Heartz DB, RT Eddison, eds. 1964. Progress in Operation Research. Vol 2. New York: Wiley, pp 110–113.

Juel H. 1975. Properties of location models. PhD Dessertation, University of Wisconsin–Madison.

Juel H, RF Love. 1976. An efficient computational procedure for solving multifacility rectilinear facilities location problems. Operat Res Q 27:697–703.

Katz IN. 1969. On the convergence of a numerical scheme for solving same location equilibrium problem. SIAM J Appl Math 17:1224–1231.

Katz IN, L Cooper. 1974. An always convergent numerical scheme for a random equation locational equilibrium problem. SIAM J Numerical Anal 17:683–693.

Kaufmann L, F Broeckx. 1978. An algorithm for quadratic assignment problem using Benders decomposition. Eur J Operat Res 2:204–211.

Keeney RL. 1980. Siting Energy Facilities. New York: Academic Press.

Kermack KA, JBS Haldane. 1950. Organic correlation and allometry. Biometrika 37:30–41.

Kolen AJW. 1986. Tree Network and Planar Rectilinear Location Theory. Amsterdam: CWI tract 25, CWI.

Krarup J, PM Pruzan. 1983. The simple plant location problem: survey and synthesis. Eur J Operat Res 12:36–81.

Land AH. 1963. A problem of assignment with interrelated costs. Operat Res Q 14:185–198.

Laporte G, Y Nobert, P Pelletier. 1983. Hamiltonian location problems. Eur J Operat Res 12:82–89.

Larson RC. 1974. A hypercube queuing model for facility location and redistricting in urban emergency services. Comput Operat Res 1:67–95.

Larson RC, AR Odoni. 1981. Urban Operations Research. Englewood Cliffs, NJ: Prentice-Hall.

Lawler EL. 1963. The quadratic assignment problem. Manage Sci 9:586–599.

Lawson CL. 1965. The smallest covering cone or sphere. SIAM Rev 7:415–417.

Leamer EE. 1968. Locational equilibria. J Region Sci 8:229–242.

Litwhiler DW. 1977. Large region location problems. PhD Dissertation, University of Oklahoma.

Litwhiler DW, AA Aly. 1979. Large region location problems. Comput Operat Res 6:1–12.

Love RF. 1967. The location of single facilities in three dimensional space by nonlinear programming. J Can Operat Res Soc 5:136–143.

Love RF. 1967. A note on the convexity of sitting depots. Int J Product Res 6:153–154.

Love RF. 1969. Locating facilities in three dimensional space by convex programming. Nav Res Logist Q 16:503–516.

Love RF. 1972. A computational procedure for optimally locating a facility with respect to several rectangular regions. J Reg Sci 21:22–33.

Love RF. 1974. The dual of a hyperbolic approximation to the generalized constrained multifacility location problem with lp distances. Manag Sci 21:22–23.

Love RF. 1976. One dimensional facility location allocation using dynamic programming. Managet Sci 22:614–617.

Love RF, PD Dowling. 1985. Optional weighted lp norm parameters for facilities layout distance characteristics. Managet Sci 31:200–206.

Love RF, PD Dowling. 1986. A generalized bounding method for facilities location models. Research and Working Paper Series No. 250, Faculty of Business, McMaster University.

Love RF, H Juel. 1982. Properties and solution methods for large location allocation problems. Operat Res Soc 33:443–452.

Love RF, H Juel. 1983. Hull properties in location problems. Eur J Operat Res 12:262–265.

Love RF, SA Kraemer. 1973. A dual decomposition method for minimizing transportation costs in multifacility location problems. Transport Sci 7:297–316.

Love RF, JG Morris. 1972. Modelling intercity road distances by mathematical functions. Operat Res Q 23:61–71.

Love RF, JG Morris. 1975. A computational procedure for the exact solution of location allocation problems with rectangular distances. Nav Res Logist Q 22:441–453.

Love RF, JG Morris. 1975. Solving constrained multifacility location problems involving lp distances using convex programming. Operat Res 23:581–587.

Love RF, JG Morris. 1979. Mathematical models of road travel distances. Managet Sci 25:130–139.

Love RF, JG Morris, GO Wesolowsky. 1988. Facilities location. Models and Methods. Amsterdam: North-Holland.

Love RF, WG Truscott, JH Walker. 1985. Terminal location problem: a case study supporting the status quo. Operat Res Soc 36:131–136.

Love RF, GO Weslowsky, SA Kraemer. 1973. A multifacility minimax location method for euclidian distances. Int J Product Res 11:37–45.
Love RF, JY Wong. 1976. Solving quadratic assignment problems with rectangular distances and integer programming. Nav Res Logist Q 23:623–627.
Love RF, JY Wong. 1976. On solving a one dimensional space allocation problem with integer programming. INFOR 14:139–143.
Love RF, WY Yeong. 1981. A stopping rule for facilities location algorithms. AIIE Trans 13:357–362.
Love RF, L Yerex. 1976. Application of a facilities location model in the prestressed concrete industry. Interfaces 6/4:45–49.
Mairs, TG, GW Wakefield, EL Johnson, K Spielberg. 1978. On a production allocation and distribution problem. Manage Sci 24:1622–1630.
Manne AS. 1964. Plant location under economics of scale-decentralization and computation. Manage Sci 11:213–235.
Marucheck AS, AA Aly. 1981. An efficient algorithm for the location allocation problem with rectangular regions. Nav Res Logist Q 28:309–323.
Mavrides LP. 1979. An indirect method for the generalized k-median problem applied to lock-box location. Manage Sci 24:1622–1630.
Miehle W. 1958. Link-length minimization in networks. Operat Res 6:232–243.
Minieka E. 1970. The m-center problem. SIAM Rev 12:138–139.
Mirchandani PB, Francis RL. 1990. Discrete Location Theory. New York: Wiley–Interscience Series in Discrete Mathematics and Optimization.
Morris JG. 1973. A linear programming approach to the solution of constrained multifacility minimax location problems where distances are rectangular. Operat Res Q 24:419–435.
Morris JG. 1978. On the extent to which certain fixed charge depot location problem can be solved by L.P. J Oper Res Soc 29:71–76.
Morris JG. 1981. Convergence of the Weiszfeld algorithm for Weber problems using a generalized "distance" function. Operat Res 29:37–48.
Morris JG. 1982. Lawson's algorithm for p-norm minimax facility problems. Paper presented at the Joint National Meeting of ORSA/TIMS, San Diego, CA.
Morris JG, JP Norback. 1980. A simple approach to linear facility location. Transport Sci 14:1–8.
Morris JG, WA Verdini. 1979. A simple iterative scheme for solving minisum facility location problems involving lp distances. Operat Res 27:1180–1188.
Nair KPK, R Chandrasekaran. 1971. Optimal location of a single service centre of certain types. Nav Res Logist Q 18:503–510.
Nauss RM, RE Markland. 1981. Optimizing procedure for lock-box analysis. Manage Sci 27:855–865.
Nugent CE, TE Vollmann, J Ruml. 1968. An experimental comparison of techniques for the assignment of facilities to locations. Operat Res 16:150–173.
O'Kelly ME. 1986. The location of interacting hub facilities. Transport Sci 20:92–106.
Ostresh LM, Jr. 1973. TWAIN-exact solution to the two source location allocation problem. In: G Rushton, MF Godchild, LM Ostresh, JR, eds. Computer programs for the allocation problems. Iowa City: Monograph No. 6, Dept Geography, University of Iowa.

Ozgen-Mehmet-Tankut; Demirbas-Kerim. 1998. Cohens bilinear class of shift-invariant space/spatial frequency signal representations for particle-location analysis of in line Fresnel holograms. Opt Soc Am, Vol 25, 2117–2137, 1975. An efficient algorithm for solving the two centre location allocation problem. J Reg Sci 15:209–216.

1977. The multifacility location problem: applications and descent theorems. J Reg Sci 17:409–419.

Pardalos PM, JB Rosen. 1986. Methods for concave minimization: a bibliographic survey. SIAM Rev 28:367–379.

Pearson K. 1901. On lines and planes of closest fit to systems of points in space. Philos Mag J Sci, Sixth Ser 2:559–572.

Perreur J, J Thisse. 1974. Central metrics and optional location. J Reg Sci 14: 411–421.

Plane DR, TE Hendrick. 1977. Mathematical programming and the location of fire companies for the Denver Fire Dept. Operat Res 25:563–578.

Prakash M, K Rajeev, T Arie. 1996. Capacitated location problem on a line. Transport Sci 30:75–80.

Prakash M, K Rajeev, T Arie. 1978. On the convergence of class of iterative methods for solving the Weber location problem. Operat Res 26:597–609.

Rao MR. 1973. On the direct search approach to the rectilinear facilities location problem. AIIE Trans 5:256–264.

ReVelle C, R Swain. 1970. Central facilities location. Geogr Anal 2:30–42.

Ritzman LP. 1972. The efficiency of computer algorithms form plant layout. Manage Sci 18:240–248.

Rodman GM, LB Schwarz. 1977. Extensions of the multi period facility phase out model. New procedures and applications to a phase in/phase out problem. AIIE Trans 9: 103–107.

Sarker BR, WE Wilhelm, GL Hogg, MH Han. 1995. Backtracking of jobs in one-dimensional machine location problems. Eur J Operat Res 85:593–609.

Sarker BR, WE Wilhelm, GL Hogg. 1994. Backtracking and its amoebic properties in one-dimensional machine location problems. J Operat Res Soc 45:1024–1039.

Sarker BR, WE Wilhelm, GL Hogg. 1994. Measures of backtracking in one-dimensional machine location problems. Product Planning Control 5:282–291.

Schilling D, DJ Elzinga, J Cohon, R Church, C ReVelle. 1979. The team/fleet models for simultaneous facility and equipment citing. Transport Sci 13:163–175.

Schrage L. 1975. Implicit representation of variable upper bounds in linear programming. Math Progr Study 4:118–132.

Scott AJ. 1971. Combinatorial Programming, Spatial Analysis and Planning. London: Methuen.

Sherali AD, CM Shetty. 1977. The rectilinear distance location allocation problem. AIIE Trans 9:136–143.

Spielberg K. 1969. An algorithm for the simple plant location problem with some side conditions. Operat Res 17:85–111.

Steinberg L. 1961. The background wiring problem: a placement algorithm. SIAM Rev 3:37–50.

Sule DR. 1988. Manufacturing Facilities Location Planning and Design. PWS Publishing Company.

Thisse JF, JE Ward, RE Wendel. 1984. Some properties of location problems with block and round norms. Operat Res 32:1309–1327.
Ting S–S. 0000. A linear-time algorithm for maxsum facility location problem on tree networks. Transport Sci 18:76–84.
Toregas CR, R Swain, C ReVelle, L Bergman. 1971. The location of emergency service facilities. Operat Res 19:1363–1373.
Toregas C, C ReVelle. 1973. Binary logic solutions to a class of location problems. Geogr Anal 5:145–155.
Urquhart M. 1977. Pipe fabrication shop layout. Undergraduate thesis, Dept of Mechanical Eng, University of Waterloo, Spring Semester.
Van Roy TJ, D Erlenkotter. 1982. Dual based procedure for dynamic facility location. Manage Sci 28:1091–1105.
Vergin RC, JD Rogers. 1967. An algorithm and computational procedure for locating economic facilities. Manage Sci 13:256–264.
Walker W. 1974. Using the set covering problem to assign fire companies to fire houses. Operat Res 22:275–277.
Walker W, JM Chaiken, EJ Ignall, eds. 1980. Fire Department Deployment Analysis. A Public Policy Analysis Case Study. New York: Elsevier/North-Holland.
Ward JE, RE Wendell. 1980. A new norm for measuring distance which yields linear location problems. Operat Res 28:836–844.
Weber A. 1909. Uber den Standort der Industrien. Tubingen (English Translation: Friedrich, C.J (translator) 1929 [Theory of the Location of Industries.] Chicago: University of Chicago Press.
Wendell RE, A Phurter. 1973. Location theory, dominance and convexity. Operat Res 21:314–320.
Wendell RE, A Phurter, TJ Lowe. 1977. Efficient points in location problems. AIIE Trans 9:338–346.
Wersan SJ, JE Quon, A Charnes. 1962. Systems analysis of refuse collection and disposable practices. American Public Works Association, Year-Book, pp 195–211.
Wesolowsky GO. 1970. Facilities location using rectangular distances. PhD Dissertation, University of Wisconsin–Madison.
Wesolowsky GO, WG Truscott. 1975. The multiperiod location allocation problem with relocation of facilities. Manage Sci 22:57–65.
Westwood JB 1977. A transport planning model for primary distribution. Interfaces 8/1:1–10.
White JA. 1971. A note on the quadratic facility location problem. AIIE Trans 3:156–157.
Witzgall C. 1964. Optimal location of a central facility: mathematical models and concepts. National Bureau of Standards Report 8388, Gaithersberg.
Wolfe P. 1961. A duality theorem for non linear programming. Q J Appl Math 19:239–244.
Woolsey RED. 1986. The fifth column: on the minimization of need for new facilities or space wars, lack of presence, and Delphi. Interfaces 16/5:53–55.
Wyman SD. III, LG Callahan. 1975. Evaluation of computerized layout algorithms for use in design of control panel layouts. Proceedings, Fourteenth Annual U.S Army Operations Research Symposium. Fort Lee, VA, November, vol. 2, pp 993–1002. Aberdeen Proving Ground, MD: Director, U.S. Army Material Systems Analysis Activity.

Index

* 9 7 8 0 3 6 7 3 9 7 5 1 7 *

T - #0023 - 111024 - C0 - 229/152/27 - PB - 9780367397517 - Gloss Lamination